# NATO ASI Series

## Advanced Science Institutes Series

*A series presenting the results of activities sponsored by the NATO Science Committee, which aims at the dissemination of advanced scientific and technological knowledge, with a view to strengthening links between scientific communities.*

The Series is published by an international board of publishers in conjunction with the NATO Scientific Affairs Division

| | |
|---|---|
| A Life Sciences<br>B Physics | Plenum Publishing Corporation<br>London and New York |
| C Mathematical and Physical Sciences<br>D Behavioural and Social Sciences<br>E Applied Sciences | Kluwer Academic Publishers<br>Dordrecht, Boston and London |
| F Computer and Systems Sciences<br>G Ecological Sciences<br>H Cell Biology<br>I Global Environmental Change | Springer-Verlag<br>Berlin Heidelberg New York<br>London Paris Tokyo Hong Kong<br>Barcelona Budapest |

PARTNERSHIP SUB-SERIES

| | |
|---|---|
| 1. Disarmament Technologies | Kluwer Academic Publishers |
| 2. Environment | Springer-Verlag/Kluwer Academic Publishers |
| 3. High Technology | Kluwer Academic Publishers |
| 4. Science and Technology Policy | Kluwer Academic Publishers |
| 5. Computer Networking | Kluwer Academic Publishers |

*The Partnership Sub-Series incorporates activities undertaken in collaboration with NATO's Cooperation Partners, the countries of the CIS and Central and Eastern Europe, in Priority Areas of concern to those countries.*

## NATO-PCO DATABASE

The electronic index to the NATO ASI Series provides full bibliographical references (with keywords and/or abstracts) to about 50000 contributions from international scientists published in all sections of the NATO ASI Series. Access to the NATO-PCO DATABASE compiled by the NATO Publication Coordination Office is possible in two ways:

- via online FILE 128 (NATO-PCO DATABASE) hosted by ESRIN, Via Galileo Galilei, I-00044 Frascati, Italy.
- via CD-ROM "NATO Science & Technology Disk" with user-friendly retrieval software in English, French and German (© WTV GmbH and DATAWARE Technologies Inc. 1992).

The CD-ROM can be ordered through any member of the Board of Publishers or through NATO-PCO, Overijse, Belgium.

Series H: Cell Biology, Vol. 102

Springer
*Berlin*
*Heidelberg*
*New York*
*Barcelona*
*Budapest*
*Hong Kong*
*London*
*Milan*
*Paris*
*Santa Clara*
*Singapore*
*Tokyo*

# Interacting Protein Domains

## Their Role in Signal and Energy Transduction

Edited by

Ludwig Heilmeyer

Ruhr-Universität Bochum
Institut für Physiologische Chemie
Abt. Biochemie Supramolekularer Systeme
D-44780 Bochum, Germany

With 119 Figures

Springer

Published in cooperation with NATO Scientific Affairs Division

Proceedings of the NATO Advanced Study Institute on Structure and Function of Interacting Protein Domains in Signal and Energy Transduction, held at Acquafredda di Maratea, Italy, September 10–19, 1996

Library of Congress Cataloging-in-Publication Data applied for

Die Deutsche Bibliothek - CIP-Einheitsaufnahme

**Interacting protein domains** : their role in signal and energy transduction ; [proceedings of the NATO Advanced Study Institute on Structure and Function of Interacting Protein Domains in Signal and Energy Transduction, held at Acquafredda di Maratea, Italy, September 10 - 19, 1996] / ed. by Ludwig Heilmeyer. Publ. in cooperation with NATO Scientific Affairs Division. - Berlin ; Heidelberg ; New York ; Barcelona ; Budapest ; Hong Kong ; London ; Milan ; Paris ; Santa Clara ; Singapore ; Tokyo : Springer, 1997
(NATO ASI series : Ser. H, Cell biology ; Vol. 102)
ISBN-13: 978-3-642-64583-9 e-ISBN-13: 978-3-642-60848-3
DOI: 10.1007/978-3-642-60848-3

ISBN-13: 978-3-642-64583-9 Springer-Verlag Berlin Heidelberg New York



Softcover reprint of the hardcover 1st edition 1997
Typesetting: Camera ready by authors/editors
Printed on acid-free paper
SPIN 10525515 31/3137 - 5 4 3 2 1 0

# Preface

This is now the fourth time that protein phosphorylation has been the focus of a NATO Advanced Study Institute. The first meeting with the topic „Signal Transduction and Protein Phosphorylation" was held on the island of Spezai, Greece, in September 1986. The second one took place in Chateau La Londe, France, in September 1989 on „Cellular Regulation by Protein Phosphorylation", the third one on „Tyrosine Phosphorylation/Dephosphorylation and Downstream Signaling" was in September 1992 in Maratea, Italy.

The titles of these books clearly mirror the developments that have taken place in the last decade. Beginning with the recognition that protein phosphorylation is at the center of signaling - clearly established in 1990 - it became apparent that many cellular processes are regulated by this mode. A new focus then emerged when it was recognized that growth factors are bound to corresponding receptors trigger protein tyrosine phosphorylation which controls cell proliferation. This was the topic of the third meeting in this series. It is now evident that further progress depends on understanding the three dimensional structure of the proteins involved. It goes without saying, for example, that understanding the location of proteins by adaptor proteins is only possible on the basis of the three dimensional protein structure.

Therefore, the fourth meeting in this series concentrated on the protein structure of signaling molecules as well as on the elucidation of the principles of protein domain interactions. Sessions, therefore, on muscle structure were included, demonstrating principles in the domain interactions that generate force. This volume covers the presentations that were made during this fourth NATO Advanced Study Institute which was co-sponsored by the Federation of European Biochemical Societies. It was especially fruitful since a center project at the Ruhr-University of Bochum, Germany (Sonderforschungsbereich 394: „Strukturelemente und Molekulare Mechanismen von Proteinen bei Energieübertragung und Signalvermittlung") is actually involved in this research area. This NATO Advanced Study Institute therefore reflects some of the ongoing research in this center project. The importance of this area of research has just begun to be recognized. Much more effort to determine three dimensional structures of proteins is still needed. I hope that these proceedings will further promote this research area.

The organizers would like to express their thanks to their co-workers for their help in organizing the meeting and in preparing the manuscript. We are especially grateful to Mrs. Humuza who organized the meeting and prepared the contributions in the camera ready form.

Bochum, 3rd March, 1997 — Ludwig Heilmeyer

# CONTENTS

## Part I: Regulation by Reversible Protein Phosphorylation



## Part II: Methodology

## Part III: Phospholipid Signaling



## Part IV: Tyrosine Phosphorylation and Downstream Signaling

## Part V: Regulatory Cascades

## Part VI: Regulation of Muscle Contraction

# Part I

# Regulation by Reversible Protein Phosphorylation

# Control of Cellular Processes by Reversible Protein Phosphorylation

Edmond H. Fischer
Department of Biochemistry, University of Washington, Seattle, WA 98195

The regulation of cellular processes requires a myriad of commands, positive and negative, that must be tightly coordinated to keep all the reactions that take place under control. In particular, to make sure that no crucial event will occur at inappropriate times or out of phase. We know today that most of the signals that are used to orchestrate these reactions, the switches that must be turned on or off, rely on reversible protein tyrosine phosphorylation. This first article will provide a rapid historical overview of the regulation of cellular processes by reversible protein phosphorylation. A second article (see below) will cover some more recent aspects of this field, particularly as they relate to tyrosine phosphatases.

Fifty years ago, very little was known about cellular regulation and nothing at all, of course, about a possible involvement of protein phosphorylation/dephosphorylation in these processes. But in the mid '50's, it was shown that protein phosphorylation could serve as a means to regulate the activity of glycogen phosphorylase, discovered by Parnas in Poland and Carl and Gerty Cori in the U.S. in the mid 30's (for review, 1). The muscle enzyme was thought to have an absolute requirement for adenylic acid for activity until, in 1943, Arda Green in Cori's lab crystallized it in a form that was active without added AMP. They called this form phosphorylase a, and very logically, assumed that it contained covalently-bound AMP. They further thought that it had to be the native form of the enzyme because, when left standing in crude extracts, it was rapidly converted to the earlier species which they called phosphorylase b. However, if that hypothesis were correct, AMP would have to be released in the reaction, but they found none. Furthermore, no AMP or adenine or ribose could be detected in the "native" enzyme, using the most sensitive microbiological assays available at that time. They knew that the enzyme existed in two forms but did not know how these two forms differed and, strangely, actually dropped the problem.

It is approximately ten years later that in Seattle, we would show with Ed Krebs, that the conversion of phosphorylase b to a involved a phosphorylation of the protein. The reaction had to be enzymatic, requiring a kinase which we called phosphorylase kinase; the reverse reaction, then, had to be catalyzed by a phosphorylase phosphatase.

In crude extracts, there was an absolute requirement for $Ca^{2+}$ or $Mn^{2+}$ which suggested that phosphorylase kinase itself would also exist in an inactive and active form and that $Ca^{2+}$ would be required for this conversion. That was indeed

NATO ASI Series, Vol. H 102
Interacting Protein Domains
Their Role in Signal and Energy Transduction
Edited by Ludwig Heilmeyer

the case so that, at this point, it was clear that we were dealing with a cascade of two enzymes acting successively on one another.

During the same period of time, Sutherland, Rall and Wosilait had arrived at a similar conclusion, working on liver phosphorylase (2,3). An epochal finding that grew out of these studies was the discovery of cAMP by Sutherland and Rall which, by an unknown mechanism, shifted the balance between non-activated and activated liver phosphorylase toward the latter form (4). They generously supplied us with samples of cAMP which allowed us to determine that its action was directed towards the activation of phosphorylase kinase by either accelerating its rate of autophosphorylation or allowing another kinase (initially called a kinase kinase) to do so. This latter hypothesis was proven by the isolation of the cAMP-dependent protein kinase 6-7 years later by Ed Krebs and Don Walsh, allowing the entire glycogenolytic cascade to be established (5).

The finding that calcium could trigger glycogenolysis was quite exciting. Physiologists had known for many years that $Ca^{2+}$ released in response to a nerve stimulus could trigger muscle contraction. The finding that precisely the same concentrations of $Ca^{2+}$ would initiate glycogenolysis and ultimately provide for the synthesis of ATP needed to maintain contraction explained how these two physiological processes could be regulated in concert.

In retrospect, we were very lucky to have selected this particular enzyme to work on (for review, 6). First, it is extremely abundant in skeletal muscle. Second, the phosphorylation reaction was unambiguous, converting a totally inactive enzyme into a fully active species. Third, it was stoichiometric, introducing 1 mole of phosphate per mole of enzyme subunit. Finally, the site phosphorylated occurred within a loose N-terminal arm of the molecule that could be easily cleaved by limited proteolysis, leaving behind the bulk of the enzyme intact. This made it easy to separate the phosphopeptide involved and determine its structure; we could show that a single seryl residue had been phosphorylated (6).

The phosphorylation reaction was so straightforward that there was no doubt in our minds that it would represent the prototype for such kinds of interconversions. As it turned out, it was really the exception. Six years went by before Joe Larner identified the next glycogenolytic enzyme to be regulated by phosphorylation-dephosphorylation, namely, glycogen synthase (7), which was inactivated rather than activated by phosphorylation (8). However, Larner rapidly found that about six phosphates were introduced during the interconversion (9). At that time, the idea that a single phosphorylation event was all that was needed to alter the state of activity of an enzyme was so ingrained that we thought that glycogen synthase was probably made up of 6 identical subunits, each of approximately 15,000 MW. Of course, we know today from Larner's work as well as that of Phil Cohen (10), Peter Roach (11) and others, that the enzyme is phosphorylated on no less than seven sites by seven different protein kinases, all totally unknown at that time. Furthermore, some of these phosphorylation events must follow a most complicated program of successive reactions that have to proceed in a strictly prescribed order (12). Had we decided to work on this enzyme rather than on phosphorylase, we would have never been able to solve the problem.

We did not know at that time whether the phosphorylation reaction was a rare event restricted to the control of these two enzymes or perhaps to carbohydrate metabolism. Once again, as luck would have it, reversible protein phosphorylation turned out to be one of the most prevalent mechanisms by which cellular processes can be regulated, being involved in the regulation of most metabolic pathways, hormone action, gene expression, the immune response, etc., and many pathological conditions including cancer.

In contrast to allosteric regulation in which the enzymes involved undergo changes in conformation in response to effectors that are generated during the normal maintenance of the cell and reflect its overall internal condition, covalent regulation (and this is perhaps one of the major lessons we have learned over the last 30 years), responds essentially only to extracellular signals. For instance, catecholamines, various peptide hormones and neurotransmitter and other stimuli such as drugs, light, odorants, etc. will act on specific 7-transmembrane (serpentine) receptors. These, in G-protein-regulated reactions, bring about the production or mobilization of second messengers such as cAMP, $Ca^{2+}$, DAG, $IP_3$, etc. which, in turn, will act on modulator proteins (mostly Ser/Thr protein kinases or phosphatases), then on target enzymes to finally elicit a physiological response. The resulting serine/theonine phosphorylations are generally quite stable and abundant and many of these pathways have been elucidated. Several other types of receptors have been identified: some have catalytic activity of their own, such as the receptors for TGFβ or activin having Ser/Thr kinase activity, or the large family of growth factor receptors with tyrosine kinase activity. Receptors for cytokines such as interferon-α or γ, IL-3, IL-4 or IL-6 have no catalytic activity of their own but are associated with intracellular tyrosine kinases of the JAK or TYK family and signal mainly by phosphorylating STAT proteins that then translocate to the nucleus and induce the transcription of susceptible genes. Other receptors respond to stress, nitric oxide, osmolarity or pheromones in yeast, and so forth.

One of the most exciting developments in this field has been the discovery some 18 years ago that protein tyrosine phosphorylation was intimately implicated in cell signaling, growth, differentiation and transformation - by bringing into play tyrosine kinases of cellular or viral origin or linked to GF-receptors.

Growth factor receptors have a similar general architecture: a single transmembrane segment separating a catalytic domain with tyrosine kinase activity inside and a considerable diversity of motifs outside recognizing different ligands:

cystein-rich segments; FNIII repeats; immunoglobulin-like loops; EGF-, Factor VIII-, or cadherin-like domains, Kringles, etc. (13). The ligands for these receptors are circulating molecules, mostly mitogenic hormones or growth factors, except for the Eph-family of receptors that recognize membrane-bound structures. As expected, they are neither mitogenic nor involved in cell differentiation like the other growth factor receptors; rather, their primary role is to help developing axons reach their proper target (14). During embryonic development, nerve growth cones must explore their local environment and choose the proper paths in order to go where they are supposed to to. Pathfinding involves highly sophisticated mechanisms with soluble and membrane-bound molecules in the target area acting as attractants or repellents (15).

We finally begin to understand how signal is transduced from these receptors: phosphotyrosyl residues generated on the receptors following their stimulation serve as docking sites for adaptor or linker proteins having SH2 (for src-homology 2) domains. This engenders some sort of tinker-toy system of successive protein-protein interactions involving SH2, SH3, PH and other domains that ultimately initiate a number of signaling pathways (16,17). But we still do not understand how selectivity is introduced in such complex signaling systems. The PDGF receptor, for example, can interact through at least 8 of its tyrosyl residues and trigger the initiation of a number of signaling pathways (18).

In a very over-simplified way, receptors can be compared to the old hand-operated telephone switchboard in which the operator linked a call from the outside to the proper recipient by inserting the plug in the appropriate hole. Paul Valery once said: "Ce qui est simple est faux, et ce qui est compliqué est incompréhensible"; the above analogy is a perfect example of something simple and wrong since, obviously, a combinatorial system of many different receptors must act in concert to modulate the response and select which process is to become operative.

Mutations of these receptors may lead to pathological conditions of varying degrees of severity. Among the most thoroughly studied are those affecting the insulin receptor and leading to various forms of NIDDM or insulin-resistant diabetes; more than two dozens are known (19). Some prevent the expression of the receptor or its transport from the ER/Golgi to the plasma membrane, or increase its rate of degradation following internalization; others interfere with ligand binding or its catalytic function. The most severe of these may lead to leprechaunism resulting in early death. As expected, the subjects are severely glucose intolerant in spite of huge amounts of circulating insulin. In fact, in a futile and desperate attempt to overcome the defect, the organism can produce insulin at concentrations 100 times above normal.

But the best known and most dramatic receptor alterations are those that lead to oncogenicity (20). Indeed, receptor or cytoplasmic tyrosine kinases represent the largest group of oncoproteins (more than 80%) which underscores the essential role tyrosine phosphorylation plays in regulating cell function. Oncogenicity results either from the overexpression of the normal gene, or mutations that would render the encoded enzymes constitutively active. They could result from a truncation of the molecule, as observed with the external domain of v-erb b, related to the EGF receptor. A single mutation of a valyl to a glutamyl residue in the transmembrane domain of the neu/her2 protooncogene also related to the EGF-receptor is responsible for neuroblastomas in rats. Others involve fusion with segments of other genes, particularly those that encode proteins that easily undergo dimerization: many have leucine repeats that predict a coiled-coil configuration. This is the case, for example, of the oncogenic forms of the Trk family of nerve growth factor receptors that undergo fusion with a segment of non-muscle tropomysin. In fact, most oncogenic mutations are those that allow these structures to undergo dimerization and, therefore, spontaneous and unrestricted transphosphorylation and activation.

With accumulating evidence implicating tyrosine phosphorylation in cell proliferation and transformation, it is hardly surprising that many groups, including our own, became interested in the enzymes that would catalyze the reverse reaction, namely, the protein tyrosine phosphatases (PTPases). This will be the subject of the second article.

**References**


1. Krebs, E. G. (1993) *Angew Chem* **32, No. 8**, 1122-1120

2. Sutherland. E. W. and Wosilait, W. D. (1955) *Nature* **175**, 169

3. Rall, T. W., Sutherland. E. W., and Wosilait. W. D. (1956) *J. Biol. Chem.* **218**, 483

4. Sutherland, E. W. and Rall, T. W. (1958) *J. Biol. Chem.* **233**, 1077-1091

5. Walsh. D. A., Perkins. J. P., and Krebs. E. G. (1968) *J. Biol. Chem.* **243**, 3763-3765

6. Fischer. E. H. (1993) *Angew Chem* **32, No. 8**, 1130-1137

7. Rosell-Perez, M.. Villar-Palasi, C., and Larner. J. (1962) *Biochemistry* **1**, 763-768

8. Friedman, D. L. and Larner, J. (1963) *Biochemistry* **2**, 669-675

9. Larner, J. and Villar-Palasi, C. (1971) *Curr. Top. Cell Regul.* **3**, 195-236

10. Cohen. P. (1986) *The Enzymes* **17A**. 461-497

11. Roach. P. J. (1986) in *Enzymes* (Boyer. P. D. and Krebs, E. G. eds) pp. 499-539, Academic Press. Orlando

12. Roach. P. J. (1990) *FASEB* **4(12)**. 2961-2968

13. Schlessinger, J. and Ullrich. A. (1992) *Neuron* **9**, 383-391

14. Tessier-Lavigne. M. (1995) *Cell* **82**. 345-348

15. Barinaga, M. (1995) *Science* **269**. 1668-1670

16. Pawson, T. and Schlessinger. J. (1993) *Curr. Biol.* **3**, 434-442

17. Cohen. C. B., Ren, R., and Baltimore. D. (1995) *Cell* **80**. 237-248

18. Heldin, C.-H. (1995) *Cell* **80**, 213-223

19. Taylor, S. I. (1992) *Diabetes.* **41**. 1473-1490

20. Cantley, L. C.. Auger. K. R.. Carpenter. C.. Duckworth. B.. Graziani. A.. Kapeller. R.. and Soltoff. S. (1991) *Cell* **64**. 281-302

# From Phosphorylase to Phosphorylase Kinase

L. N. Johnson

Laboratory of Molecular Biophysics, University of Oxford, South Parks Road, Oxford OX1 3QU, UK.

Glycogen phosphorylase was the first protein shown to be controlled by reversible phosphorylation (Fischer and Krebs 1955). Control by protein phosphorylation is now known to be a ubiquitous process for intracellular signalling mechanisms. There is a growing number of protein kinase and protein phosphatase structures solved by X-ray diffraction methods (Table 1) that is providing insight into some of these control mechanisms.

| Ser/Thr Kinases | Ser/Thr Phosphatases |
|---|---|
| cAPK_P (active)<br>CDK2 (inactive)<br>CDK2/cyclin A (partially active)<br>CDK2_P/cyclin A (fully active)<br>MAPK (inactive)<br>Twitchin kinase (inactive)<br>Casein kinase 1 (active)<br>Phosphorylase kinase (active)<br>Calmodulin dependent kinase 1 (inactive) | Calcineurin (PP2B)<br>Protein phosphatase 1A<br>Protein phosphatase 2C |
| **Tyr kinases** | **Tyr Phosphatases** |
| Insulin receptor kinase (inactive)<br>FGF receptor kinase (inactive)<br>Lck_P kinase (active) | PTP1B<br>Yersinia PTP<br>Low MW PTP<br>Dual specificity PTP<br>Receptor-like PTP |

Table 1. Summary of protein kinases and protein phosphatases whose crystal structures are known to date. The active or inactive state of the kinases is indicated. _P indicates the kinase is phosphorylated on the activation segment in the crystal structure.

How can the addition of a phosphate group control the activity of a protein? The response of a protein on phosphorylation is dictated by the special properties of the phosphate group that distinguish it from the naturally occurring amino acids.

NATO ASI Series, Vol. H 102
Interacting Protein Domains
Their Role in Signal and Energy Transduction
Edited by Ludwig Heilmeyer

The phosphate group with 4 oxygen atoms can participate in extensive hydrogen bond interactions and these can link different parts of the polypeptide chain. The phosphate group (approximate pK 6.7) is likely to be dianionic at physiological pH. The property of a double negative charge is a property that is not available to the naturally occurring amino acids and electrostatic effects are important in control by phosphorylation. Analysis of protein phosphate interactions in existing protein structures has shown that the most common interaction is between the phosphate oxygens and the main chain nitrogens at the start of a helix where the most frequently found residue is glycine. In non-helix interactions, phosphate groups most commonly interact with arginine residues. The guanidinium group is suited for interactions with phosphate by virtue of its planar structure and its ability to form multiple hydrogen bonds. Because of its resonance stabilisation, the guanidinium group is a poor proton donor ($pK_a > 12$) and cannot function as a general acid catalyst in the hydrolysis of phosphorylated amino acids. Electrostatic interactions between arginine and phosphate groups provide tight binding sites that appear to play a dominant role in recognition and stabilisation of protein conformations.

## 1. Phosphorylase

Structural studies on glycogen phosphorylase have shown that the N-terminal 20 amino acids are relatively mobile in the unphosphorylated form of the enzyme (Johnson 1992; Johnson et al., 1992; Johnson and Barford 1993) On phosphorylation and conversion of the inactive to the active protein, the N-terminal 20 amino acids shift so that the serine (Ser14-P) moves by about 50 Å and changes its contacts from intrasubunit to intersubunit contacts (Figure 1). In its new position the Ser14-phosphate contacts 2 arginine residues (Figure 2). The main incentive for the shift in the serine on phosphorylation appears to be electrostatic. The movements of the Ser14-P and its adjacent amino acids are linked with substantial changes at the subunit-subunit contacts that lead to changes in tertiary and quaternary structure. These changes provide a route of communication from the allosteric and phosphorylation sites to the catalytic site which is over 30 Å away.

To a first approximation phosphorylase can be understood in terms of the Monod-Wyman Changeux theory for allosteric proteins in which the enzyme exist in equilibrium between a less active T state and a more active R state. Phosphorylation of Ser14 shifts the equilibrium in favour of the R state.

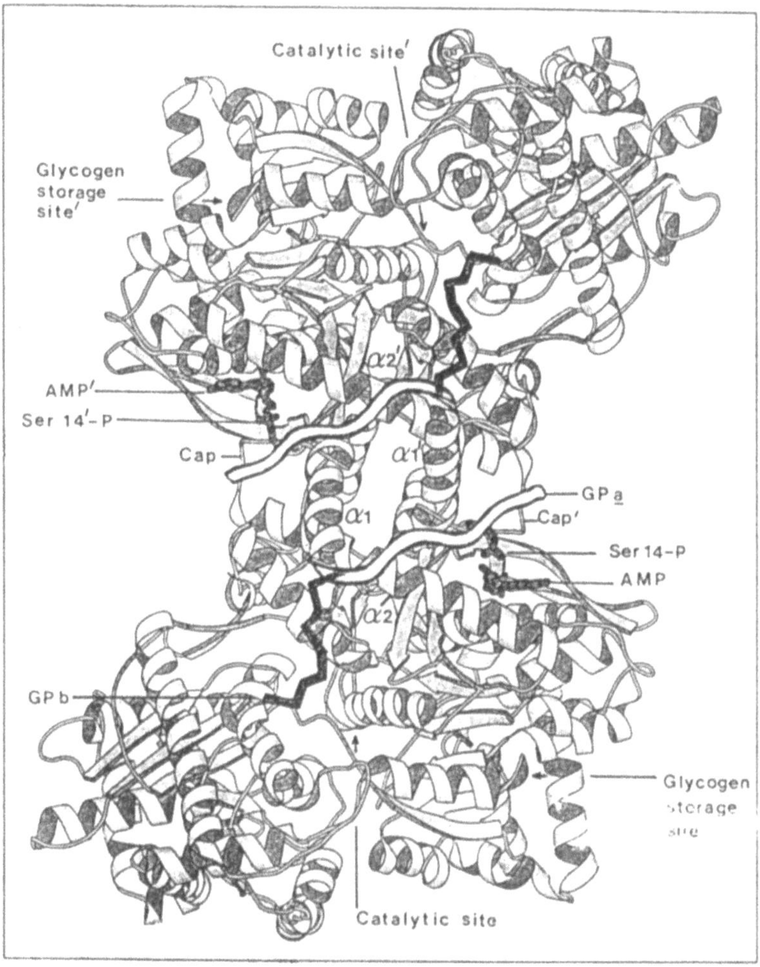


Figure 1. Schematic diagram of the phosphorylase dimer viewed down the 2-fold axis with the Ser14-P and AMP allosteric sites towards the viewer. Access to the catalytic site is from the far side of the molecule. The position of residues 14 to 23 in phosphorylase b (the less active non-phosphorylated form of the enzyme) are shown as a black line. The positions of residues 10-23 in phosphorylase a (the active phosphorylated form ) are shown as thick white line.

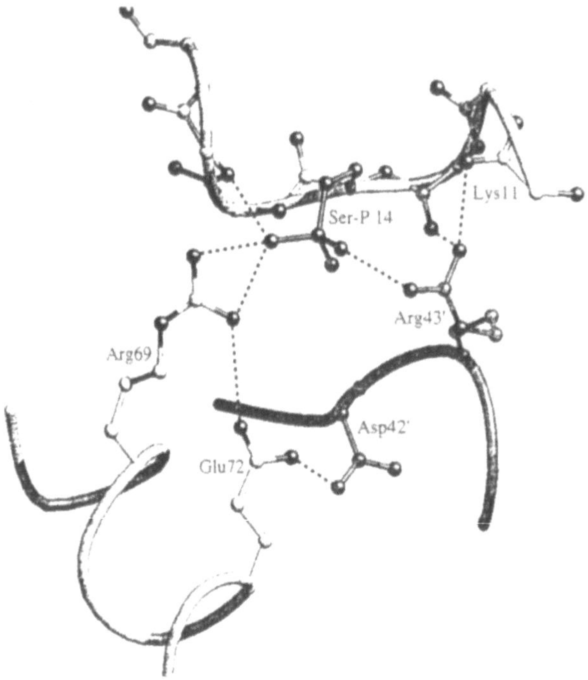


Figure 2. Details of the Ser14-P interactions in glycogen phosphorylase a.

We have recently determined the crystal structure of *E. coli* maltodextrin phosphorylase (Watson et al., 1996) . The bacterial enzyme exhibits no regulatory properties and its activity is controlled at the level of gene expression. Analysis of the structure shows that, despite 40% similarity in sequence, none of the regulatory sites exist and the structure is locked in an active conformation. As a result of changes in the subunit-subunit contacts, the 280s loop, which in T state mammalian phosphorylase acts as a gate to control access to the catalytic site, is held in an open conformation in the bacterial phosphorylase. The catalytic sites of R state mammalian and *E. coli* phosphorylase are essentially identical, including the environments of the essential cofactor, pyridoxal phosphate.

## 2. Phosphorylase kinase

Phosphorylase kinase was the second phospho-regulatory protein to be discovered. It is a large protein with subunit composition $(\alpha\beta\gamma\delta)_4$. The $\alpha$ and $\beta$ subunits are the regulatory subunits and convey activation following phosphorylation by cyclic AMP dependent protein kinase. The δ subunit is essentially identical to calmodulin and conveys control by calcium. The γ subunit is the catalytic subunit and is composed of a kinase domain which has 36% sequence identity to cyclic AMP dependent protein kinase and a calmodulin binding domain. The catalytic domain of phosphorylase kinase is constitutively active and requires no post-translational modification for activity. This is in contrast to many other protein kinases, such as cyclic AMP dependent protein kinase, MAP kinase, CDK2, that require phosphorylation on a threonine or tyrosine in the activation segment for activity.

An important interaction made by the phospho-amino acid in these kinases is an interaction with an arginine residue which is adjacent in sequence to an essential aspartate. Determination of the X-ray crystal structure of phosphorylase kinase domain (Owen et al., 1995) has shown that a glutamate has taken the place of the phospho-amino acid in the activation segment and contacts the arginine (Figure 3).

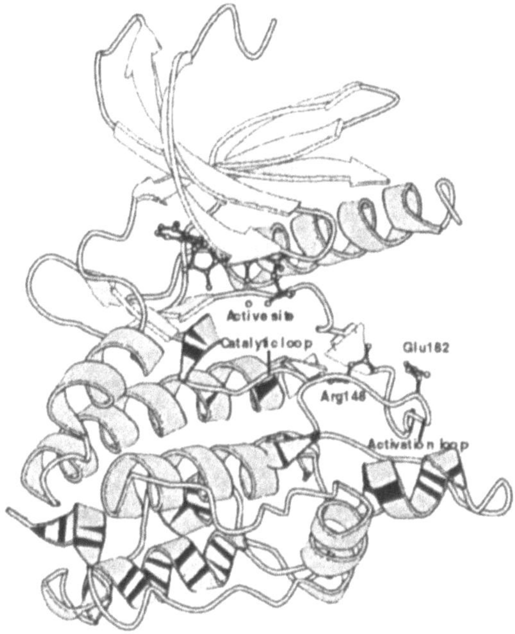


Figure 3. The kinase domain of phosphorylase kinase showing AMPPNP bound at the catalytic site and the activation and catalytic segments.

In the other kinases there is invariably a charge cluster whose neutralisation can best be effected by a dianionic phosphate group. In phosphorylase kinase there is no charge cluster and the active conformation of the activation segment can be promoted by a carboxylate group. Some of these results on active and inactive protein kinases and their control by phosphorylation on the activation segment have been reviewed (Johnson et al., 1996; Johnson and O'Reilly 1996) . Phosphorylation on a threonine or a tyrosine residue in the activation segment appears to be important in order to correctly align the catalytic site residues, especially those that are involved in substrate recognition.

**Acknowledgements.** I thank my colleagues, especially David Barford, Martin Noble , David Owen and Kim Watson, whose work formed the basis for the results described in this review.

**References**

Fischer, E. H. and Krebs, E. G. (1955). Conversion of phosphorylase b to phosphorylase a in muscle extracts. J. Biol. Chem. 216, 121-132.

Johnson, L. N. (1992). Glycogen phosphorlase: control by phosphorylation and allosteric effectors. FASEB J. 6, 2274-2282.

Johnson, L. N. and Barford, D. (1993). The effects of phosphorylation on the structure and function of proteins. Annu. Rev. Biophys. Biomol. Struct. 22, 199-232.

Johnson, L. N., Noble, M. E. M. and Owen, D. J. (1996). Active and inactive protein kinases. Cell 85, 149-158.

Johnson, L. N. and O'Reilly, M. (1996). Control by phosphorylation. Current Opinion in Structural Biology 6, 762-769.

Owen, D. J., Noble, M. E., Garman, E. F., Papageorgiou, A. C. and Johnson, L. N. (1995). Two structures of the catalytic domain of phosphorylase kinase: an active protein kinase complexed with substrate analogue and product. Structure 3, 467-482.

Watson, K. A., Schinzel, R., Palm, D. and Johnson, L. N. (1996). The crystal structure of E. coli maltodextrin phosphorylase. EMBO J. In press,

# PROTEIN KINASE X :
# A Novel Human Protein Kinase closely related to the Catalytic Subunit of cAMP-dependent Protein Kinase

B. Zimmermann*, A. Klink°, G. Rappold°, F.W. Herberg*
*Institut für Physiologische Chemie I, Ruhr Universität Bochum, 44780 Bochum, Germany
°Institut für Humangenetik, Universität Heidelberg, 69120 Heidelberg, Germany

## INTRODUCTION

The recently isolated gene PRKX encodes a novel type of human protein kinase closely related to the cAMP-dependent protein kinase (cAPK). The hydrophilic protein of about 41 kDa is called Protein Kinase X (Prkx) because the gene is located on the short arm of the human X-chromosome (Klink et al. 1995). It has a striking sequence similarity to the DC2 protein kinase from *Drosophila melanogaster* (62.3% identity in the conserved catalytic core region) and has a lower similarity to the catalytic subunits of lower organisms like that of *Ascaris suum* (53.5% identity) and *Caenorhabditis elegans* (51.1%). These protein kinases differ from the „classical" cAMP-dependent protein kinases (cAPKs) by their isoelectric points and the lengths of their branches in the phylogenetic tree (table 1, figure 1). Most residues important for substrate recognition and binding of the regulatory subunits RI and RII are identical comparing the catalytic subunit of cAMP-dependent protein kinase alpha (Adams and Taylor 1993; Grant et al. 1996) and Protein Kinase X. The core region of the kinase is also highly conserved whereas the N-terminus is totally different. The N-terminus of Protein Kinase X contains a putative binding motif for WW-domains, which is a proline-rich region called PY-motif (Macias et al. 1996), that might be important for regulation or subcellular localization. Northern blot analysis with different subfragments of cDNA clones indicates a widespread expression with the highest level of expression observed in fetal and adult brain, kidney and lung and low levels of expression in all other tested tissues (Klink et al. 1995). Prkx was expressed in *E.coli* using several different expression systems. The purified protein was identified, tested for activity and tested as substrate for the catalytic subunit of cAMP-dependent protein kinase alpha.

| „classical" cAPKs | | cAPKs in yeast/fungi | | novel putative subgroups | |
|---|---|---|---|---|---|
| Ancylostoma | 8.99 | Blastocladiella | 8.86 | human Prkx | 6.37 |
| C. eleg. 1 | 8.93 | Ustilago | 7.79 | C. eleg. 2 | 6.25 |
| mam. cAPKα | 8.84 | Magnaportha | 7.79 | Ascaris suum | 5.29 |
| mam. cAPKβ | 8.84 | Schizosac. | 6.72 | Drosophila (DC2) | 4.87 |
| Aplysia | 8.79 | Sacch. cer.(tpk2) | 6.64 | | |
| Drosophila (DC0) | 8.78 | Sacch. cer.(tpk3) | 5.93 | | |
| mam. cAPKγ | 8.50 | Sacch. cer.(tpk1) | 5.32 | | |

Table 1: isoelectric points (pIs) of cAMP-dependent protein kinases based on theoretical calculations (ExPASy WWW-server, compute pI/Mw tool), abbreviations refer to figure 1

NATO ASI Series, Vol. H 102
Interacting Protein Domains
Their Role in Signal and Energy Transduction
Edited by Ludwig Heilmeyer

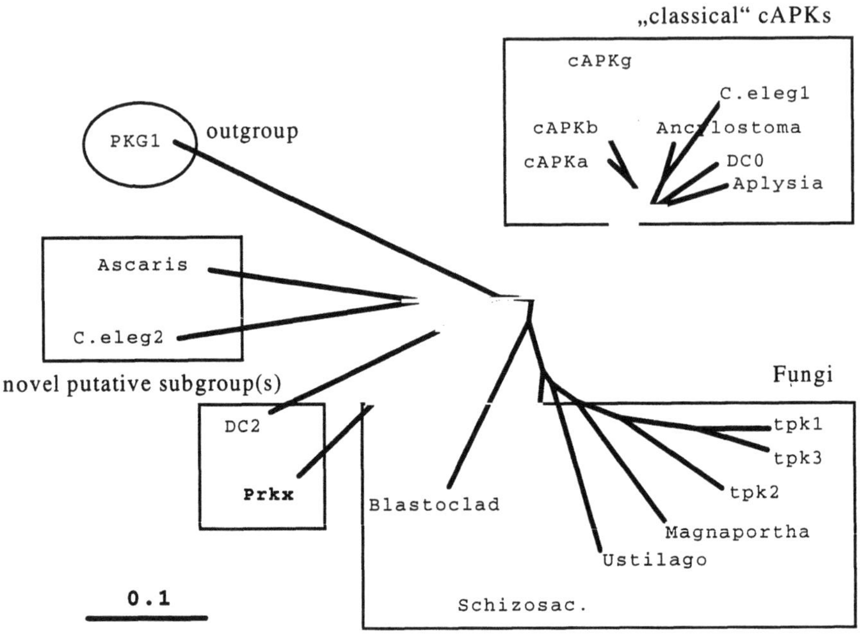


**Figure 1: Phylogenetic tree of cAMP-dependent and related protein kinases**

The phylogenetic tree includes the cAMP-dependent protein kinases which are part of the AGC1-subfamily of cyclic nucleotide regulated protein kinases (Hanks & Hunter 1995) and their closest relatives based on sequence alignment of the catalytic core. It is rooted by including the cGMP-dependent protein kinase.
Abbreviations : **cAPKa, cAPKb, cAPKg** mammalian cAMP-dependent protein kinase α, β, and γ-subunit; **C.eleg.** nematode *Caenorhabditis elegans*; **Ancylostoma** dog hookworm *Ancylostoma caninum,* **DC0** cAPK fruit fly *Drosophila melanogaster;* **Aplysia** sea hare (mollusca) *Aplysia californica;* **Blastoclad.** water mold (aquatic fungus) *Blastocladiella emerssonii*; **Schizosac.** *Schizosaccharomyces pombe*; **Ustilago** fungus *Ustilago maydis;* **Magnaportha** fungus *Magnaportha grisea;* **tpk1-3** cAPK genes from *Saccharomyces cerevisiae*; **DC2** DC2 kinase from *Drosophila melanogaster;* **Prkx** human protein kinase Prkx; **Ascaris** nematode *Ascaris suum;* **PKG** cGMP-dependent protein kinase (α-subunit) from *bos taurus*

## METHODS

The cDNA of Prkx was cloned into the pGEX-4T2-vector (Pharmacia) and the pGEX-KG-vector (Guan & Dixon 1991). The pGEX-KG-vector includes a linker between the coding sequence of glutathione-S-transferase (GST) and the Prkx sequence to improve the cleavage of the fusion protein with thrombin. Additionally the PRKX sequence was cloned into the pT7-vector pRSET5B (Invitrogen). The DNA sequences were verified with an automated ABI-sequencer. GST-Prkx and GST-KG-Prkx were expressed in *E.coli* BL21.DE3 cells. The cells were grown up to OD=0.6 at 37°C and after induction with 0.4 mM IPTG grown for additional 3 hours at room temperature. After harvesting by centrifugation the cells were stored at -80°C. In the coexpression system we used BL21.DE3 cells containing the GST-KG-PRKX vector under kanamycin resistance and a pT7-vector coding for the catalytic subunit of cAMP-dependent protein kinase under ampicillin resistance. The cells were resuspended in lysis buffer containing 50 mM Tris-Cl (pH 7.5), 150 mM NaCl, 5 mM DTT, 5 mM EDTA and 5 mM EGTA, and were lysed in a French pressure cell (Aminco). After centrifugation (20000xg, 30 min., 4°C) the supernatant was incubated with freshly equilibrated glutathione-agarose for 2 hours at 4°C. The glutathione-agarose was transferred into a Ø12mm column and washed with lysis buffer containing 500 mM NaCl. The GST fusion protein was eluted in Tris buffer containing 150 mM NaCl, 1 mM DTT and 15 mM glutathione (pH 7.5).

The recombinant proteins were identified by SDS polyacrylamide gel electrophoresis (SDS-PAGE) and by Western Blotting with specific GST antibodies. The purified fusion protein was cleaved with thrombin (Guan & Dixon 1991). Following cleavage the putative Prkx protein was verified by Edman sequencing. Kinase activity was either checked spectrophotometrically using Kemptide as substrate in a coupled assay (Cook et al. 1982) or with a more sensitive radioactive assay. Additionally, to test for autophosphorylation and phosphorylation by cAMP-dependent protein kinase, SDS-PAGE was used after phosphorylation with $^{32}$P-ATP, the gel was dried and the radioactive bands were visualized by autoradiography.

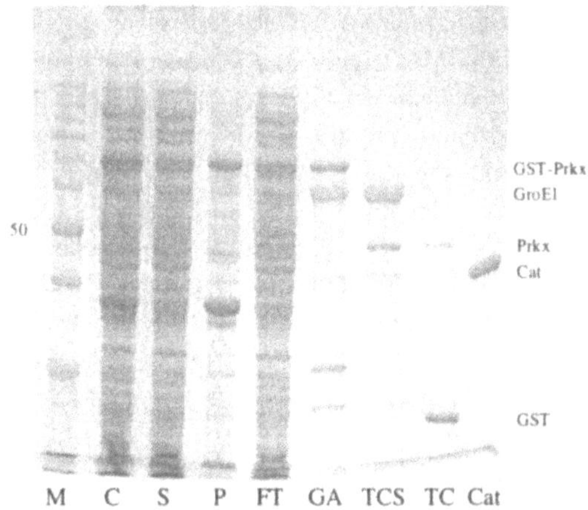


Figure 2: SDS-PAGE of the purification and thrombin-cleavageof GST-Prkx

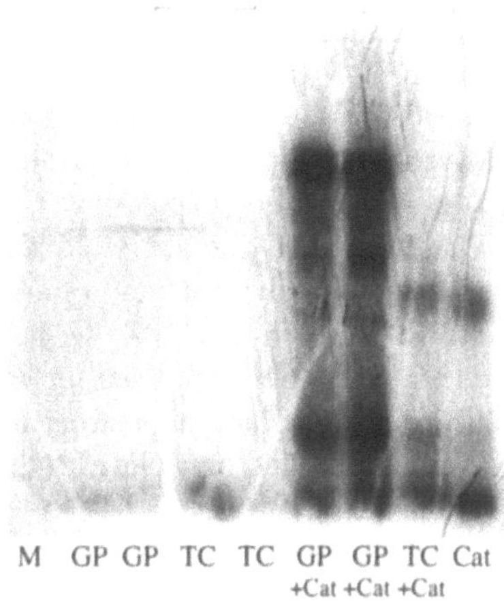


Figure 3: Autoradiogramm of the phosphorylation of GST-Prkx by cAPK

Abbreviations: **M** 10 kDa-Marker (GIBCO BRL), **C** crude lysate, **S** supernatant, **P** pellet fraction, **FT** flow through, **GA** glutathione-agarose bound fraction, **TCS** supernatant after thrombin cleavage, **TC** thrombin cleaved fraction, **Cat** catalytic subunit of cAPK (alpha subunit), **GP** GST-Prkx

## RESULTS AND DISCUSSION

Primary sequence analysis identified Protein Kinase X (Prkx) as a member of the protein kinase superfamily. The alignment of Prkx with different protein kinases revealed the highest similarity with the catalytic subunit of cAMP-dependent protein kinase especially the DC2-kinase of *Drosophila melanogaster*. Prkx and DC2 differ from the „classical“ cAMP-dependent protein kinases by their isoelectric points (table 1) and the position of their branches in the phylogenetic tree (figure 1) indicating a new subfamily of protein kinases in the AGC1-family of cyclic nucleotide regulated protein kinases (Hanks & Hunter 1995). The N-terminus of Prkx does not show any similarity to cAPK. However, we found a putative binding motif for a WW-domain [PPVY] whose consensus was shown to be PPxY (Macias et al. 1996). The expression of Prkx in *E.coli* as a GST-fusion protein yielded up to 70% soluble but inactive enzyme determined by a coupled spectrophotometric or a more sensitive radioactive assay using the most commonly used substrate for cAPKs, the heptapeptide Kemptide. The GST-fusion protein copurified with the heat shock protein 60 (hsp60) homologue GroEl from *E. coli* (60kDa-band, figure 2) that was identified by Edman sequencing. The GroEl could not be separated from Prkx by washing with high salt concentrations (1M NaCl), reducing agents (100mM DTT),

2mM ATP/10mM $MgCl_2$ or detergents (1% Triton X-100). Following cleavage of GST-Prkx bound to glutathione-agarose with thrombin the GroEl coelutes with the cleaved Prkx (figure 2) suggesting a tight binding to the kinase.

Prkx might be inactive because of the potential lack of important posttranslational modifications (phosphorylation on Thr197 is essential for activity in mammalian cAPKs) or the inhibitory effect of the copurified heat shock protein 60 (hsp60) homologue GroEl from *E. coli* that prevents binding of substrates and the regulatory subunits RI and RII of cAMP-dependent protein kinase.

In vitro phosphorylation experiments demonstrated that cAMP-dependent protein kinase is able to phosphorylate the inactive GST-Prkx (figure 3). Control experiments showed that the phosphorylation site is not located in the GST fusion part since the cleaved Prkx without GST is also phosphorylated. Finally, a coexpression system using two vectors expressing both Prkx and the catalytic subunit of cAPK in BL21.DE3 cells did not yield active enzyme. Because of the striking homology between the catalytic subunit of cAPK and Prkx an inactive protein was not expected, since in the case of cAPK we have a fully phosphorylated and catalytically active protein when overexpressed in *E. coli* (Herberg et al. 1993). It still has to be elucidated if the lack of posttranslational modification or the interaction of Prkx with the bacterial chaperonin GroEl is the reason for the inactive protein.

## LITERATURE


Adams JA & Taylor SS (1993) Effects of pH on the Phosphorylation of Peptide Substrates for the Catalytic Subunit of cAMP-dependent Protein Kinase. J. Biol. Chem. 268: 7747-7752

Cook PF, Neville ME, Vrana KE, Hartl FT & Roskoski J R. (1982) Adenosine Cyclic 3',5'-Monophosphate Dependent Protein Kinase: Kinetic Mechanism for the Bovine Skeletal Muscle Catalytic Subunit. Biochemistry. 21: 5794-5799

ExPASy World Wide Web (WWW) molecular biology server from the Geneva University Hospital and the University of Geneva, Compute pI/Mw tool

Grant BD, Tsigelny I, Adams JA & Taylor SS (1996) Examination of an Active-Site Electrostatic Node in the cAMP-dependent Protein Kinase Catalytic Subunit. Protein Science. 5: 1316-1324

Guan KL & Dixon JE (1991) Eukaryotic Proteins Expressed in Escherichia coli: an Improved Thrombin Cleavage and Purification Procedure of Fusion Proteins with Glutathione S-transferase. Anal. Biochem. 192: 262-267

Hanks SK & Hunter T (1995) Protein Kinases 6. The Eukaryotic Protein Kinase Superfamily - Kinase (catalytic) Domain Structure and Classification [Review]. Faseb Journal. 9: 576-596

Herberg FW, Bell SM & Taylor SS (1993) Protein Eng. 6, 771-777

Klink A, Schiebel K, Winkelmann M, Rao E, Horsthemke B, Luedecke H-J, Claussen U, Scherer G & Rappold G (1995) The human protein kinase gene PKX1 on Xp22.3 displays Xp/Yp homology and is a site of chromosomal instability. Hum. Mol. Genet. 4: 869-878


## ACKNOWLEDGEMENTS


We thank Dr. A. Beyer (Bochum) for an introduction into sequence alignment and building phylogenetic trees, E. Baraldi (Heidelberg) for helpful discussions concerning the WW-domain and K. Schiebel (Heidelberg) for discussions on the Prkx sequence.

# INTERACTION OF PROTEIN KINASE $A_c$ FROM *ASCARIS SUUM* WITH PROTEINS AND PEPTIDES: COMPARISON WITH THE MAMMALIAN ENZYME

S. Jung[1], R. Hoffmann[1], T. Treptau[1], P. H. Rodriguez[2], B. G. Harris[3], P.F. Cook[3], and H. W. Hofer[1,*]

[1]University of Konstanz, Germany, [2]Universidad de Chile, Santiago, Chile,[3]NTSU Health Science Center, Fort Worth, U.S.A.

Protein kinases belong to a single large family of proteins (Hanks *et al.*, 1988) and exhibit considerable similarity in primary and tertiary structure. While they are highly specific for ATP as phosphate donor (only a small minority uses GTP as an alternative), their limited *in vitro* specificity for protein substrates is remarkable with regard to the specialized functions of protein kinases in signal transduction, and cellular and metabolic regulation. Most obvious specificities attributed to certain subfamilies concern their ability to catalyze the formation of phosphate esters of either aliphatic or aromatic amino acid residues (or both) and the requirement for given recognition sites. Typically, these recognition sites are degenerate and comprise only a few amino acids in the vicinity of a serine, threonine or tyrosine residue, thus hardly providing elaborate tools of molecular recognition.

Elucidation of the three-dimensional structure of a ternary complex between the catalytic subunit of cAMP-dependent protein kinase (PKA), ATP, and the pseudo substrate PKI(5-24) (Knighton *et al.* 1991) provided a superb basis for the understanding of the enzyme substrate interactions within the catalytic groove of PkA and also suggested interactions with the polypeptide substrate exerted by a "hyrophobic pocket" in addition to those of the catalytic centre. Taking into account that the size of protein substrates is large compared to PKI(5-24), also other regions of the protein kinase may confer substrate specificity. The experimental detection of such regions by a functional approach using mammalian PkA, however, is hampered by extremely low variability in the primary structure natural kinases. Consequently, functional influences on substrate binding are below the limits of detection of kinetic assays.

In the course of a study on regulation of phosphofructokinase (PFK) in the muscle of the intestinal parasite *Ascaris suum* we had previously identified and characterized a form of PkA that exhibited clear differences in its specificity for protein substrates compared to mammalian PkA (Hofer *et al.* 1982, Daum *et al.*, 1986, Thalhofer *et al.* 1988). There is also a substantial difference in primary structure: The amino acid sequence derived from a cDNA clone is only ~48% identical with the enzyme from mammalian sources (Jung *et al.*, 1995).

* Author for correspondence.

NATO ASI Series, Vol. H 102
Interacting Protein Domains
Their Role in Signal and Energy Transduction
Edited by Ludwig Heilmeyer

Remarkably, the PkA from *Caenorhabditis elegans*, also a nematode, and therefore belonging to the same subfamily of helminths as *Ascaris*, was 89% identical with the mammalian enzyme. The enzyme from *Ascaris* is apparently a member of a subfamily different from that comprising the PkA of mammals and *C. elegans*. Based on the similarity of the mitochondrial genome (Okimoto *et al.*, 1992), the period of separate evolution of *Ascaris* and *C. elegans* was estimated to $7 \times 10^7$ years and, therefore, considerably less than the period separating nematodes and mammals. The apparent discrepancy would be resolved by the hypothetical assumption that PkA forms occurred during some stages in evolution which belonged to different subfamilies. Persisting examples may still occur as DC0, DC2 (and perhaps DC1) in *Drosophila*, and Apl-C, and Apl-B in *Aplysi*a. However, there is no indication that a PkA structurally belonging to the "mammalian" subfamily is present in *Ascaris* (Jung *et al.*, 1995), since unique fragments of restricted genomic DNA hybridized with a degenerated oligonucleotide encoding the hallmark sequence DFGFAK of PkA but were unreactive with mammalian cDNA. Moreover, the oligonucleotide probe also yielded a single signal of mRNA on northern blots and a single peak of protein was found in eluates from a DEAE-cellulose column reacting with an antiserum raised against the bacterially expressed protein. This peak was eluted in parallel with PkA activity and was used for purification of the enzyme.

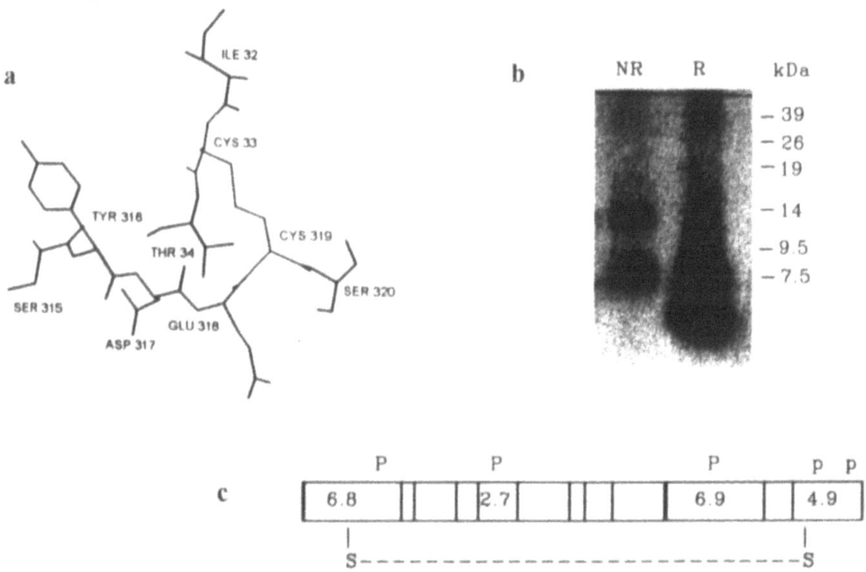


Fig. 1.
a. Energy minimized structures of the region formed by $Cys^{33}$ and $Cys^{319}$ of the *Ascaris* C-subunit in oxidized state. b. Separation of CNBr fragments of auto-phosphorylated PkA from *Ascaris* under reducing (R) and non-reducing (NR) conditions. c: Schematic representation of the location of methionins in the sequence of Ascaris PkA and the fragments resulting from CNBr cleavage. The suggested S-S bridge is indicated. P, presumptive autophosphorylation sites. The theoretical size of the fragments of the enzyme is 11.7, 6.9, and 2.7 kDa in the non-reduced (NR) and 6.9, 6.8 4.9, and 2.7 kDa in the reduced (R) state.

There are several important structural distinctions between *Ascaris* and mammalian PkA. The N-terminal 14 amino acids and hence the myristoylation site and the stabilizing A-helix of mammalian PkA are not contained in the enzyme from *Ascaris*. Two cysteins (Cys33 and Cys319) replace a basic residue in the N-terminal region and an acidic amino acid near the C-terminus which are conserved in all known C-subunits from other sources. The substitutions provide the possibility of disulfide bridge formation between the N-terminal and C-terminal parts of the protein (Fig. 1a). SDS-PAGE of BrCN fragments of non-reduced Ascaris PkA showed a ~11.9 kDa fragment which may represent the N-terminal 6.8 kDa peptide connected to the C-terminal 4.9 kDa peptide (Fig. 1 b). In fact, the 11.9 kDa fragment disappeared when the experiment was performed under reducing conditions and gave rise to a ~4.9 kDa peptide (Fig. 1c).

Modelling of the sequence into the coordinates of the X-ray structure of the mammalian enzyme (Knighton *et al.*, 1991) suggested a high degree of conservation in three-dimensional structure. The catalytic loop appeared exactly positioned as in mammalian protein kinase, although tethered by two non-conserved ß-strands. The coordinates of the amino acid side chains essentially conserved in all protein kinases, as derived from the X-ray structure of mouse PkA, neeeded no change in the model of *Ascaris* PkA. Distinct variations in the model structure, however, turned out at the surface of the protein near the catalytic cleft and are likely to account for variations in substrate specificity between the purified protein kinases from *Ascaris* and mammals.

Table 1: $K_m$ and V/K Values of PkA from *Ascaris* and Bovine Heart

| Substrate | $K_m$ [µM] | | relativeV/K | |
|---|---|---|---|---|
| | Bovine | Ascaris | Bovine | Ascaris |
| L R R A S L G ("Kemptide") | 20 | 19 | 1 | 1 |
| A K G R S D S* I V P T (derived from Ascaris PFK) | 55 | 54 | 0.4 | 0.4 |
| *Ascaris* PFK | 58 | 4.6 | 0.15 | 1.1 |

Phosphofructokinase is a predominant physiological substrate for PkA in the muscle of A. suum whose energy metabolism entirely depends on anaerobic glycolysis. Phosphorylation of a single serine residue lowers the $K_m$ for fructose 6-phosphate by a factor of 100 (Payne *et al.*, 1991). The serine is probably located close to the N-terminus and the encompassing sequence is conserved in other non-mammalian PFK whose activity is regulated by phosphorylation-dependent interconversion (Schaloske *et al.*, 1996). In contrast, phosphorylation occurring near the C-terminus in mammalian PFK does not significantly change the kinetic behaviour in solution. The phosphorylation of the mammalian enzyme was not catalyzed by Ascaris PkA, whereas Ascaris PFK was a better substrate for the protein kinase from the nematode in comparison to the mammalian protein kinase due to a 10-fold lower Michaelis constant (Table 1). Selective phosphorylation by the two kinases was also observed with some in vitro substrates. In addition, quantitative differences in the interactions between R- and C-subunits from Ascaris and bovine heart were observed. These differences were not obvious when short synthetic peptides were assessed as substrates. Peptides whose sequences reflected the phosphorylation site of Ascaris suum phosphofructokinase (AKGRSDS*IV), or variations of it, were phosphorylated with the same efficiency by both protein kinases (Treptau *et al.*, 1996).

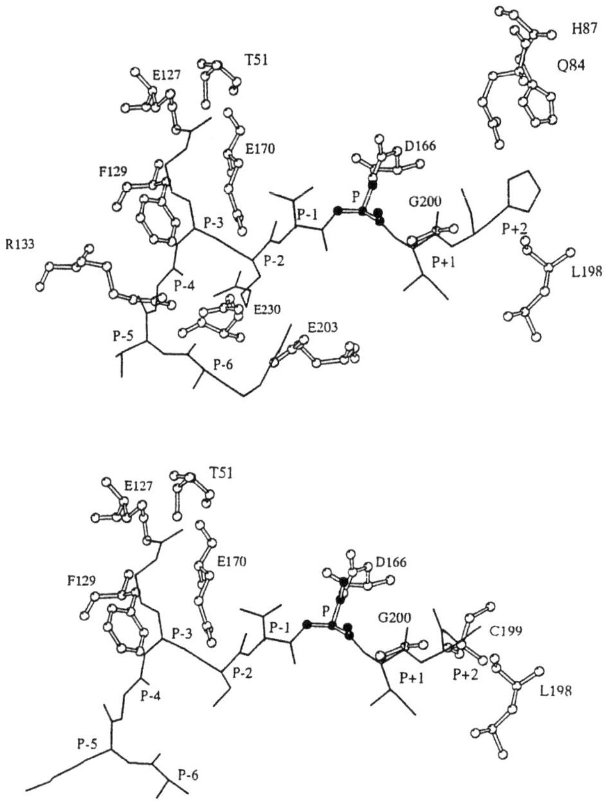


Fig. 2
Amino acid side chains of protein kinase A (stick and ball representation) located within less than 0.35 nm and likely to interact with LRRASLG (Top) and PFK-peptide (Lower Panel). Only the carbon skeleton of the substrate peptides is represented.

As shown by Fig. 2a, the binding of the peptide LRRASLG is stabilized by ionic interactions of the P-2 and P-3 positions with E230 and E127 and by hydrophobic interactions of the P+1 site with L198 and G200. These amino acids as well as their three-dimensional location are fully conserved in the *Ascaris* protein kinase. Fig. 2b outlines an energy-minimized model for the binding of the peptide representing the PFK phosphorylation site. The P+1 residue is also able to interact with L198 and G200 and the arginine at the P-3 site is close to E127 and E170. Yet, the ionic interaction of the arginine at the P-2 site of LRRASLG with E230 is lacking with the

PFK peptide thus explaining the reduced affinity and higher $K_m$. The geometry of the P site relative to the catalytic base (D166 in mouse and D152 in Ascaris PkA) and ATP (not shown), however, appears the same in mouse and *Ascaris* PkA. This fact may account for the almost identical $k_{cat}$ for LRRASLG and the PFK peptide.

The presence of a hydrophobic amino acid at the P+1 position in addition to the arginine at the P-3 position is an absolute requirement to confer substrate properties to the PFK peptide (Treprau *et al.*, 1996). The absence of the interaction at P-2 for the PFK peptide leads to a modest 2.5-fold increase in Km in comparison to the LRRASLG peptide. Evidently, two strongly interacting side chains of the substrate are sufficient to keep the serine in the proper position and interactions at different positions can replace one another at least to some extent.

The identical behaviour of *Ascaris* and mammalian protein kinases towards short peptides rules out the possibility that the difference in the catalytic efficiency for phosphofructokinase is caused by a variation in recognition at the catalytic groove. The X-ray structure of the PkA-PkI binary complex, however, also demonstrated interactions of a hydrophobic pocket of the protein kinase with the P-11 side chain. The sequence within this pocket is $Y^{235}$PPFF in the mammalian but $I^{221}$PPFR in the *Ascaris* enzyme. Structural differences, however, also exist at the surface next to the catalytic cleft and may cause steric effects or provide secondary attachment points for

**(a)** **(b)**

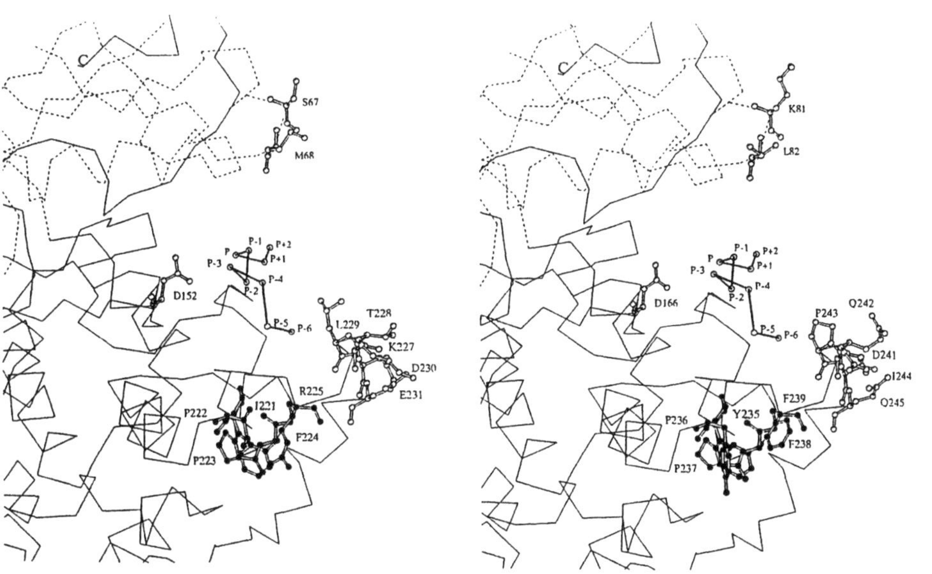


Fig. 3

Schematic representation of the location of peptide substrates in the catalytic cleft of protein kinase A from *Ascaris* (a) and mouse (b). The backbone of the protein kinases is partly shown by dashed lines (N-terminal part) and full lines (C-terminal part). Only the backbone of the P-6 to P+2 region of the substrate is shown. The catalytic base ($D^{152}$ in *Ascaris* and $D^{166}$ in the mouse PkA) is shown at the bottom of the catalytic cleft.

larger peptides and proteins. For example, $K^{81}L^{82}$ at the small lobe of the bovine and mouse protein kinases is substituted by $S^{67}M^{68}$ in the *Ascaris* protein kinase and the sequence $Q^{242}P^{243}I^{244}$ forming the surface of the catalytic cleft of the mammalian enzyme at the big lobe is $T^{228}L^{229}D^{230}$ in Ascaris (Fig. 3). For physiological substrates such differences in charge and hydrophobicity may create differential selectivity.

The functional importance of PkA, like that of other protein kinases, is stressed by conservation during evolution of their fundamental properties affecting the overall molecular architecture, recognition, and binding of substrates in the catalytic centre, and supposedly also the details of the catalytic process itself. On the other hand, rising complexity of organisms led to an increase in the number of protein kinases and their isozymes present in a cell concommitant with the demand for more subtle regulation. It is likely that such an evolution has occurred for protein kinases and their substrates in an interdependent manner. The amino acids on the surface of the molecule are probably subjected to less stringent constraints required by molecular architecture and basic catalytic function and substitutions are more liberally allowed. PkA and PFK in *Ascaris* present themself as proteins that have undergone mutual functional optimization by development: PFK has gained the profit of regulation in response to cAMP (whose concentration is regulated by the neuromodulator serotonin and possibly other extracellular stimuli) and PkA improved its catalytic efficency for this substrate, apparently without forfeit in its capacity to regulate other processes.

References

Daum G., Thalhofer, H.P., Harris, B.G., Hofer, H.W. (1986) . Biochem. Biophys. Res. Commun. 139, 215-221.

Hanks, S.K., Quinn, A.M., Hunter, T. (1988) Science 241, 42-52.

Hofer, H.W., Allen, B.L., Kaeini, M.R., Harris, B.G. (1982), J.Biol.Chem. 257, 3807-3810.

Jung, S., Hoffmann, Rodriguez. P., Mutzel,R. (1995), European J. Biochem. 232, 111-117.

Knighton, D.R., Zheng, J., Ten Eyck, L.G., Ashford, V.A., Xuong, N., Taylor, S.S., Sowadski, J.M. (1991) . Science 253, 407-420.

Knighton, D.R., Zheng, J., Ten Eyck, L.G., Xuong, N., Taylor, S.S., Sowadski, J.M. (1991) Science 253, 414-420.

Kulkarni, G., Rao, G.S.J., Srinivasan, N.G., Hofer, H.W., Yuan, P.M., Harris, B.G. (1987) J.Biol.Chem. 262, 32-34.

Okimoto, R., Macfarlane, J.L., Clary, D.O., Wolstenholme, D.R. (1992) Genetics 130, 471-498.

Payne, M.A., Rao, G.S.J., Harris, B.G., Cook, P.F. (1991) J. Biol. Chem. 266, 8891-8896.

Schaloske, R., Biethinger, M., Fothergill-Gilmore, L.A., Hofer, H.W. (1996) Biochem. J. 317, 377-383.

Thalhofer, H.P., Daum, G., Harris, B.G., Hofer, H.W. (1988) J.Biol.Chem. 263, 952-957.

Treptau, T., Piram, P., Cook, P.F., Rodriguez, P.H., Hoffmann, R., Jung, S., Thalhofer, H.P. Harris, B.G., Hofer, H.W. (1996) Biol. Chem. Hoppe-Seyler 377, 203-209.

# Part II

# Methodology

# Interaction Studies Using Biosensors

Ariane Maleszka and Friedrich W. Herberg

Ruhr Universität Bochum, Physiologische Chemie, MA 2/40, 44780 Bochum, Germany

## Introduction

Novel technologies employing surface plasmon resonance (BIAcore, Pharmacia/ Biosensor) or resonance mirrors (IAsys, Fisons Instruments) can be used for the characterization of protein-protein, protein-DNA and ligand-receptor interactions (for review see Szabo, *et al*, 1995; Raghavan *et al.*, 1995) by measuring directly the binding of an analyte to a ligand immobilized on a sensor surface. The chemistries for coupling molecules to the surface via amine, sulfhudryl, carboxyl, and other groups are well defined and reproducible. Non covalent, site directed immobilization can be achieved by avidin-biotin interaction, by the use of antibodies or by binding fusion proteins to specific surfaces (Gershon and Khilko, 1995). Surfaces containing lipid monolayers are available, too (Soulages *et al.*, 1995).

The BIAcore system consists out of a light source emitting polarized light, a sensor microchip, an automated liquid handling system (constant flow) and a diode array position-sensitive detector. When light shines on an interface between two transparent media in an angle of total internal reflection, an electromagnetic field component called the evanescent wave penetrates a short distance into the medium of lower refractive index. If the interface between the media is coated with a thin layer of metal a phenomena, referred as surface plasmon resonance (SPR) occurs, resulting in an intensity dip in the reflected light. The angle in which SPR occurs is dependent on the refractive index of the media of lower refractive index, which is directly proportional to mass changes close to the metal layer.

Changes in mass on the surface of a biosensor chip/cuvette are proportional to changes in the refractive index. These are detected by the biosensor, which makes the system suitable to follow ligand-analyte interaction. The functional characteristics of biological interactions, such as kinetics, affinity and binding topology (epitopes, antibodies), are analyzed by monitoring separately the association and dissociation rate constants. Principly the method can also be used qualitatively: for example to determine relative binding affinities, for ligand fishing or to detect expression of recombinant proteins.

The binding of an analyte to a ligand under constant flow can be regarded as a 'pseudo' first-order reaction since the concentration of analyte is constant in the flow cell. Using a series of different concentrations of analyte, the association rate constants can be calculated. All interactions in the matrix of the sensor surface occur in a diffusion limited environment resembling closely in vivo interactions. Applications of this technology include probing for the importance of dimerization for function (Cunningham and Wells, 1993) unwinding signal transduction pathways and analyzing macromolecular complexes (Schuster *et al.*, 1993; Raghavan *et al.*, 1995) or

NATO ASI Series, Vol. H 102
Interacting Protein Domains
Their Role in Signal and Energy Transduction
Edited by Ludwig Heilmeyer

measuring binding constants for protein domain interactions (Ladbury *et al.*, 1995; Felder *et al.*, 1993)

The cAMP dependent protein kinase (cAPK) offers an excellent system to apply Surface Plasmon Resonance. cAPK is a heterotetramer containing two regulatory (R) and two catalytic (C) subunits. Each R- subunit contains two tandem cAMP-binding domains, and the dissociation of the inactive holoenzyme complex is mediated by the cooperative, high affinity binding of cAMP to these two domains (Herberg *et al.*, 1996) promoting dissociation of the complex into a dimeric R-subunit and two monomeric, active C-subunits. Several forms of the R-subunit exist and are classified as type I and type II (Beebe and Corbin, 1986; Taylor *et al.*, 1990). The C - and the R -subunits could be immobilized readily to the carboxy methyl dextran of the surface. The association and dissociation rates of the other subunit could then be monitored. After each binding event the surface with the bound R or C-subunit was regenerated physiologically, simply by injecting cAMP onto the surface (Herberg *et al.*, 1994). This test system was used to investigate the interaction between mutant forms of the R- and/or C-subunit and between ligands and protein domains of this protein kinase. The work presented here focuses on the application of different immobilization strategies also employing site directed immobilization of the subunits of cAPK via GST- and poly-his fusion parts or via specific ligand, here cAMP.

## Material and Methods

**Protein Purification:** Following overexpression in *E. coli* BL21/DE3 (Taylor *et al.*, 1989) the cells were lysed in buffer A (30 mM MES, 1 mM EDTA, 50 mM KCl, 5 mM b-mercaptoethanol, pH 6.5). The C-subunits were purified by phosphocellulose chromatography (P11 Whatman) as described previously (Yonemoto *et al.*, 1991) and were stored at 4°C in buffer A or further purified using Mono S chromatography as described previously (Herberg *et al.*, 1993). Typical yields were 10 mg/l of cell culture.

Recombinant $R^I$-subunits were overexpressed in *E. coli* 222, purified as described previously (Saraswat *et al.*, 1986), and stored at -20$^o$C. To obtain cAMP-free R-subunit, the R-subunit was unfolded with 8 M urea and refolded in buffer B (20mM MOPS, 150mM KCl, 1mM DTT, pH 7.0) as described by Buechler *et al.*, 1993.

GST-fusion proteins, overexpressed in *E. coli* BL21/DE3 (Herberg *et al.*, in press 1996), were purified according to the procedure of Smith and Johnson (Smith and Johnson, 1988). Following overexpression in *E. coli* BL21/DE3 poly-his proteins were bound to a $NI^{++}$-NTA fast flow Sepharose (Quiagen) according to the procedure described by the manufacturer and eluted with 120mM imidazole.

**Surface Plasmon Resonance:** All experiments were performed on a BIAcore 2000 system (Pharmacia/Biosensor) using different immobilization chemistries as described in Results. To obtain association and dissociation rate constants and to calculate apparent $K_D$'s, the interactions between analyte and ligand were measured at concentrations between 25 and 500 nM protein in buffer B. Experiments were performed in buffer B + 0.005% detergent P20 in the presence or absence of 1mM ATP 5mM $MgCl_2$ as indicated using a flow of 5 or 10 µl/min. In the association phase the interactions were monitored for 350 seconds. The dissociation phase was then monitored for 15 minutes with a concentration of 500 nM protein. The following sensor chips were used: CM 5 research grade, SA 5 with covalently immobilized streptavidin, and NTA, research grade.

For the use of NTA chips the BIAcore system was modified according to the instructions using as an eluent buffer 10mM HEPES, 0.15 mM KCl, 50μM EDTA, 0.005% P20, pH 7.4 or in buffer B +1mM ATP, 5 mM MgCl2. As a dispenser buffer 10 mM HEPES, 0.15 mM NaCl, 3.4 mM EDTA, 0.005% P20, pH 7.4. The surface was prepared by injecting 20 μl of 500 μM NiCl2 in eluent buffer.

8-AHA-cAMP was immobilized by amine coupling to a CM 5 chip. Therefore 20 μl of 3 mM 8-AHA-cAMP in 10 mM HEPES pH 8 were injected over a surface activated for 7 minutes. This injection was repeated twice and a total of 91 RU's was immobilized after injection of 1 M ethanolamine for 7 minutes corresponding to a surface concentration of 4 mM 8-AHA-cAMP. This calculation was based on the assumption that 800 RU corresponded on a surface concentration of $1 ng/mm^2$.

Calculations: Kinetic constants were calculated by linear regression of data using the pseudo first order rate equation, dR/dt = kaCRmax-(kaC+kd)Rt ka is the association rate, kd is the dissociation rate C is the concentration of the injected analyte and R is the response. Plots of dR/dt vs Rt have a slope of ks. When ks is plotted against C the resulting slope is equal to the ka. kd was calculated by integrating the rate equation when C = 0 yielding ln(Rt1/Rtn = kd (tn-t1). Affinity constants were calculated from the equation KD=kd/ka. Nonlinear fits were used fitting both monophasic and biphasic rate constants for the association and dissociation phases using the software Biaevalution Version 2.1.

## Results and Discussion

BIAcore allows several principle strategies of immobilization of a analyte molecules to the surface of a sensorchip: covalently to carboxymethylated dextran matrix (CM 5 chip) via i.e. primary amines, thiol groups and aldehydes, and non-covalently via a "capture molecule" (Szabo *et al.*, 1995) via a lipid monolayer (HPA-chip) or a specific ligand. The immobilization using primary amines is a simple, broadly applicable method and binds an analyte protein either via the free N-terminus or Lysines in a more or less random manner. If a defined orientation is required in some cases cysteine residues can be introduced to specific sites and thiol coupling can be applied. Glutathione-S transferase (GST) fusion proteins or poly-his tags can be utilized as capture molecules, allowing a site directed binding to a specific antibody (i.e. anti-GST) covalently immobilized to a sensor chip or a specific surface ($Ni^{++}$-NTA). A novel approach is the "capture" of the R- subunit of cAPK on covalently immobilized 8-AHA-cAMP (Biolog, Bremen) using a CM5 chip.

**Direct coupling** Direct coupling of the rC-subunit via primary amines to the CM dextran was achieved by activating the surface of the chip for 2 minutes with NHS/EDC according to the manufacturers instructions and by injection of 15 μl of protein was 7 μg/ml in 10 mM sodium phosphate (pH 6.2) containing 1 mM ATP and 2 mM $MgCl_2$ and 425 RU's were immobilized (1000 RU = 1 $ng/mm^2$). Residual NHS-esters on the sensorchip surface were reacted with ethanolamine (1 M in water, pH 8.5). Non-covalently bound protein was washed off with 500 mM NaCl. All binding interactions were performed at 20° C in buffer B ±1 mM ATP, 5 mM $MgCl_2$. To determine unspecific binding and bulk refractive index changes blank runs were performed with 500 nM R-subunit using a non-activated sensor chip surface. After injections of the R-subunits the C-subunit surface was regenerated by injection 10 µl of 100 µM cAMP, 2.5 mM EDTA in buffer B. To immobilize the C-subunit in a physiologically active conformation the addition of MgATP during the immobilization process appeared to be important. Surface activity of the rC-subunit (=stoichiometry of binding of the R-subunit) was increased from 30% to 88% (311 RU's) for the binding of a R-subunit monomer (mutation Δ1-91$R^I$, (Herberg *et al*,. 1994)) assuming a stoichiometric relationship of 1:1 in the absence or presence of 1 mM ATP and 2 mM $MgCl_2$, respectively. This may be due to blocking of the active site Lysine, Lys72, (Zoller and Taylor, 1979) from coupling to the matrix. R-subunit dimer bound 525 RU's corresponding to 55% of the calculated maximum for the surface assuming 1:1 binding. If the dimer is binding monovalently so that only one R protomer is free to bind on immobilized rC-subunit, then this may correspond to about 100%. Surface activity was calculated using the equation: $S = MW_L R_A / MW_A R_L$ where S is the stoichiometry, $_L$= ligand(immobilized protein), $_A$= analyte (injected protein), R= response in RU's, MW= molecular weight.

To immobilize the R-subunit, a surface was activated for 6 minutes, protein was injected at 10µg/ml in 10mM acetate buffer (pH 3.8). A total of 500 RU's of a mutant of the $R^I$-subunit, $R^I$R209K was immobilized (Herberg *et al.*, 1994). In contrast to the wild type R-subunit this mutant R-subunit forms holoenzyme instantaneously and is dissociated by micromolar concentrations of cAMP (Herberg *et al.*, 1996) Figure 1 shows the interaction of the wild-type recombinant C-subunit with immobilized R209K $R^I$-subunit (for apparent binding constants see also table 1).

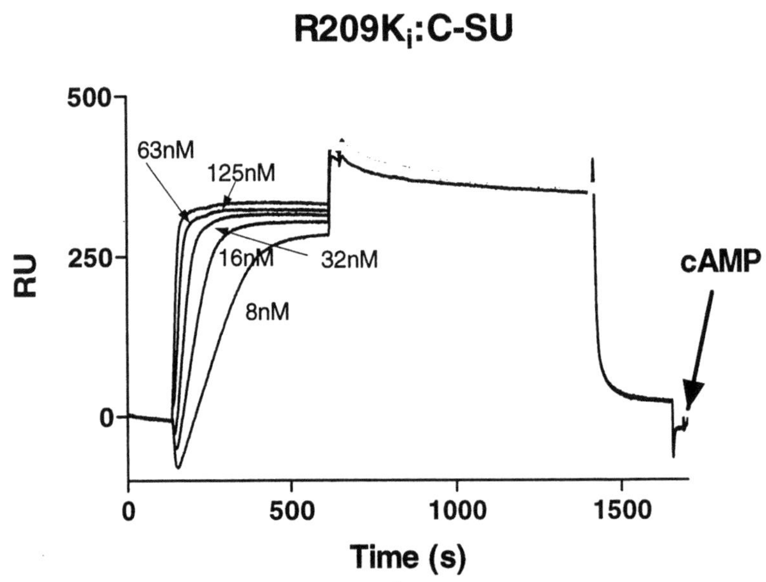


**Figure 1** A mutant form of the $R^I$-subunit (R209K) was immobilized by amine coupling and the interaction of recombinant C-subunit was monitored. Regeneration was performed by addition of 100µM cAMP, 2.5mM EDTA in running buffer.

In some cases (acetic proteins) a low pH is required to concentrate proteins on the CM 5 sensor chip, however, because of the decreased efficiency of the coupling chemistry (amine- or thiol coupling) only low amounts of protein can be immobilized. These chemistries can be performed before coupling to the sensor chip with biotinylated agents at higher pH followed by binding to a streptavidin-coated sensor chip as demonstrated for cardiac troponin C (Reiffert *et al.*, 1996 ).

**Site directed coupling**

**1. GST-fusion proteins**: Immobilization of the αGST-antibody was performed as recommended by the Biosensor GST-coupling kit. A total of 6000 RU of the antibody were immobilized. A different mutant form of the R-subunit (GSTΔ1-45/R209K) was bound by injecting 20 μl of 5 μg/ml 900 RU's were bound to the surface. With a baseline drift 0.6 RU/min this surface was relatively stable, allowing to investigate the binding of several concentrations of C-subunit and regeneration with cAMP, as described before, without adding again GSTΔ1-45/R209K. However, the stoichiometry of binding was only 20%. Figure 2 shows the interaction of C-subunit between 12 and 250 nM with the immobilized GSTΔ1-45/R209K after normalization of the baseline. The inset in figure 2 shows the drift after each cycle of 15 minutes. Apparent binding constants are listed in table 1.

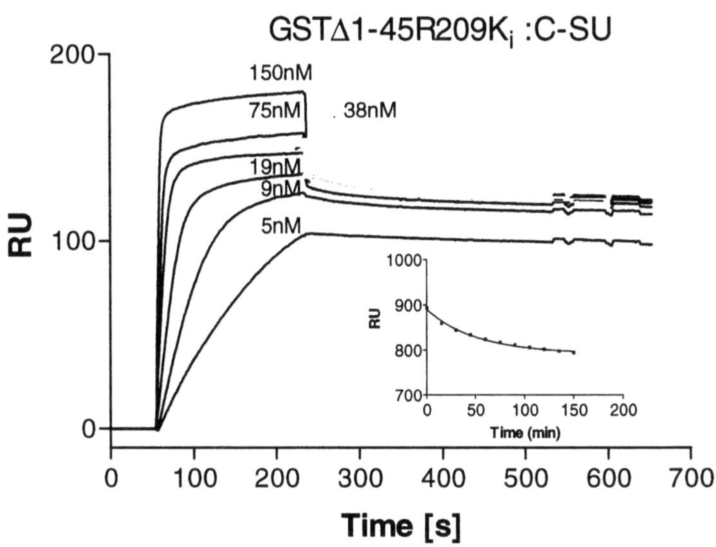


**Figure 2** Interaction of a fusion captured $R^I$-subunit (GSTΔ1-45/R209K) with recombinant C-subunit. The baselines are normalized; the inset shows the amount of $R^I$-subunit after each run.

**2. Poly-his fusion proteins**: A NTA sensor chip was used to couple poly-his R-subunit ($R^{II}$R213K, an analog mutant protein to $R^I$R209K). After preparation of the surface, $R^{II}$R213K corresponding to 800 RU's were immobilized. After 2 pulses of 1 min injections of 500μM NiCl2 330 RU's remained on the surface. C-subunit was injected on this surface between concentration from 18 nM to 300 nM (figure 3) However, the stability of the surface was rather poor (- 0.5 - 1 RU /min ). Although the on rates were comparable to those of the $R^I$R209K mutant protein, the off-rates were one order of magnitude faster. If this faster off-rate is due to an accelerated kinetic of the $R^{II}R213K$ mutant protein or due to a faster dissociation of the poly-his fusion complex remains to be clarified. Figure 3 shows the interaction of the C-subunit with the immobilized $R^{II}R213K$ protein before (inset) and after normalization of the baseline.

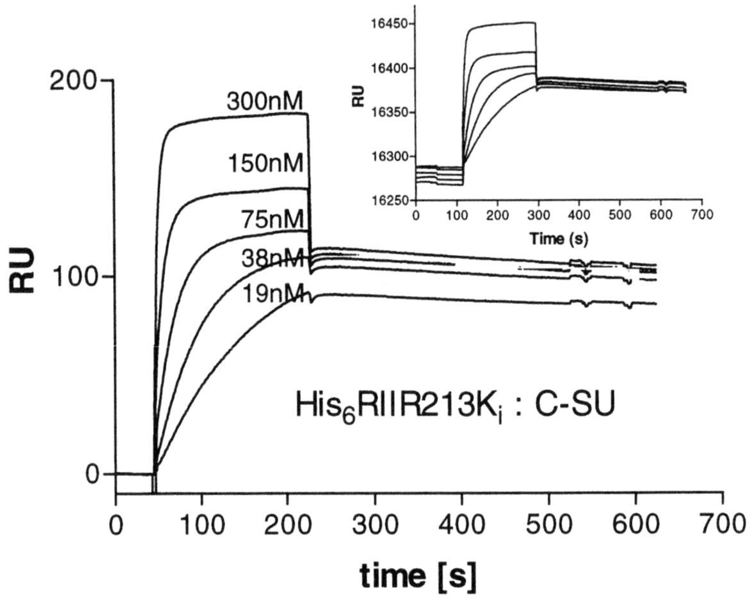


**Figure 3** Interaction of C-subunit of indicated concentrations with $His_6R^{II}/R213K$ captured on a NI++-NTA chip. The data were normalized on one baseline; the inset shows the same data not normalized.

| Immobilized ($_{im}$) | $_{app}K_D$ (nM) | $_{app}k_{ass}$ ($M^{-1}s^{-1}$) ± SE | $_{app}k_{diss}$ ($s^{-1}$) ± SE |
|---|---|---|---|
| $R^IR209K_{im}$/C-SU | 0.073 | $2.6 \times 10^6 \pm 3.4 \times 10^5$ | $1.9 \times 10^{-4} \pm 5.9 \times 10^{-6}$ |
| $His_6$-$R^{II}R213K_{im}$/C-SU | 0.75 | $5.7 \times 10^5 \pm 8.1 \times 10^4$ | $4.3 \times 10^{-4} \pm 9.2 \times 10^{-5}$ |
| GST-$R^IR209K_{im}$/C-SU | 0.055 | $2.2 \times 10^6 \pm 3.8 \times 10^5$ | $1.2 \times 10^{-4} \pm 4.2 \times 10^{-5}$ |
| $C$-$SU_{im}/R^IR209K$ | 0.036 | $1.4 \times 10^6$ | $5.1 \times 10^{-5}$ |

**Table 1** Apparent rate- and binding constants for the interaction of type I and type II R-subunits with the C-subunit.

**3. cAMP**: 8-AHA-cAMP was immobilized by amine coupling to a CM 5 chip as described in Material and Methods. Surface activity was tested by injecting 30 µl of 400 nM cAMP stripped $R^I$-subunit at a flow rate of 5 µl/min over the surface yielding 3249 RU's with a standard error of the mean of ±3.4 RU's (n=7). This number corresponds to 20 mg/ml and to 74 % stoichiometry assuming the binding of 2 cAMP molecules per dimer R-subunit. As shown in figure 4 the interaction of the R-subunit with the cAMP matrix was strictly mass transport controlled. This allowed to quantify binding of different amounts of R-subunit as shown in figure 4. In contrast to the $Ni^{++}$-NTA- and the αGST-surface the stability of the surface was extremely high with an off rate in the range of $10^{-7}sec^{-1}$. Even after 20 hours of running time still 80 % of the original amount of $R^{I}$-subunit was present (data not shown). This allowed interaction studies with A-Kinase binding proteins (AKAP's), which bind the type II R-subunit (Tasken and Herberg, in preparation). Surprisingly the R-subunit could not be competed away readily with cAMP. Even an injection for 4 minutes using 10 mM of cAMP would displace only 10 % of the R-subunit from the cAMP-surface.

Even more surprisingly, the binding of C-subunit did not increase the response due to the additional mass added, but instantaneously took off the R-subunit from the surface. This effect was amplified by the addition of MgATP and cAMP suggesting a physiological competition of the C-subunit with the bound R-subunit (Herberg *et al.*, in preparation). However, this regeneration was not completely efficient. Finally 5 µl of 0.05% SDS removed all bound R-subunit.

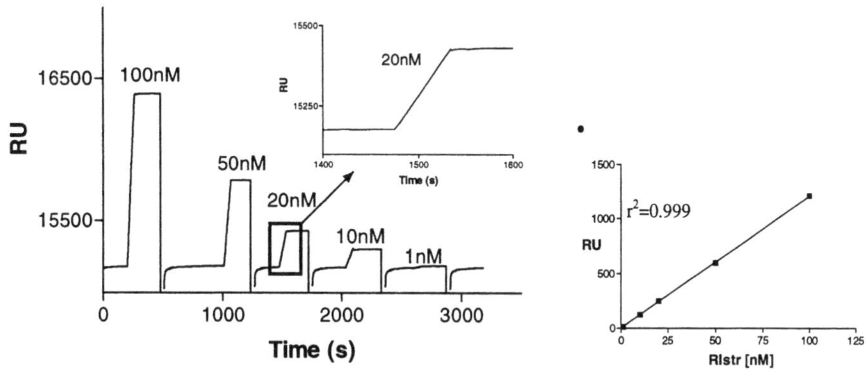


**Figure 4** Injection of cAMP-stripped RI-subunit (concentrations as indicated) on a chip with immobilized 8-AHA-cAMP. The inset (middle) clearly demonstrates that the interaction is purely mass transport controlled. The plot on the right shows the absolute response taken after each injection plotted against the concentration of the $R^{I}$-subunit.

Although the reproducibility of each single interaction study is extremely good as demonstrated above, the absolute numbers derived from the association and dissociation phases have to be evaluated carefully. This becomes apparent when data of "vice/versa"-experiments are compared. Here first one of the interacting partners is immobilized (R-subunit) and the binding of the other partner (C-subunit) is monitored and then other way around. By comparing the numbers in table 1 it becomes clear that, although the association and dissociation rate constants differ significantly, the apparent KD is very similar. The differences in the rate constants might be due to different degrees of freedom for the immobilized partner or for several other reasons as discussed below. SPR should be accompanied with other methods preferably in free solution to check if the apparent numbers reflect a realistic interaction in vivo. This has been done for the R/C-interaction using analytical gel filtration (Herberg *et al.*, 1993), activity assays (Hofmann, 1980), calorimetric studies (S. Cox, personal communication), or fluorescense studies (SS. Taylor, personal communication). Analytical ultracentrifugation is another technique, which could be employed. Resonance mirrors have been used to characterize the interaction of peptide motifs of the C-subunit in interaction with the R-subunit (Sahara *et al.*, 1996).

There are several factors which could have a major impact on the data observed by SPR. Aside of the immobilization chemistry, this could be mass transfer, fluid dynamic and volumetric effects, „crowding" on the surface of the chip, as well as the valence and inhomogeneity of the ligand or the analyte. Therefore an important factor is the surface concentration of the ligand. Principally lower immobilization levels are preferable yielding more reproducible results, less mass transport effects and generally lower tendency of rebinding. Often higher affinities are obtained at lower immobilization levels (Ladbury *et al.*, 1995) and the association and dissociation phases of interaction studies performed can more readily explained with simple models of interaction. However, the number of possible binding sites and the choice of a model for these interactions have to be chosen carefully. A lot of these problems could be avoided by using solution competition assays (Nieba *et al.*, 1996).

One main problem of Biosensors often in the lack of proper regeneration conditions. As demonstrated above the non-covalent immobilization methods offer a solution because after each binding event the whole complex is removed from the surface a sensor chip. This allows repetitive measurements with variable concentrations and different interacting partners on the same surface.

SPR has been prooven to be a powerful technique for the determination of relative association and dissociation rate constants by evaluating the effects of single or multiple amino acid substitution introduced by site directed mutagenesis. Here the focus lies on the combination of the effects of site directed mutagenesis on protein-protein interaction. Even with the know three dimensional structure of the C-subunit (Knighton *et al.*, 1993) and the R-subunit (Su *et al.*, 1995) the interaction sites have to be mapped out (R. Gibson and SS Taylor in preparation). Other experiments focus on the roles of nucleotides and divalent cations on subunit interaction.

Acknowledgments: We like to thank Bastian Zimmermann and Marco De Stefano for technical support and LMG Heilmeyer, SS Taylor and C. Schulz for helpful discussions.

**Abbreviations:** AKAP, A-kinase anchor protein; ATP, Adenosine triphosphate; C, catalytic subunit of cAPK, cAMP, adenosine 3'-,5'- cyclic monophosphate; 8-AHA-cAMP, 8-(6-Aminohexyl)aminoadenosine-3'-,5'- cyclic monophosphate; CAP, catabolic gene activating protein; cAPK, cAMP-dependent protein kinase; DTT, dithioerythritol; EDC, N-ethyl-N'-(3-diethylaminopropyl)-carbodiimide; EDTA, N,N,N',N'-Ethylenediamine tetra acetic acid; HEPES ,N-[2-Hydroxyethyl]-piperazine-N'-[2-ethanesulfonic acid]; MOPS, 3[N-Morpholino]propane sulfonic acid; NHS N-hydroxysuccinimide; PKI, heat stable protein kinase inhibitor; R, regulatory subunit of cAPK; $R^I$, $R^{II}$, type I and type II, respectively, of the R-subunit of cAPK; RU, response units (1000 RU = 1 ng/mm$^2$); SDS, sodium dodecyl sulfate; SPR, surface plasmon resonance.

**References:**

Beebe SJ & Corbin JD (1986) Cyclic Nucleotide-Dependent Protein Kinases. In *The Enzymes: Control by Phosphorylation Part A* (Krebs EG and Boyer PD, eds) pp 43-111, Academic Press, Inc. New York

Buechler YJ, Herberg FW & Taylor SS (1993) Regulation Defective Mutants of Type I cAMP-dependent Protein Kinase: Consequences of Replacing Arg94 and Arg95. J. Biol. Chem. 268: 16495-16503

Cunningham BC & Wells J (1993) Comparison of a Structural and Functional Epitope. J. Mol Biol. 234: 554-563

Felder S, Zhou P, Hu J, Urena J, Ullrich A, Chaudhuri M, White M, Shoelson S & Schlessinger J (1993) SH2 Domains Exhibit High-Affinity Binding to Tyrosine-Phosphorylated Peptides Yet also Exhibit Rapid Dissociation and Exchange. Mol. Cell. Biol. 13: 1449-1455

Gershon PD & Khilko S (1995) Stable chelating linkage for reversible immobilization of oligohistidine tagged proteins in the biacore surface plasmon resonance detector. Journal of Immunological Methods. 183: 65-76

Herberg F, Taylor S & Dostmann W (1996) Active site mutations define the pathway for the cooperative activation of cAMP-dependent protein kinase. Biochemistry. 35: 2934-2942

Herberg FW, Bell SM & Taylor SS (1993) Expression of the Catalytic Subunit of cAMP-dependent Protein Kinase in *E. coli*: Multiple Isozymes Reflect Different Phosphorylation States. Prot. Eng. 6: 771-777

Herberg FW, Dostmann WRG, Zorn M, Davis SJ & Taylor SS (1994) Crosstalk between Domains in the Regulatory Subunit of cAMP-dependent Protein Kinase: Influence of the Amino-Terminus of cAMP-binding and Holoenzyme Formation. Biochemistry. 33: 7485-7494

Herberg FW & Taylor SS (1993) Physiological Inhibitors of the Catalytic Subunit of cAMP-dependent Protein Kinase: Effect of MgATP on Protein/Protein Interaction. Biochem. 32: 14015 - 14033

Herberg FW, Zimmermann B, McGlone MM & Taylor SS (in press 1996) Importance of the A-helix of the Catalytic Subunit of cAMP-dependent Protein Kinase for Stability and for Orienting Subdomains at the Cleft Interface. Protein Science.

Hofmann F (1980) Apparent Constants for the Interaction of Regulatory and Catalytic Subunit of cAMP-Dependent Protein Kinase I and II. J. Biol. Chem. 255: 1559-1564

Knighton DR, Zheng J, Ten Eyck LF, Ashford VA, Xuong N-h, Taylor SS & Sowadski JM (1991) Crystal Structure of the Catalytic Subunit of cAMP-dependent Protein Kinase. Science. 253: 407-414

Ladbury JE, Lemmon MA, Zhou M, Green J, Botfield MC & Schlessinger J (1995) Measurement of the binding of tyrosyl phosphopeptides to sh2 domains - a reappraisal. Proceedings of the National Academy of Sciences of the United States of America. 92: 3199-3203

O'Shannessy DJ & Brigham-Burke M (1993) Determination of rate and equilibrium constants for macromolecular interactions using surface plasmon resonance: Use of nonlinear least square analysis methods. Anal. Biochem. 212: 457-468

Raghavan M, Bjorkman P, Bjorkman PJ, Hughes P & Hughes H (1995) Biacore - a microchip-based system for analyzing the formation of macromolecular complexes. Structure. 3: 331-333

Reiffert S, Jaquet K, Heilmeyer LMG jr, & Herberg FW (in preparation, 1996)

Sahara S, Sato K, Kaise H, Mori K, Sato A, Aoto M, Tokmakov AA & Fukami Y (1996) Biochemical evidence for the interaction of regulatory subunit of camp-dependent protein kinase with ida (inter-dfg-ape) region of catalytic subunit. Febs Letters. 384: 138-142

Saraswat LD, Filutowics M & Taylor SS (1986) Expression of the Type I Regulatory Subunit of cAMP-Dependent Protein Kinase in Escherichia coli. J. Biol. Chem. 261: 11091-11096

Schuster SC, Swanson RV, Alex LA, Bourret RB & Simon MI (1993) Assembly and Function of a Quartternary Signal Transduction Complex Monitored by Surface Plasmon Resonance. Nature. 365: 343-347

Smith DB & Johnson KS (1988) Single-Step Purification of Polypeptides Expressed in Escherichia coli as Fusions with Glutathione S-transferase. Gene. 67: 31-40

Soulages JL, Salamon Z, Wells MA & Tollin G (1995) Low concentrations of diacylglycerol promote the binding of apolipophorin iii to a phospholipid bilayer - a surface plasmon resonance spectroscopy study. Proceedings of the National Academy of Sciences of the United States of America. 92: 5650-5654

Su Y, Dostmann W, Herberg FW, Durick K, Xuong NH, TenEyck L, Taylor SS & Varughese KI (1995) Regulatory subunit of protein kinase A: structure of a deletion mutant with cAMP binding domains. Science. 269: 807-813

Szabo A, Stolz L & Granzow R (1995) Surface plasmon resonance and its use in biomolecular interaction analysis (bia). Current Opinion in Structural Biology. 5: 699-705

Taylor SS, Buechler J, Slice L, Knighton D, Durgerian S, Ringheim G, Neitzel J, Yonemoto W, Dostmann W & Sowadski J (1989) cAMP-dependent Protein Kinase: A Framework for a Diverse Family of Enzymes. Cold Spring Harbor Symposium. 53: 121-130

Yonemoto W, McGlone ML, Slice LW & Taylor SS (1991) Prokaryotic Expression of the Catalytic Subunit of cAMP-dependent Protein Kinase. In *Protein Phosphorylation (Part A)* (Hunter T and Sefton BM, eds) pp 581-596, Academic Press, Inc. San Diego

Zoller MJ & Taylor SS (1979) Affinity Labeling of the Nucleotide Binding Site of the Catalytic Subunit of cAMP-dependent Protein Kinase Using *p*-Fluorosulfonyl-[$^{14}$C]benzoyl 5'-Adenosine: Identification of a Modified Lysine Residue. J. Biol. Chem. 254: 8363-8368

# Advances in Determination of Protein Structure by X-ray Diffraction Methods

L. N. Johnson

Laboratory of Molecular Biophysics, University of Oxford, South Parks Road, Oxford OX1 3QU, UK.

X-ray analysis of protein crystals provides a wealth of detailed information on molecular structure and intermolecular interactions. Such information is essential for understanding the biological function of macromolecules. In the early days, protein crystallography was laborious and uncertain. The situation is different today. In the last decade the technique has seen many spectacular technical advances that allows a new structure to be determined in a matter of weeks or, if the protein has a fold similar to that which already exists in the Protein Data Bank, the structure can be solved within days. Of course this is the best case scenario. There are still some proteins that present special problems. In 1986 there were less than 300 protein structures in the Protein Data Bank; by the end of 1995 there were more than 4000 and in 1995 alone 1269 entries were received for deposition. This lecture identifies some of the advances that has led to this increased efficiency in the solution of protein crystal structures. The steps involved in a protein crystal structure determination are summarised in Figure 1. The basics of protein crystallography have been defined in the early textbook of Blundell & Johnson (Blundell and Johnson 1976). A more up to date account is given by Drenth (Drenth 1994) and a simplified account in Branden & Tooze (Branden and Tooze 1991).

1. Crystallisation: Systematic screening
2. Data collection: Synchrotron radiation sources; cryocrystallography
3. Phase calculation: MAD
4. Electron density map interpretation: O
5. Refinement: XPLOR

Total time: 10 days to 10 years

Figure 1. Steps in protein crystal structure determination

NATO ASI Series, Vol. H 102
Interacting Protein Domains
Their Role in Signal and Energy Transduction
Edited by Ludwig Heilmeyer

## 1. Crystallisation

Protein crystallisation was once considered an art but is now more scientific. There are a number of standard screening protocols available which have some physical-chemical justification and which are known to have worked for other proteins. Improved quantities and quality of proteins, made possible through expression of recombinant DNA, and better techniques for protein purification have yielded more homogeneous preparations than were available in the past. The commercial screening kits produced by Hampton Research (25431 Cabot Road, Suite 205, Laguna Hills, CA 92653-5527, USA; http://www.hamptonresearch.com) provide flexible screening kits for a variety of macromolecules. The Hampton Research catalogue includes much useful advice.

Modification of proteins has on occasion led to a product that crystallised when the native protein had proved intractable. Recognition of the domain construction of many large proteins from sequence analysis and other predictive methods may suggest a suitable break point in the chain that can be made either by limited proteolysis or by site directed mutagenesis and subsequent expression to yield a smaller product that crystallises. For example in our work with phosphorylase kinase the engineering of a stop codon between the kinase domain and the calmodulin binding domain led to a kinase domain that crystallised well (Owen et al., 1995). In work with cyclin A, deletion of the N-terminal domain led to the C-terminal active cyclin box fold which was able to activate the cyclin dependent protein kinase. This product crystallised (Brown et al., 1995) whereas the full length cyclins, which carry the N-terminal region with the destruction box to target the protein for ubiquitination and subsequent proteolysis, have not yet crystallised. On occasions, mutation of a single amino acid has resulted in a product that crystallised. For example in HIV integrase, the catalytic core, residues 50-212, had very low solubility. In a heroic set of experiments, hydrophobic residues were replaced in turn with either alanine or lysine. Out of 29 mutants, one was successful (Phe185Lys) which gave rise to a crystallisable protein that led to the structure (Jenkins et al., 1995). Other examples have included human ferritin and GroEL.

## 2. Data Collection

Efficiency and precision of X-ray data collection have improved significantly in recent years with increased access to bright synchrotron radiation sources and the introduction of more sensitive and automatic X-ray detectors (Helliwell 1992). In Europe synchrotron radiation sources are available with protein crystallography stations at SRS, Daresbury, UK; LURE, Paris, France; DESY, Hamburg, Germany; ESRF, Grenoble, France; and ELETTRA, Trieste, Italy. In the US there are sources at CHESS, Ithaca, NY; SSRL, Stanford, Ca.; NSLS,

Brookhaven National Laboratory, Upton, NY; and APS, Argonne, Il. The Photon Factory at Tsukuba, Japan also provides a unique facility to the international community. Development of techniques which allow protein crystals to be frozen

1. Bragg's Law (1913) $2\, d \sin \theta = n\, \lambda$

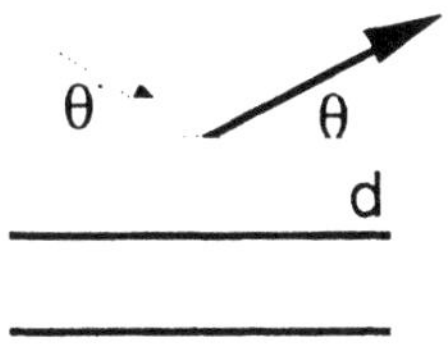


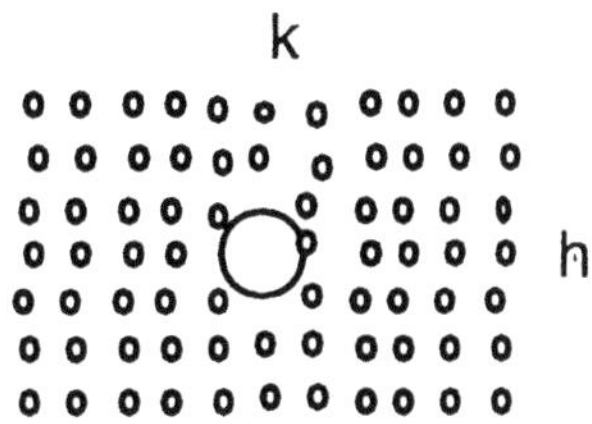


d is interplanar spacing of atoms in crystal, θ is the Bragg scattering angle, n is an integer (0,1,2,...) and corresponds to the order of diffraction, λ is wavelength of X-rays.

2. Structure Factor equation

$$F(hkl) = \Sigma_j\, f_j \exp 2\pi i\, (hx_j + ky_j + lz_j)$$

$I(hkl) = |F(hkl)|^2$

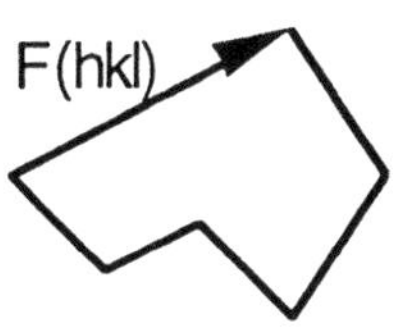


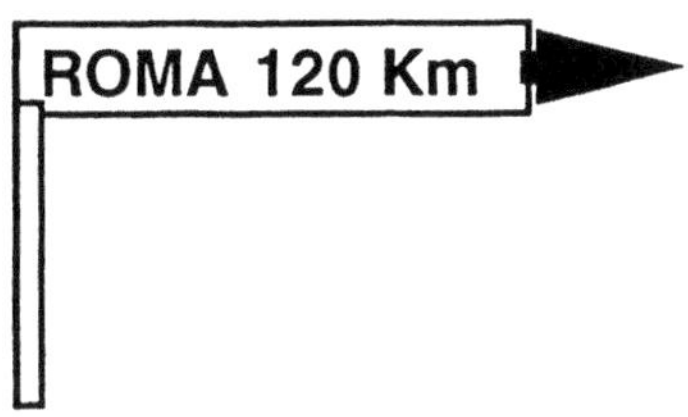


Where $f_j$ is the atomic scattering factor and $x_j$, $y_j$ and $z_j$ are the positional coordinates for the jth atom

3. Electron Density equation

$$\rho(xyz) = 1/V\ \Sigma\Sigma\Sigma_{hkl}\, F(hkl) \exp -2\pi i\, (hx + ky + lz)$$

Where ρ(xyz) is the electron density at point x, y, z, V is the volume of the unit cell, the summation is over all h, k, l of the diffraction pattern and F(hkl) is the structure factor for reflection hkl.

Figure 2. Fundamental equations of X-ray diffraction

to liquid nitrogen temperatures has helped to alleviate radiation damage and provided considerable advantages when crystals are bombarded with high intensity radiation (Teng 1990; Mitchell and Garman 1994). In the past crystals of at least 0.1 mm were required before measurements could be made; with cryocrystallographic methods crystals of dimensions as small as 10 m can successfully be used for data collection. High speed computers with large memory have taken the drudgery out of data processing and new software (such as DENZO (Otwinowski 1993)) has allowed efficient and reliable intensity measurements.

### 3. Phase determination

Figure 2 summarises the 3 fundamental equations for X-ray diffraction and Figure 3 gives a definition of resolution. Bragg's law tells us the conditions and the direction of the diffracted beam and allows a strategy for data collection of all the diffracted beams that arise when the crystal is irradiated from all different angles. The structure factor equation gives the algebraic relationship that allows the intensity of the diffraction beam to be computed from knowledge of the structure of the molecule in the crystal. The structure factor has both amplitude and phase (magnitude and direction). Both these quantities are required in order to compute the electron density from the diffraction pattern as shown in equation 3. The amplitudes may be measured directly from the square root of the intensities of the diffracted beams but all information of the relative phases of the diffracted beams is lost when these beams interact with the detector. The phase problem was for many years the most serious problem which presented a rate limiting step to crystal structure determination.

$2\,d\,\sin\theta = \lambda$ Resolution: $d = \lambda / 2 \sin\theta$

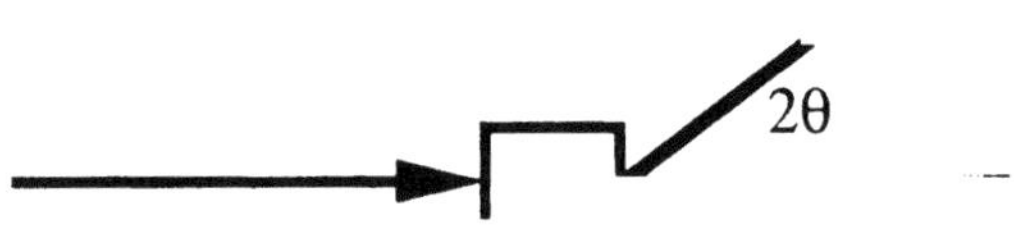


| | |
|---|---|
| 3 Å resolution: | Identify polypeptide fold and side chains with the help of sequence |
| 2Å resolution: | Identify main chain and side chains with greater precision. Interpretation possible without sequence. Identify water molecules. |
| 1Å resolution: | Resolve individual atoms. Describe intermolecular contacts with high precision. Detect anisotropic motion of atoms. Possibly identify deviations from standard geometry. |

Figure 3. Definition of resolution

Anomalous scattering occurs when the wavelength of the incident radiation is close to the natural absorption edge of the target atom (corresponding to K or L edge). Away from the absorption edge, the atom electrons can be considered as free electrons: close to the edge, the inner electrons which are tightly attached to the nucleus, produce an additional component.

$$F_H\,(\text{anom}) = f + f' + if''$$

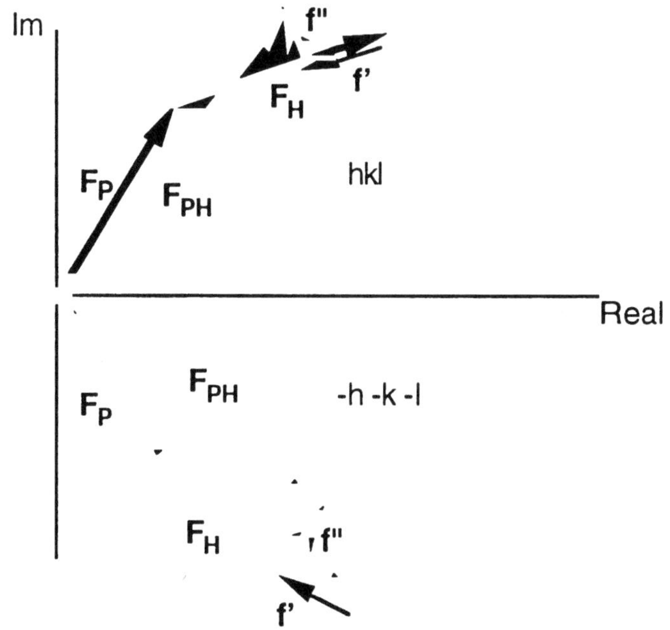


$|F(hkl)| \neq |F(-h-k-l)|$

Example: Tungstate in protein phosphatase 1A (Egloff et al., 1995)

| Wavelength | | f' | f" |
|---|---|---|---|
| λ1 | 1.2123Å | -12.86 | 18.54 |
| λ2 | 1.2136 | -24.39 | 9.948 |
| λ3 | 1.1350 | -6.85 | 9.124 |

Some absorption edges for frequently used atoms in anomalous scattering:

Zinc (30) K abs edge 1.28340 Å Selenium (34) Kabs edge 0.97974 Å

Terbium (65) LIII abs edge 1.64970 Å Tungsten (74) LIII abs edge 1.21550Å

Figure 4. Multiwavelength Anomalous Dispersion (MAD) method for phase determination

Phase determination has now become more routine with a number of standard heavy atom reagents available for heavy atom isomorphous replacement methods (Blundell & Johnson 1976) and the development of the multi-wavelength anomalous dispersion (MAD) methods (Hendrickson 1991) (Figure 4).

Molecular replacement methods have become most powerful when the fold of the protein is close to a fold has been previously determined. A search model is constructed from the likely homologous structure and this object rotated and translated within the unit cell of the unknown structure until a good correlation between observed and calculated intensities is reached.

## 4. Electron density map interpretation

Computer graphics have had an impact on the ease with which structures can be interpreted with user friendly programmes such as O (Jones et al., 1991). The programme allows simultaneous display of the electron density contoured at a suitable level (often equivalent to 1 or 2 times the standard deviation of the electron density map) with the model of the protein structure. Initially this may be a trace of the continuous part of the electron density to represent the course of the main chain. Side chains may then be added with known standard geometry. Atoms, groups of atoms and parts of the structure can be shifted individually or as a whole to achieve the best fit to the experimental electron density. This model is then refined against
the experimental data to provide an objective assessment of its correctness. Iterative cycles of model building and refinement are usually required to produce the final structure.

## 5. Refinement of protein structures.

The refinement of protein structures is routinely achieved with programmes such as XPLOR (Brunger 1992). This results in a structure that accounts for the experimental X-ray data within the restraints of standard stereochemistry. The programme is based on an energy function approach: arbitrary combinations of different energy functions that describe the experimental data are minimised by gradient descent, simulated annealing and conformational search procedures. The protocols provide an objective criteria of the correctness and reliability of the structure. The functions minimised are summarised in Figure 5.

XPLOR : Includes stereochemical restraints to increase the number of "observations": include simulated annealing to allow structure to escape from false energy minimum.

Minimise an pseudo energy term E:

$E = E_{xray} + E_{energy}$

where

$E_{xray} = \Sigma_{hkl}\, w(hkl)\, \{|F_{obs}(hkl)| - |F_{calc}(hkl)|\}^2$

and

$E_{energy} = \Sigma\, E_{bond} + E_{angle} + E_{dihedral} + E_{improper} + E_{vdw} + \text{etc.}$

where $E_{bond} = \Sigma_{bonds}\, k_{bond}(d_o - d)^2$
and $d_o$ ideal bond, d observed bond distance and $k_{bond}$ is energy constant

Assess progress with:

$R_{cryst} = \Sigma_{hkl}\, |F_{obs}(hkl) - F_{calc}(hkl)| \,/\, \Sigma_{hkl}\, F_{obs}(hkl)$

and

$R_{free}$ as for R but calculated with 5% of data not included in the refinement

Examples:

cd1 nitrite reductase 1.55 Å : R= 0.176; $R_{free}$ = 0.193 (Fulop et al., 1995)

Cyclin A 2.0 Å R = 0.216; $R_{free}$ = 0.299 (Brown et al. 1995)

Figure 5. Outline of refinement strategy with XPLOR

**References**

*(Only a few key references are given.)*

Blundell, T. L. and Johnson, L. N. (1976). Protein Crystallography. London, Academic Press.

Branden, C.-I. and Tooze, J. (1991). Introduction to Protein Structure. New York, Garland.

Brown, N. R., Noble, M. E. M., Endicott, J. A., Garman, E. F., Wakatsuki, S., Mitchell, E. P., Rasmussen, B., Hunt, T. and Johnson, L. N. (1995). The crystal structure of cyclin A. Structure 3, 1235-1247.

Brunger, A. T. (1992). X-PLOR: Version 3.1; a system for protein crystallography and NMR. New Haven, Yale University Press.

Drenth, J. (1994). Principles of Protein Crystallography. New York, Springer-Verlag.

Egloff, M.-P., Cohen, P. T. W., Reinemer, P. and Barford, D. (1995). Crystal sreucture of the catalytic subunit of human protein phosphatase 1 and its complex with tungstate. J. Mol. Biol. 254, 942-959.

Fulop, V., Moir, B., Ferguson, S. J. and Hajdu, J. (1995). The anatomy of a bifunctional enzyme: structural basis for reduction of oxygen to water and synthesis of nitric oxide by cytochrome cd1. Cell In press.

Helliwell, J. R. (1992). Macromolecular Crystallography with Synchrotron Radiation. Cambridge, Cambridge University Press.

Hendrickson, W. A. (1991). The detrmination of macromolecular structure from anomalous diffraction of synchrotron radiation. Science 254, 51-58.

Jenkins, T. M., Hickman, A. B., Dyda, F., Ghirlando, R., Davies, D. R. and Craigie, R. (1995). Catalytic domain of human immunodeficiency virus type 1 integrase: identification of a soluble mutant by systematic replacement of hydrophobic residues. Proc. Natl. Acad. Sci. USA 92, 6057-6061.

Jones, T. A., Zou, J. Y., Cowan, S. W. and Kjeldgaard, M. (1991). Improved method for building models in electron density maps and the location of errors in these models. Acta Crystallogr. A47, 110-119.

Mitchell, E. P. and Garman, E. F. (1994). Flash freezing of protein crystals: investigation of mosaic spread and diffraction limit with variation in cryoprotectant concentration. J. Appl. Cryst. 27, 1070-1074.

Otwinowski, Z. (1993). DENZO. Data Collection and Processing., SERC Laboratory, Daresbury, Warrington, UK. DL/SC1/R34,

Owen, D. J., Papageorgiou, A. C., Garman, E. F., Noble, M. E. M. and Johnson, L. N. (1995). Expression, purification and crystallisation of phosphorylase kinase catalytic domain. J. Mol. Biol. 246, 376-383.

Teng, T.-Y. (1990). Mounting of crystals for macromolecular crystallography in a free standing thin film. J. Appl. Crystallogr. 23, 387-391.

# Spectroscopical studies on the interaction between ww domain and proline-rich peptides

**Elena Baraldi, Maria Macias, Marko Hyvonen, Hartmut Oschkinat and Matti Saraste**

EMBL, Meyerhofstrassse 1, 69117 Heidelberg, Germany.

The WW domain is a new member of the family of protein modules, like the SH2, SH3, PH and PTB domains , regulatory domains for protein-protein and protein-lipid interaction.
It consist of 38 aminoacids with a high content of hydrophobic and aromatic residues and two highly conserved tryptophanes (hence the WW domain ) and one invariant proline.
It is found in different cytoskeletal and signal-transducing proteins ( for example dystrophin, Yap 65, ubiquitin-protein ligases, the transcriptional activator FE65 and proteins binding to formin) (Fig.3) . It occurs in single or multiple tandem copies and its function is not yet totally understood; in different cases it has been shown to associate to proline-rich regions of proteins and to compete with the SH3 domain for the same sequence motif.

| | | | |
|---|---|---|---|
| Yap/Mouse-1 | 143 | SSFEIPDDVPLPAGWEMAKTSS.GQRYFLNHNDQTTTWQDPRKAMLS | X80508 |
| Yap/Mouse-2 | 210 | QTLMNSASGPLPDGWEQAMTQD.GEVYYINHKNKTTSWLDPRLDPRF | X80508 |
| Rsp5/Yeast-1 | 220 | YSSFEDQYGRLPPGWERRTDNF.GRTYYVDHNTRTTTWKRPTLDQTE | L11119 |
| Rsp5/Yeast-2 | 322 | TGGTTSGLGELPSGWEQRFTPE.GRAYFVDHNTRTTTWVDPRRQQYI | L11119 |
| Rsp5/Yeast-3 | 379 | QQQPVSQLGPLPSGWEMRLTNT.ARVYFVDHNTKTTTWDDPRLPSSL | L11119 |
| Dmd/Human | 3044 | PASQHFLSTSVQGPWERAISPN.KVPYYINHETQTTCWDHPKMTELY | P11532 |
| Ykb2/Yeast-1 | 1 | MSIWKEAKDAS.GRIYYYNTLTKKSTWEKPKELISQ | P33203 |
| Ykb2/Yeast-2 | 30 | ELISQEELLLRENGWKAAKTAD.GKVYYYNPTTRETSWTIPAFEKKV | P33203 |
| Yo61/Caeel-1 | 41 | GIDESHSSPSVESDWSVHTNEK.GTPYYHNRVTKQTSWIKPDVLKTP | P34600 |
| Yo61/Caeel-2 | 86 | PLERSTSGQPQQGQWKEFMSDD.GKPYYYNTLTKKTQWVKPDGEEIT | P34600 |
| Db10/Tobac | 10 | GPSYAPEDPTLPKPWKGLVDGTTGFIYFWNPETNDTQYERPVPSSHA | D16247 |
| Yfx1/Yeast | 1 | MAQSKSNPPQVPSGWKAVFDDEYQTWYYVDLSTNSSQWEPPRGTTWP | Z46255 |
| 56G7/Caeel-1 | 220 | MEIASSSQTPPESHWKTYLDAK.KRKFYVNHVTKETRWTKPDTLNNN | Z46793 |
| 56G7/Caeel-2 | 363 | VWLFADITQPLPSGWECITMNN..RTVFLNHANKETSFYDPRIRRFE | Z46793 |

The WW domain from Yap 65 binds to proline rich sequences in which the PPPYmotif (named "the PY motif") is present. The PY motif is also present in the Epithelial $Na^{+}$ channel which bins to the WW domain of the Nedd4 protein. Mutations in the ENaC protein involving the region corresponding to the PY motif affects the control of the channel and cause the Liddle's syndrome, a hereditary form of hypertension.

NATO ASI Series, Vol. H 102
Interacting Protein Domains
Their Role in Signal and Energy Transduction
Edited by Ludwig Heilmeyer

We determine the structure of the WW domain from Hs Yap 65 in complex with proline-rich peptides GTPPPPYTVG or SPPPYTV, both containing the PY motif . These proline-rich sequences were found in two newly discovered proteins ( WBP1 and WBP2, for WW binding proteins 1 and 2) which associate with the Yap 65 protein *in vitro.*
The structure consist of a twisted and slightly bent three-stranded antiparallel beta sheet presenting a concave and a convex sides at opposite sites (Fig. 2 and 3).

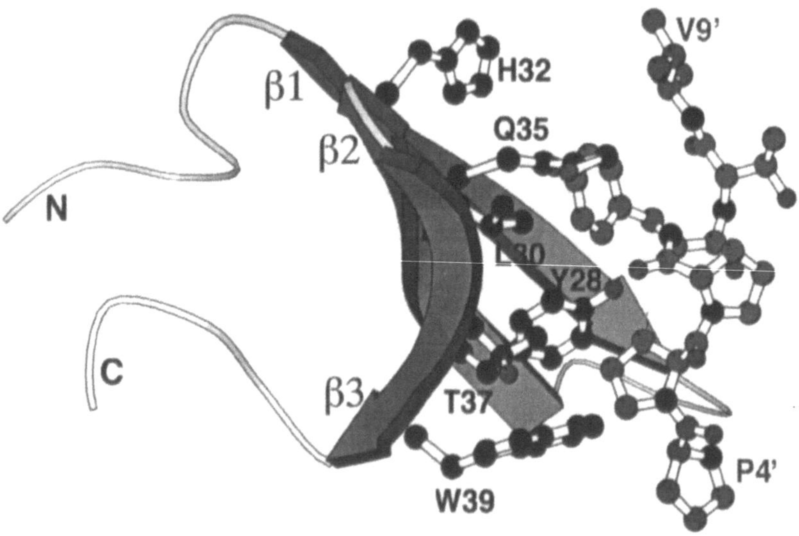


The side chain of Y28 and W39 fill its concave side to form an almost flat hydrophobic surface which represent the binding surface for the proline-rich peptide. The N and C termini of the domain meet on the convex side to form a hydrophobic buckle, in correspondence of which the side chains of P14 and P42 are located opposite each other above W17. The binding site for the proline-rich peptides correspond to the area defined by residues Y28, L30, H32 and six residues from D34 to W39 on the concave side of the domain.. This area present a very hydrophobic patch on the protein surface due to the side chains of Y28,L30 and W39 and different threonine methyl groups. The contacts between the ligand and the domain are in correspondence of residue P5' with W39, residue Y7' with L30 and H32 and residue V9' and H32.

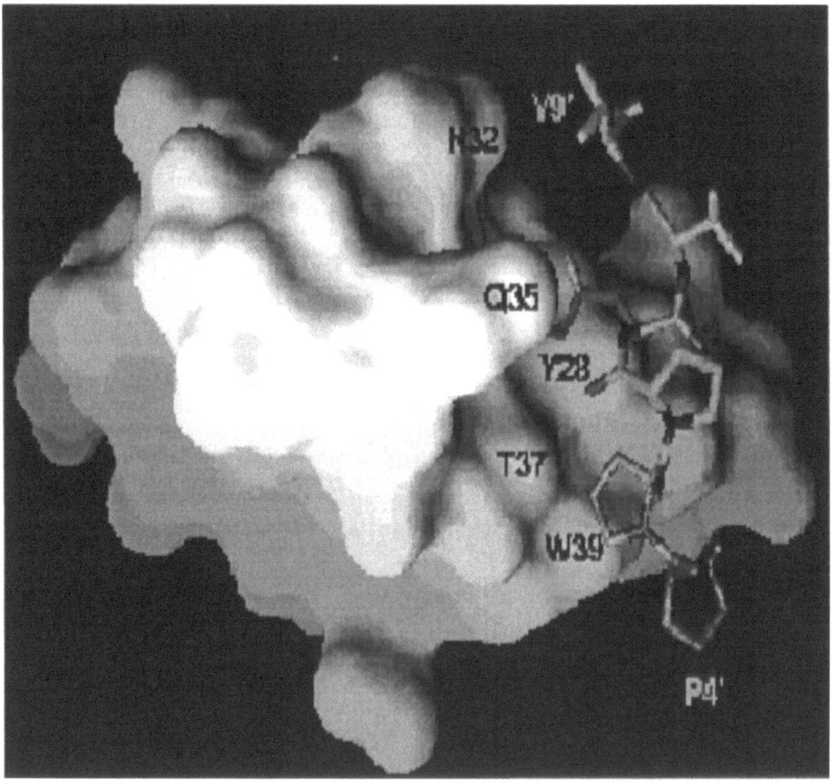


The proline-rich peptide GTPPPPYTVG was restrained into a polyproline helix type II. There are smooth contacts between t he ligand and the domain. The central prolines P4' and P5' contact W39 and the carbonyl group of P6' points towards the OH group of the conserved Y28. The side chain of Y7' is located in a hydrophobic pocket formed by L30 and H32. Fluorescence spectroscopy has been chosen for better understand the nature of the interaction between the WW domain and the proline-rich partners. The W39 fluorescence emission undergoes evident changes upon interaction with the ligand. These changes are proportional to the binding affinity. (Fig.4).We synthesized an alanine scanning peptide library using as a template the SPPPYTV peptide, and we measured the Kd's of each of these peptide with the WW domain of Yap 65. The dissociation constants measured for the WW domain and the SPPPYTV or the GTPPPPYTVG peptides are very similar ( 50 um cca). This affinity is mostly affected when the two centre prolines in the peptide are substituted by alanines, whereas it is not changed when this substitution involves the first proline. When the tyrosine in the PPxY motif is substituted for alanine, leucine or phenylalanine the affinity is also affected. In the first two cases the binding was completely abolished and in the latter one it was decreased dramatically.

These results confirm the importance of the PY motif in determining the binding with the WW domain from Yap 65. Furthermore the residues L30, H32, Q35 in the WW domain were mutated to alanine and these mutant were assayed for their retention of capability to bind to the proline -rich peptide SPPPYTV. These mutants were completely unable to bind the ligand, demonstrating the importance of these residues in the interaction.

**Calculated Kd's**

1) Wild type and mutant forms of the protein

| Protein: Yap 65 WW domain | Peptide | Kd (um) |
|---|---|---|
| WILD TYPE | GTPPPPYTVG | 52+/-6 |
| | SPPPYTV | 47+/-6 |
| MUTANT L30K | GTPPPPYTVG | 40+/-5 (*) |
| | SPPPYTV | NB |
| MUTANT Q35A | GTPPPPYTVG | 160+/-18 |
| | SPPPYTV | NB |
| MUTANT H32A | GTPPPPYTVG | NB |
| | SPPPYTV | NB |

2) Wild type protein and alanine scanning mutated peptides

| Protein | Peptides | Kd (um) |
|---|---|---|
| WILD TYPE Yap 65 WW domain | SPPPPATV | NB |
| | SPPPPLTV | NB |
| | SPPPPFTV | 118+/-17 |
| | SAPPPYTV | 53+/-17 |
| | SPAPPYTV | NB |
| | SPPAPYTV | NB |
| | SPPPAYTV | 63+/-17 |
| | SPPPPYAV | 110+/-13 |
| | SPPPPYTA | 150+/-15 |

This study confirm the hypothesis previously supported by *in vitro* data that the WW domain is a modular domain involved in the regulation of interactions to proline rich regions of proteins, representing one alternative to the SH3 domain.

# Novel Microscope-Based Approaches for the Investigation of Protein-Protein Interactions in Signal Transduction

György Vereb[1,2], Christoph K. Meyer[1], and Thomas M. Jovin[1]

[1]Department of Molecular Biology, Max Planck Institute for Biophysical Chemistry, D-37018 Göttingen, Germany
[2]Department of Biophysics, University Medical School of Debrecen, H-4012 Debrecen, Hungary

## 1 Introduction

In addition to the biochemical characterization of interacting proteins, the spatio-temporal localisation *in situ* or *in vivo* of the transiently associating participants of a cascade mechanism can greatly enhance our understanding of the given signal transduction process. However, the mere determination of colocalization based on fluorescent tags that are compatible with the *in vivo* observation does not generally suffice due to the limitation in resolution imposed by optical diffraction in conventional light microscopy.

We resorted to two approaches in order to overcome this barrier. The first exploited fluorescence resonance energy transfer (FRET), a phenomenon capable of reporting molecular proximities over distances of a few nanometers. FRET was successfully implemented both in full field and confocal microscopy. In addition, scanning force microscopy (SFM) and scanning near-field optical microscopy (SNOM) were employed to achieve spatial resolution in the 10-100 nanometer range, using immunogold and fluorescent tags, respectively.

## 2 Fluorescence Resonance Energy Transfer (FRET)

### 2.1 Measuring Energy Transfer by Photobleaching the Fluorescence Donor

Fluorescence resonance energy transfer (FRET) is based on non-radiative dipole-dipole interaction, the efficiency of which is inversely proportional to the 6$^{th}$ power of the distance between the fluorescence energy donor and acceptor [1,2]. After suitably labeling the proteins of interest (directly, or with tagged antibodies), their molecular proximity in the nm range can be inferred from the presence of energy transfer [3,4]. One method for measuring FRET efficiency is based on the photobleaching of the donor, the rate of which is slower in the presence of an acceptor due to the introduction by FRET of an additional deactivation path, thus shortening the lifetime of the excited state [5,6]. An example is provided in Fig. 1, in which $\beta_2$-microglobulin ($\beta_2$m), an

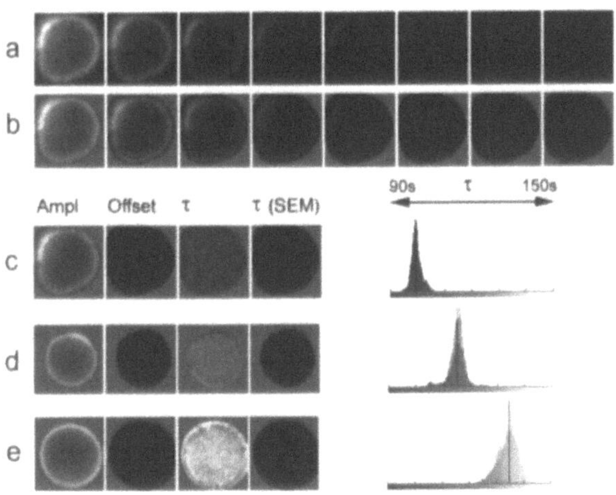


Fig. 1. Photobleaching FRET shows MHC class I aggregation on lymphoblastoid cells

NATO ASI Series, Vol. H 102
Interacting Protein Domains
Their Role in Signal and Energy Transduction
Edited by Ludwig Heilmeyer

intramolecular constituent of the MHC class I antigens, is visualized on the surface of JY lymphoid cells by FITC-conjugated specific Fab fragments. Row *a* shows the loss of fluorescence from photobleaching during successive exposures, and row *b* presents the corresponding images generated from pixel-by-pixel monoexponential fits to the data; the fluorescence intensity is shown on a grey scale. In row *c* the decay amplitude and offset (constant background) are depicted together with an image of the photobleaching constant $\tau$, its standard error, and its distribution histogram. Rows *d* and *e* display similar results, but in these cases a fluorescent acceptor (TRITC) was bound via conjugation to an Fab directed either to $\beta_2$m (*d*) or another epitope of the MHC class I molecules (*e*). FRET efficiency (*E*) can be calculated from the expression $E = 1 - (\tau_D/\tau_{D+A})$, where the indices *D* and *A* refer to the presence of the donor and acceptor labels, respectively. The photobleaching constant $\tau$ was prolonged to different extents in the two experiments (rows *d* and *e*). $\beta_2$m, an integral component of the MHC class I complex, was apparently positioned very close to the other detected epitope (row *e*), judging from the high transfer efficiency of 26%. The presence of significant FRET ($E$ = 13%) between $\beta_2$m subunits (row *d*) was indicative of a clustering of the MHC-I complexes, which is a potential means for achieving high avidity binding during T-cell activation [7].

## 2.2 FRET Measured by Photodestructing the Fluorescence Acceptor

When assessing FRET by donor photobleaching, one has to rely on population averages with cells labeled separately with the donor alone, i.e. to provide a value for $\tau_D$. However, with a confocal microscope it is possible to investigate a cell population uniformly labeled with both donor and acceptor, by carrying out the selective photodestruction of the acceptor in a part or all of the field of observation [8,9]. In this case the FRET efficiency is calculated from $E = 1 - (F_{D+A}/F_D)$, where $F_{D+A}$ is donor fluorescence in the presence of acceptor, corresponding to a "pre-bleach" image, and $F_D$ is measured after bleaching the acceptor. An example is provided by A431 epidermoid carcinoma cells that were stimulated with the growth factor EGF and then permeabilized so as to allow the labeling of the activated EGF receptors with specific Cy5-conjugated antibodies. For demonstration purposes, the antibodies were further tagged with a Cy3-conjugated secondary antibody, serving as an efficient donor for Cy5 as acceptor. Panel *a* shows the latter signal, which was mostly localised to the membranes in this confocal section. Panel *b* depicts the same region, but after photobleaching the acceptor in the upper half of the image. The lower half of the image served as a control. The donor (Cy3) fluorescence before and after bleaching the acceptor is given in panels *c* and *d*, respectively. A slight increase in fluorescence was seen, corresponding to a mean E of 22%. The grey scale bar represents *E* values of 0 - 43% in panel *e*.

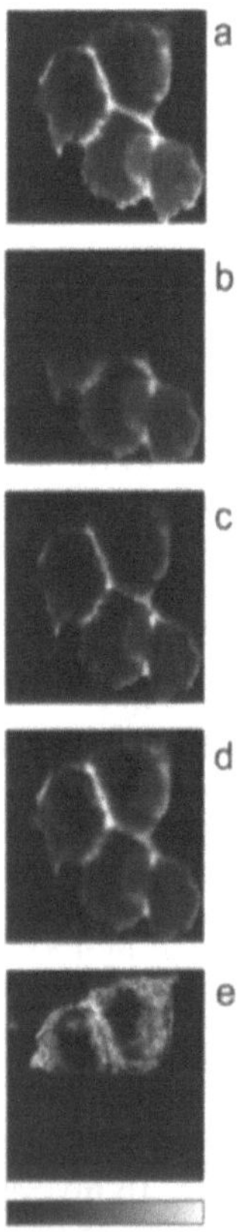


Fig. 2. Assessing FRET from acceptor bleaching

## 2.3 Combination of Donor and Acceptor Photobleaching FRET

FRET evaluated from acceptor photodestruction is not very sensitive, inasmuch as the observed intensity changes are proportional to *E*. In practice, values > 5-10% are required for unambiguous interpretations. The extent of FRET is obviously dictated by the selected dyes and the nature of the proteins and protein complexes under investigation. In cases leading to low *E*, the more sensitive donor photobleaching method can be used after photodestructing the acceptor in a region-of-interest of the sample. The donor photobleaching rates in the double labeled and the acceptor-destructed regions of the same sample can then be compared directly without a separate reference sample [10].

# 3 Scanning Probe Microscopies (SPM)

## 3.1 MHC-I Clusters on Lymphoid Cells as Seen by Scanning Force Microscopy

While FRET reveals molecular proximities on the scale of 1-10 nm, it is not a good measure of protein aggregation in the 10-100 nm range lying beyond the diffraction limit of light microscopes. For such situations scanning probe microscopies, capable of very fine motions of the sample or of the probe using piezo scanners, can be of great utility, especially for studies of cell surfaces. In the scanning (atomic) force microscope (AFM, SFM), a very fine tip of (generally) Si or $Si_3N_4$ is attached to a flexible cantilever with a spring constant of < 0.1 N/m. A split photodiode monitors a laser beam reflected from the top of the cantilever. As the tip touches the sample, the cantilever is deflected, leading to a corresponding displacement of the beam on the photodiode. The ensuing signal provides a very sensitive measure of changes in height above a reference surface [11,12]. By raster scanning along the *x* and *y* axes, a 3 dimensional topographic map can be produced, as shown in the central image of Fig. 3. The SFM image is of the external surface of a HUT-102B2 T cell labeled for MHC class I antigens with specific antibodies and then tagged with secondary antibodies conjugated with colloidal gold. Clustered gold labels are clearly visible, indicating a higher order of MHC class I association than that perceived by the shorter-range FRET determination [7]. Such a distribution could be significant in the activation or viral transformation of cells [13]. Figure 3 also demonstrates an apparently uniform labeling of cell surface antigens by conventional fluorescence microscopy (lower left corner), but beyond this at least two hierarchical levels of antigen association can be perceived. One is in the range of a few nm, revealed by FRET and modeled by MHC class I molecules (upper right) situated in the membrane (dashed lines). The primary, fluorescently labeled antibodies (white symbols) that generate the FRET signals are shown attached to the MHC. The secondary antibodies conjugated to gold microspheres (represented by the large sphere) reveal the second level of association in the SFM image: small clusters of gold beads covering some tens of MHC-I antigens over a 10-100 nm scale. This visualisation technique in the SFM is also appropriate for studying live cells under buffer [7,14].

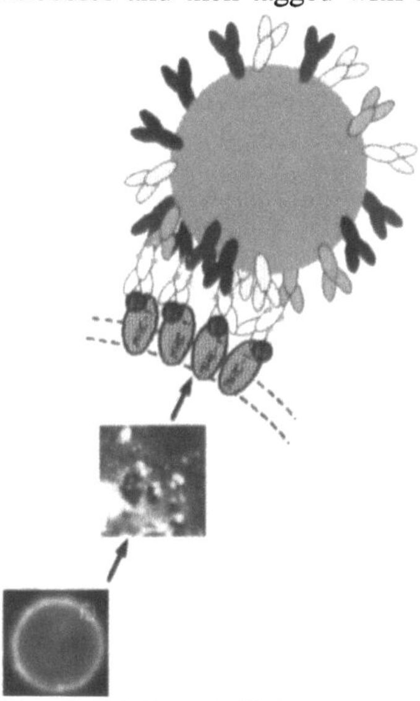

Fig. 3. FRET and SFM reveal two hierarchical levels of MHC-I clustering

## 3.2 Clustering of PDGF Receptors on T98G Glioblastoma Cells Revealed by Scanning Near-field optical microscopy (SNOM)

Scanning near-field optical microscopy (SNOM) is a new and potent technique that employs a raster scanning technique similar to that of the SFM, but providing an optical signal in addition to height (topographic) information from every scanned point of the surface [15,16]. SNOM achieves a spatial resolution beyond the limits of the optical microscope and is thus ideally suited for scanning cell surface receptors. A pulled-out optical fibre is used as the scanning probe; this is vibrated above the sample and the damping of the vibration provides feedback for height control. The device is capable of precisely following the contours of the surface while simultaneously generating an optical (fluorescence) image. In our case, the aperture of the fibre tip was shared between the excitation and emission light using filters and dichroic mirrors similar to those of epifluorescence microscopy [17]. In Fig. 4 we demonstrate the capacities of SNOM with platelet-derived growth factor (PDGF) receptors on T98G glioblastoma

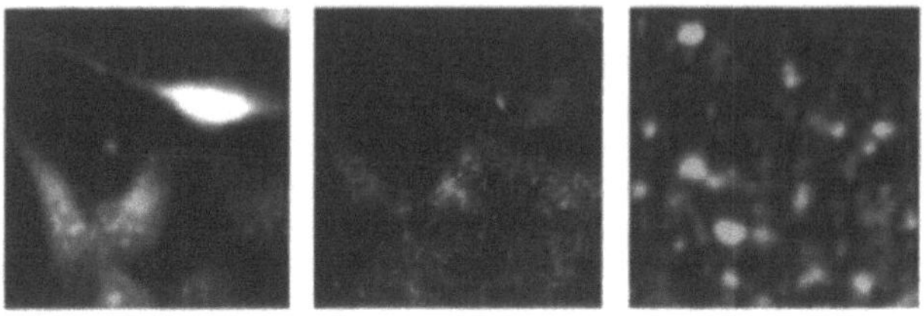


Fig. 4. PDGF receptor clustering on glioblastoma cells revealed by SNOM

cells. The cells were grown to half confluence, fixed, and labeled with anti-PDGFR antibodies on ice, followed by a secondary Cy3 conjugated $(Fab)_2$ fragment. Panel *a* shows a grey scale image of the height topography (in arbitrary units) of several cells. Panels *b* and *c* are fluorescence images obtained at equal distances from the cell surface. The lateral dimensions in panels *a*, *b*, and *c* were 120, 120, and 7 µm, respectively. The cells exhibited a patchy surface staining with cluster sizes down to 80-100 nm, likely a sign of receptor pre-aggregation.

## Acknowledgements

GV was a recipient of a postdoctoral grant of the Alexander von Humboldt Foundation and was supported by the Hungarian National Research Fund grant F-013335.

## References

1. Förster T. (1946) Energiewanderung und Fluoreszenz. Naturwissenschaften, 6, 166-175.
2. Stryer L, Haugland RP. (1967) Energy transfer: a spectroscopic ruler. Proc. Natl. Acad. Sci. USA, 58, 719-726.
3. Szöllösi J, Trón L, Damjanovich S, Helliwell SH, Arndt-Jovin D, Jovin TM. (1984) Fluorescence energy transfer measurements on cell surfaces: a critical comparison of steady-state fluorimetric and flow cytometric methods. Cytometry, 5, 210-216.
4. Jovin TM, Arndt-Jovin D. (1989) FRET Microscopy: Digital imaging of fluorescence resonance energy transfer. Application in cell biology. In: Cell Structure and Function by Microspectrofluorometry. (eds. E Kohen, JG Hirschberg). Academic Press, New York.
5. Jovin TM, Arndt-Jovin DJ. (1989) Luminescence digital imaging microscopy. Annu. Rev. Biophys. Biophys. Chem., 18, 271-308.
6. Kubitscheck U, Schweitzer-Stenner R, Arndt-Jovin DJ, Jovin TM, Pecht I. (1993) Distribution of type I $Fc_{\varepsilon}$-receptors on the surface of mast cells probed by fluorescence resonance energy transfer. Biophys. J., 64, 110-120.
7. Damjanovich S, Vereb G, Jr., Schaper A, Jenei A, Matkó J, Starink JP, Fox GQ, Arndt-Jovin DJ, Jovin TM. (1995) Structural hierarchy in the clustering of HLA class I molecules in the plasma membrane of human lymphoblastoid cells. Proc. Natl. Acad. Sci. USA, 92, 1122-6.
8. Bastiaens PIH, Jovin TM. (1996) FRET microscopy in cellular signal transduction. In: NATO Advanced Research Workshop: Analytical use of fluorescent probes in oncology. (eds. E Kohen, JG Hirschberg). Plenum Press, Miami, FL
9. Bastiaens PIH, Majoul IV, Verveer PJ, Söling HD, Jovin TM. (1996) Imaging the intracellular trafficking and state of the $AB_5$ quaternary structure of cholera toxin. EMBO J., 15, 4246-4253.
10. Bastiaens PIH, Jovin TM. (1996) Fluorescence resonance energy transfer (FRET) microscopy. In: Cell Biology: a Laboratory Handbook, 2nd ed. (ed. JE Celis) Academic Press, San Diego, CA
11. Binnig G, Quate CF, Gerber C. (1986) Atomic force microscopy. Phys. Rev. Lett., 56, 930-933.
12. Hoh JH, Hansma PK. (1992) Atomic force microscopy for high-resolution imaging in cell biology. Trends Cell. Biol., 2, 208-213.
13. Matkó J, Bushkin Y, Wei T, Edidin M. (1994) Clustering of class I MHC molecules on the surfaces of activated and transformed human cells. J. Immunol., 152, 3355-3360.
14. Vereb G, Jr., Damjanovich S, Jovin TM. (1995) Immobilization of molecules, membranes and cells for modern optical and non-optical microscopy by photo-crosslinking. J. Photochem. Photobiol. B: Biol., 27, 275-277.
15. Betzig EA, Lewis A, Harootunian A, Isaacson M, Kratschmer E. (1986) Near-field scanning optical microscopy (NSOM) - development and biophysical applications. Biophys. J., 49, 269-279.
16. Pohl DW. (1991) Scanning near-field optical microscopy (SNOM). Adv. Optical Electron Microsc., 12, 243-312.
17. Kirsch A, Meyer C, Jovin TM. (1996) Integration of optical techniques in scanning probe microscopes: the scanning near-field optical microscope (SNOM). In: NATO Advanced Research Workshop: Analytical use of fluorescent probes in oncology (eds. E Kohen, JG Hirschberg). Plenum Press, Miami, FL

# Binding studies with SH2 domains from the phosphotyrosine kinase ZAP70 using surface plasmon resonance and scintillation proximity assays.

**D.Whitaker, J.Hartley, P.Rugman, E.Keech, R.Handa , L.Price, E.Conway and K.Hilyard**

Department of Biology, Roche Research Centre, 40 Broadwater Road, Welwyn Garden City, Hertfordshire, United Kingdom, AL7 3AY

## 1.Introduction

During activation of T cells a number of proteins containing SH2 domains are found associated with the T-cell receptor (TCR) complex. There is currently much interest in the identification and characterisation of these proteins, in the expectation that alteration of these interactions mght lead to modulation of the immune response. A model of some of the proteins in the activated TCR complex is shown in Figure 1.

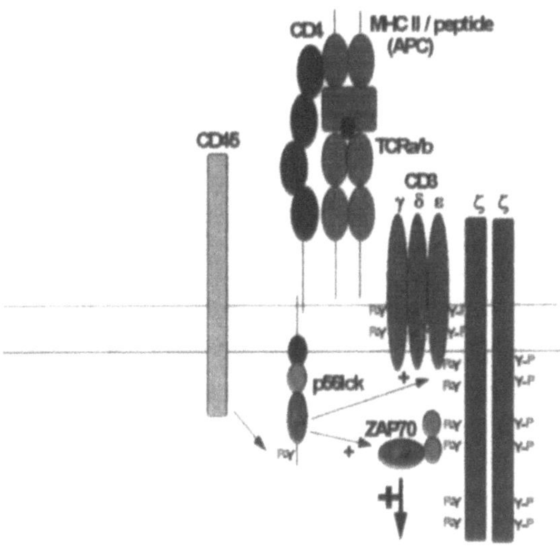


Figure 1. Schematic view of TCR/Majorhistocompatibility (MHC) complex. Binding of ZAP70 to immunoreceptor tyrosine activation motif (ITAM) of the zeta chain and subsequent phosphorylation by p56lck leads to activation of the T-cell.

A key protein in the initial response of the TCR to antigen is the phosphotyrosine kinase ZAP70. The protein is localised to the TCR complex shortly after antigen binding and is thought to be responsible for the initial events in the downstream

NATO ASI Series, Vol. H 102
Interacting Protein Domains
Their Role in Signal and Energy Transduction
Edited by Ludwig Heilmeyer

propagation of the cellular response. ZAP70 is a member of the *syk* family of phosphotyrosine kinases (PTKs), it contains two SH2 domains which direct it to the phosphorylated ITAMs on the zeta chain of the TCR complex. Expressed separately *in vivo* these SH2 domains do not bind to the TCR complex (1).
SH2 domains are protein modules of about 100 amino acids which bind, in a sequence specific manner, to phosphotyrosine containing proteins and peptides. ZAP70 contains two SH2 domains in tandem which bind diphosphopeptide motifs called ITAMs found in single copies in CD3 chains and as three copies in zeta chains (2). These have the general structure $YXXLX_{6-8)}YXXL/I$. A recent crystal structure (3) has shown that, as with other SH2 domains, the tyrosine and leucine side chains of the TAMs sit in deep pockets in the protein surface, these are known as the pY and pY+3 pockets respectively. In the case of ZAP70 one of these pockets is produced from the interface between the two SH2 domains (Figure 2).
Our experiments were aimed at evaluating the part played by the N and C phosphotyrosines in the affinity of the ITAM phosphopeptides for the tandem SH2 domains of ZAP70.

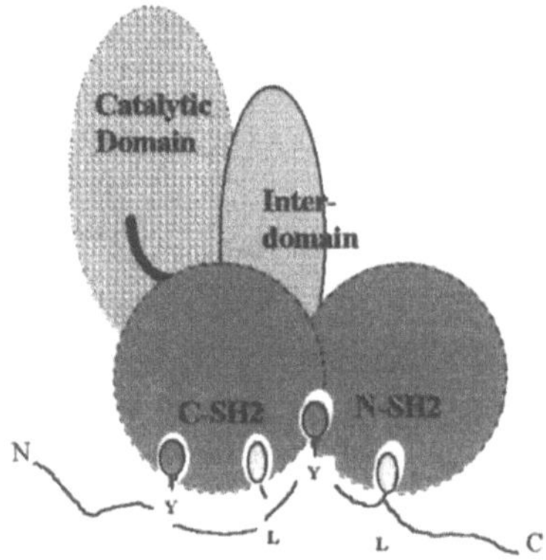


Figure 2. Schematic representation of the structure of ZAP70 based on the crystal structure showing a diphosphopeptide bound to tandem SH2 domains. Note that terminal phosphotyrosine binding site is formed from the interface of two SH2 domains.

## 2. Methods

### 2.1 Protein production

The tandem SH2 domain of ZAP70 (ZAP2SH2) was produced in *E.coli* as a fusion protein with GST. Induction of expression with isopropyl-β-D-thiogalactopyranoside caused the fusion protein to accumulate in inclusion bodies. To solubilise the protein the inclusion bodies were denatured in urea and subsequently renatured by dilution of the urea concentration. Correctly folded protein was purified on a glutathione sepharose column and finally desalted before storage at -80°C.

## 2.2 Peptide synthesis

Peptides were synthesised on a Perseptive Biosystems 9050 peptide synthesiser using PEG-PS resin/ Fmoc phosphotyrosine. Purification of the crude peptides was acheived by reverse phase HPLC.

## 2.3. SPA Assay

The assay was performed in 96 well plates. Each well contained biotinylated peptide, competing peptide and streptavidin coated SPA beads. To this was added a mixture of ZAP2SH2, rabbit anti-GST antibody, and $^{125}$I labelled anti-rabbit antibody. Plates were incubated for 16 h before counting using a Wallac Microbeta scintillation counter. (See Figure3)

**ZAP-SH2 SPA ASSAY**

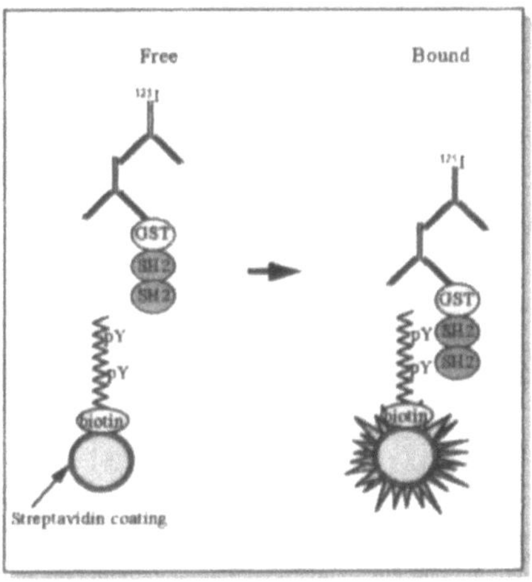


Figure 3 Schematic diagram of Scintillation Proximity Assay
Binding of iodinated antibody /GST-ZAP2SH2 to SPA beads produces signal

## 2.4. Surface plasmon resonance assay

Biotinylated peptides were bound to streptavidin coated chip at a concentration of 0.03μM in HBS buffer. GST-ZAP2SH2 was then injected over the streptavidin peptide surface. The concentration of GST-ZAP2SH2 required to give a good response varied according to the peptide used, this was determined experimentally in each case. The surface was regenerated with 0.1% SDS. Competing peptide was added to the protein solution before injection . Calculation of $IC_{50}$ was made from measurements taken at equilibrium.

## 5. Results

### 5.1 SPA Assay

| Peptide Sequence | $IC_{50}$ |
|---|---|
| KNPQEGLpYNELQKDKMAEApYSEIG | 16nM |
| KNPQEGLYNELQKDKMAEApYSEIG | 0% I at 500μM |
| KNPQEGLpYNELQKDKMAEAYSEIG | 0% I at 500μM |

Table 1. Peptide binding results from SPA assay

ZAP2SH2 exhibits a very high affinity for the doubly phosphorylated peptide ($IC_{50}$= 16nM) this is in marked contrast to the monophosphorylated peptides which failed to bind even at 500μM. At concentrations above this solubility limited measurement of affinities.

### 5.2. Surface Plasmon Resonance (Biacore) Results

| Peptide Sequence | $IC_{50}$ |
|---|---|
| KNPQEGLpYNELQKDKMAEApYSEIG | 11.9nM |
| KNPQEGLYNELQKDKMAEApYSEIG | 20μM |
| KNPQEGLpYNELQKDKMAEAYSEIG | 49μM |

Table 2: Peptide binding results from Biacore assay

In experiments using surface plasmon resonance measurements the diphosphopeptide has a far higher affinity than the monophosphates thus confirming the results from the SPA assay. Phosphorylation of the C-terminal tyrosine give a peptide with a slightly higher affinity than the N-terminal phosphorylated peptide but this may not be relevant *in vivo*.

## 6. Conclusion

The phosphorylation of tyrosine residues on ITAM sequences of zeta chains of the TCR complex by phosphotyrosine kinases produces high affinity binding sites for ZAP70. Cellular reconstitution experiments with ZAP702SH2 domains have shown that both N and C terminal SH2 domains are required for high affinity binding to ITAM motifs (1).

Our experiments have confirmed this observation by measurement of the binding of ZAPSH2 to singly and doubly phosphorylated ITAM peptides. These results are consistent with those of Chan *et al* (5) who showed that phosphorylation of both tyrosine residues in a ITAM2 peptide was required for high affinity binding of ZAPSH2 domains.
The reason why ZAP70 possesses two SH2 domains is not known but there have been suggestions that the binding of tandem SH2 domains leads to an allosteric change in ZAP kinase activity. Recent crystallographic evidence (3) has suggested a mechanism for this via the interdomain structure of ZAP70. A similar regulation of catalytic activity of other proteins containing tandem SH2 domains such as the p85 sub unit of phosphatidyl-3-OH kinase has been shown.

## 7. References


1. Wange, R.L., Malek, S.N., Desiderio, S. and Samuelson, L.E. J.Biol. Chem. 268, 19797-19801, (1993)

2. Irving, B.A., Chan ,A.C. and Weiss, A. J. Exp.Med. 177, 1093-1103 (1993).

3. Hatada, M.H.., *et al* Nature, 377, 32-38, (1995).

4. Hu, Q., *et al* Science 268 100-102, (1995).

5. Bu, Jia Ying, Shaw, A.S. and Chan, A. Proc. Nat. Acad. Sci USA 92, 5106-5110, (1995)

# Leucine Zipper mediated Homodimerization of Autoantigen L7 analyzed by Electrospray Ionization Mass Spectrometry and Yeast Two Hybrid Interactions

Stephan Witte[1,2], Frank Neumann[1], Michael Przybylski[2], and Ulrich Krawinkel[1]

[1]Fakultät für Biologie, Universität Konstanz, Postfach 5560, 78434 Konstanz, Germany
[2]Fakultät für Chemie, Universität Konstanz, Postfach 5560, 78434 Konstanz, Germany

## Introduction

The eucaryotic protein L7, originally isolated from rat liver cells, associates in the cytoplasm with the large subunit of ribosomes (1). Like other riboproteins, L7 has been identified as a potent autoantigen in systemic autoimmune diseases such as Systemic Lupus Erythematosus, Mixed Connective Tissue Disease and Progressive Systemic Scleroderma, while a defined biological function has remained as yet unknown (2). The N-terminal region of protein L7 contains a sequence motif which is similar to the "Basic-Region-Leucine-Zipper (BZIP)" domain of eucaryotic transcription factors (1). In the canonical BZIP-domain the leucine zipper region promotes homo- or heterodimerization through coiled coil formation, and the BZIP-dimer interacts with specific target sequences on DNA (3,4). We have previously shown that the basic domain of protein L7 interacts specifically with as yet uncharacterized cognate sites on mRNAs thereby inhibiting their cell-free translation (5), and that L7 is capable of forming homodimers according to cross-linking and gel electrophoresis studies (1), thus suggesting that L7 interacts with mRNA as a dimer. Upon transfection into Jurkat T-lymphoma cells L7 selectively suppresses the translation of two nuclear proteins and induces apoptotic cell death (6).
The development of "soft ionization" methods of mass spectrometry, particularly matrix-assisted laser desorption-ionization (MALDI-MS) and electrospray-ionization (ESI-MS) has enabled precise molecular weight determinations of polypeptides up to large (> 100 kDa) proteins (7,8). In addition to the already established applications to determine primary structures and post-translational modifications of proteins, the capability of ESI-MS to the direct analysis of specific non-covalent interactions and supramolecular complexes of biomacromolecules has found particular interest (9). Several types of non-covalent protein complexes have been identified by ESI-MS in recent work including dimeric and trimeric complexes of DNA-binding proteins (9). We have identified by direct ESI-mass spectrometry, the homodimer complex of the BZIP-like region of L7, and have demonstrated that the dimerization domain is the predominant structure mediating the inhibition of cell-free translation (10). Here we confirm these results by showing in the yeast Two Hybrid System, that the BZIP like region of L7 is fully sufficient to mediate dimerization in vivo.

NATO ASI Series, Vol. H 102
Interacting Protein Domains
Their Role in Signal and Energy Transduction
Edited by Ludwig Heilmeyer

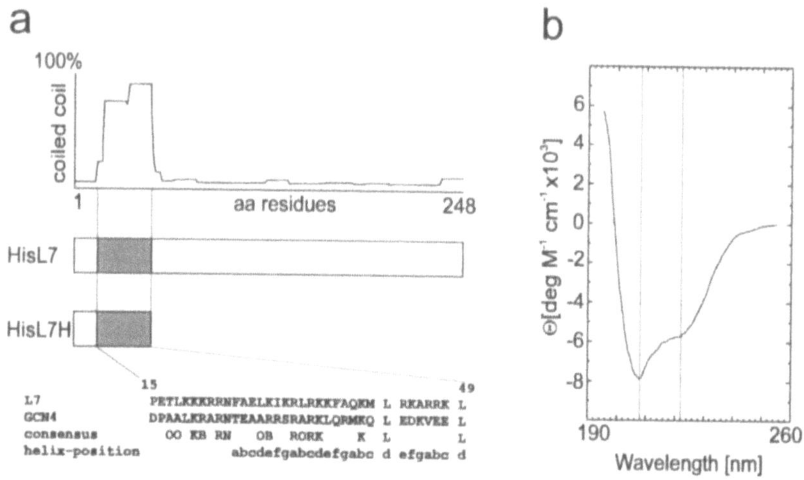


Fig. 1. Schematic depiction of sequences and the probability of helical coiled coil BZIP-like regions for proteins HisL7 and HisL7H (a) and CD spectrum of His L7 (b).

## Recombinant L7 proteins

For detailed structural analysis the entire human protein L7 and ist partial sequence comprising the BZIP-like domain were expressed in E. coli as fusion proteins with an N-terminal oligo-His-tag . (HisL7 and HisL7H; Fig. 1a) (10).

## ESI mass spectrometric identification of the HisL7H-homodimer complex

The identification of several non-covalent protein complexes including leucine zippers has been reported recently(9). In the desolvation process of charged macroions, a small (~10-100 V) potential difference between the ESI-capillary tip and a skimmer-type repeller electrode has been shown to be of critical importance for the analysis of non-covalent complexes (counter electrode-skimmer potential, DCS), in that the highest yield of non-covalent complex ions is obtained at lowest DCS. Furthermore, in the homogeneous series of multiply charge-state ions the oligomerization state of a complex of identical polypeptide chains is unequivocally defined by molecular ions with non-integer charges, when divided by the number of complex components (10); hence leucine zipper-type homodimers are characterized by all odd-charged ions.

ESI spectra of HisL7H at DCS of 10 and 20 V are shown in Figure 2. The spectrum at low declustering potential (Fig. 2a) provides the direct identification of the homodimer complex by the precise molecular weight (M2, 25072 Da) derived from the 15+, 17+, and 19+ charged ions, in addition to the multiply charged ion series of the monomer of higher abundance. The ions of the dimer are completely absent from the spectrum at DCS = 20 V, indicating facile dissociation of the complex (Fig. 2b). Relative abundances of complex ions were found increased at higher concentration (> 10 μM) and decreased at higher capillary temperature; no quantitative formation of dimers was obtained, as already observed for canonical leucine zippers due to their facile dissociation at the polar capillary surface (10). No ions indicative of higher oligomerization states were detected. Furthermore, these data were fully consistent with results of cross-linking experiments involving cell-free translated complete protein L7 (9) which did not contain a His-tag as a linker peptide. We conclude that protein L7 forms homogeneous dimers through non-covalent, coiled coil-like interactions between the regions.

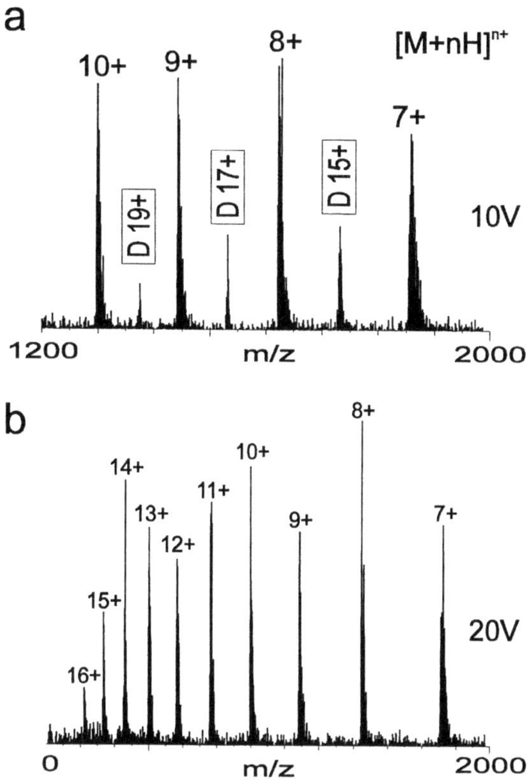


Fig. 2. ESI mass spectra of HisL7H at 10 V (a) and 20 V (b) declustering potential. $[M + nH]^{n+}$ ions of the dimer with odd charge numbers in a are boxed.

**The BZIP-like region of protein L7 is sufficient to mediate homodimerization in vivo**

To confirm dimerization of the BZIP like domain of protein L7 in vivo we utilized aYeast Two Hybrid assay (11). We constructed a series of LexA fused L7 proteins (baits) and AD-fused L7 proteins and assayed them against each other in an Interaction Trap assay.

| | AD-L7$_{1-248}$ | AD-L7$_{125-248}$ | AD-L7$_{1-124}$ |
|---|---|---|---|
| LexA-L71-248 | ++ | - | + |
| LexA-L7125-248 | - | - | - |
| LexA-L71-124 | ++ | - | ++ |

Table 1. Growth of yeast EGY48 cotransfected with plasmids indicated on leucine minus medium (11).

In this assay we show, that the N-terminal BZIP-like domain (aa 1-124) of L7 is fully sufficient to mediate homodimerization in vivo. This results were confirmed by quantitative determination of the dissociation constants of His L7H and HisL7 with wild type L7 which were 120 and 40 nM for HisL7H and HisL7, respectively). Furthermore, both proteins are indistinguishable in their ability to inhibit cell free translation of reporter mRNAs (10). In conclusion, we defined the structure-function relationship of the BZIP like region of autoantigen L7. This may be useful to elucidate the role of protein L7 in pathomechanisms of autoimmune arthritis.

**References**


1. Hemmerich, P. von Mikecz, A., Neumann, F., Sözeri, O., Wolff-Vorbeck, G., Zoebelein,R. and Krawinkel, U. (1993) *Nucl. Acids Res.* **21**, 223- 231.
2. Neu, E., Hemmerich, P., Peter, H.-H., Krawinkel, U. and von Mikecz, A. *Arthritis Rheum.* in press
3. Landschulz, W.H., Johnson, P.F., and McKnight, S. L. (1988) *Science* **240**, 1759-1764.
4. Busch, S.J., and Sassone-Corsi, P. (1991) *Trends Biochem. Sci.* **6**, 36 - 40.
5. Neumann, F., Hemmerich, P., von Mikecz, A., Peter, H.H., and Krawinkel, U. (1995) *Nucl. Acids Res.* **23**, 195-202.
6. Neumann, F. and Krawinkel, U. *Exp. Cell. Res.* in press
7. Karas, M., Bahr, U., and Gießmann, U. (1991) *Mass Spectrom. Rev.* **10**, 335-358.
8. Smith, R.D., Loo, J.A., Orgorzalek-Loo, R.R., Busmann, M., and Udseth, H.R. (1991) *Mass Spectrom. Rev.* **10**, 359-401.
9. Przybylski, M., and Glocker, M.O. (1996) *Angew. Chem Int. Ed. Engl.* **35**, 793-920
10. Witte, S., Neumann, F., Krawinkel, U., and Przybylski, M.(1996) *J.Biol. Chem.* **271**, 18171-18175
11. Estojak, J., Brent, R., and Golemis, E.A. (1995) *Mol. Cell. Biol.* **15** 5820-5829

# PROTEINS AT WORK: TIME-RESOLVED FTIR STUDIES OF BACTERIORHODOPSIN AND THE GTP-BINDING PROTEIN P21

**Klaus Gerwert**

Ruhr-Universität Bochum, Lehrstuhl für Biophysik, D-44780 Bochum, Germany,

Fax: +49 234 7094238

## INTRODUCTION

Time-resolved FT-IR difference spectroscopy has recently been established as a new biophysical tool for the investigation of molecular reaction mechanisms of proteins at the atomic level.[1] Complementary to NMR spectroscopy and x-ray structure analysis, which provide the static protein ground state structure with atomic resolution, time-resolved FT-IR difference spectroscopy monitors the dynamics of proteins at the atomic level. Although infrared spectroscopy is a classical method of analytical and structural investigations [2], it can also provide valuable insight into the mechanisms of chemical reactions. Especially timeresolved Fourier-Transform-Infrared-difference-spectroscopy has recently proved itself as a powerful new method for studies of molecular reaction mechanisms of large proteins up to 120000 Dalton and time resolution up to nanoseconds.

The infrared spectrum of a protein is dominated by its peptide backbone amide I (C=O) and amide II (C-N, NH) vibrations. In addition to the strong amide I and amide II bands water also contributes largely to the absorption (1650 $cm^{-1}$).The major problem in measuring reactions consists in selecting small absorption bands of molecular groups which undergo reactions from the large background absorption of water and of the entire protein. This difficulty is met by obtaining difference spectra between spectra of the protein in its ground state and in an

NATO ASI Series, Vol. H 102
Interacting Protein Domains
Their Role in Signal and Energy Transduction
Edited by Ludwig Heilmeyer

activated state. Such measurements require highly sensitive instrumentation. FTIR has two advantages over conventional dispersive IR spectroscopy: the multiplex and the Jaquinot advantage. This makes it possible to reliably detect such small absorption changes with a time resolution of a few nanoseconds.

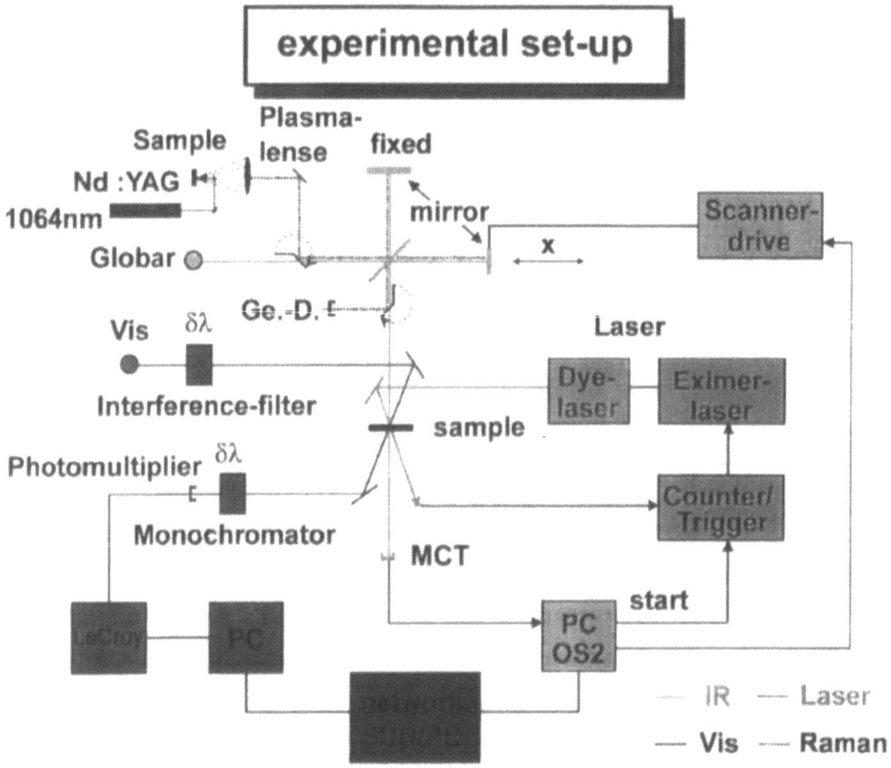


*Fig. 1. The experimental setup, consisting of an FT-IR apparatus (Bruker IFS66v) with globar, beamsplitter, mirrors, controller, detector and preamplifier connected to a 200 kHz and a 200 MHz transient recorder; a photolysis setup with light source, interference filters, monochromator, photomultiplier and transient recorder; and an excimer pumped dye laser system to activate the sample. The VIS and IR data are transfered to a workstation (SUN) network for kinetic analysis.*

In protein chemistry the macromolecule studied most intensely thus far by time-resolved FT-IR difference spectroscopy is the light-driven proton pump bacteriorhodopsin. Many research groups worldwide have contributed to the understanding of the mechanism of this membrane protein [3]. For reviews on FT-IR work on bacteriorhodopsin, see references [4,5] and [6]. Time-resolved FT-IR spectroscopy has furthermore provided a detailed picture of the light-induced electron transfer mechanisms of bacterial photosynthetic reaction centers. For reviews and recent FT-IR work, see references [7,8] and [9]. Beside photobiological proteins, time-resolved FT-IR spectroscopy can also be applied to proteins without an intrinsic chromophore by using photolabile trigger compounds.[10] Using "caged" GTP the molecular GTPase mechanism of the oncogenic protein H-ras p21, a molecule that plays a central role in the growth of cancer cells,[11] can be monitored.[12] Another example is "caged" $Ca^{2+}$, by which one can investigate the molecular mechanism of the $Ca^{2+}$-ATPase.[13,14] Furthermore "caged" electrons can be used to study redox reactions in proteins.[15]

Various time-resolved FT-IR techniques exist: the rapid-scan technique, which gives millisecond time resolution[16-18], the stroboscope technique, which gives microsecond time resolution[19, 20], and the step-scan technique, which gives nanoseconds time resolution.[21,22].In fig 1 our experimental set up is schematically drawn. Beside time resolved FTIR, simultaneously, Visible absorbance changes and FTRaman can be measured with this set up. Here two examples for the application of time-resolved FTIR spectroscopy to study molecular reaction mechanisms of proteins are given: bacteriorhodopsin and h ras p21.

## *BACTERIORHODOPSIN*

After light-excitation of bacteriorhodopsin's light-adapted ground-state $BR_{570}$ a photocycle starts with the intermediates $J_{610}$, $K_{590}$, $L_{550}$, $M_{410}$, $N_{530}$ and $O_{640}$ distinguished by different absorption maxima given by the indices. During the rise of the M-intermediate a proton is released to the extracellular side, and during the M-decay a proton is taken up from the cytoplasmic side. This creates a chemic osmotic proton gradient across the membrane.

### *The light-induced BR to K transition*

In Fig.2 a BR and K FT-Resonance-Raman spectrum is shown. The K intermediate is stabilized at 77 K. The band pattern in the 'fingerprint' region between 1300 $cm^{-1}$ and 1100 $cm^{-1}$ is typical of an all-trans configuration in BR and a 13-cis in K[4].The bands at 958 $cm^{-1}$ and 809 $cm^{-1}$ are caused by retinal Hoop (Hydrogen out of plane) vibrations. They are indicative of a strong distortion of the retinal. The BR and K Raman spectra indicate that isomerization of the chromophore induce tension in the protein that drives the following thermal reactions.

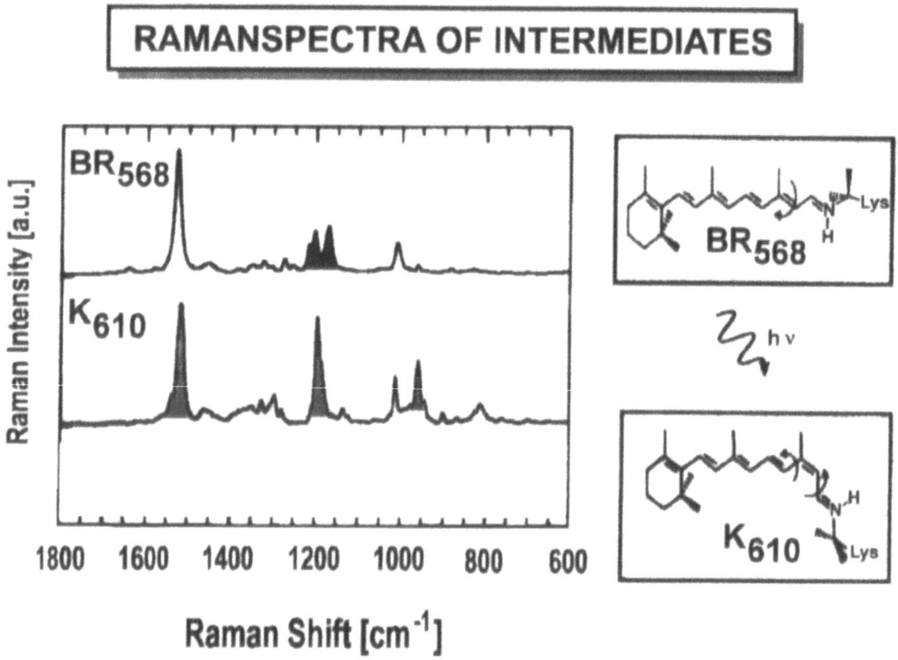


***Fig 2 In a) FT-Resonance Raman spectrum of the BR ground state and in b) of the K intermediate stabilized at 77K is shown.***

*The thermal reactions: K →L→M→N→O→BR*

The absorbance changes accompanying this reaction pathway can be monitored under physiological conditions by the step scan FTIR technique with a time resolution of 100 ns.[22]

In fig. 3 the IR absorbance changes between 1800 $cm^{-1}$ and 1000 $cm^{-1}$ during the photocycle are shown in a three-dimensional representation. Beyond a background absorbance of up to one absorbance unit, changes on the order of $10^{-3}$ to $10^{-4}$ are monitored with 3 $cm^{-1}$ spectral resolution and 100 ns time-resolution. As an example the absorbance change at 1188 $cm^{-1}$ is presented as a function of time in fig. 4.

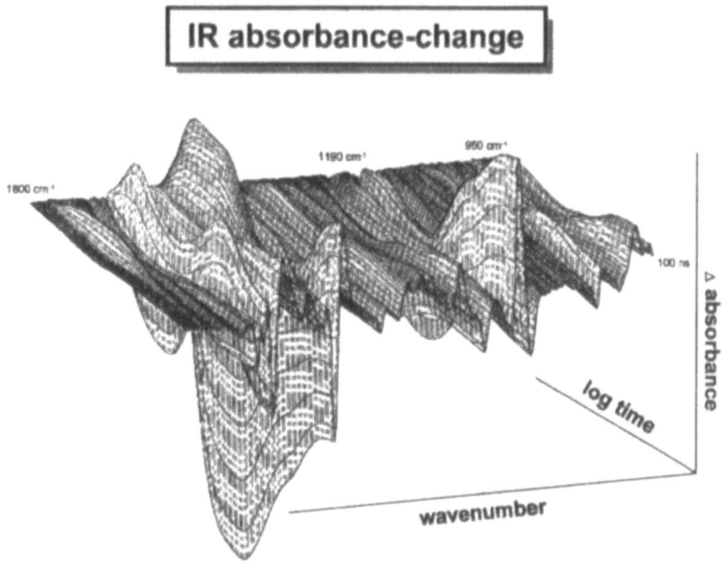


*Fig. 3. A three dimensional representation of the IR absorbance changes between 1800 $cm^{-1}$ and 1000 $cm^{-1}$ with 100 ns time-resolution and 3 $cm^{-1}$ spectral resolution accompanying bacteriorhodopsin's photocycle as revealed by a global-fit analysis. The time axis has a logarithmic scale in order to show a complete bacteriorhodopsin photocycle in one representation.*

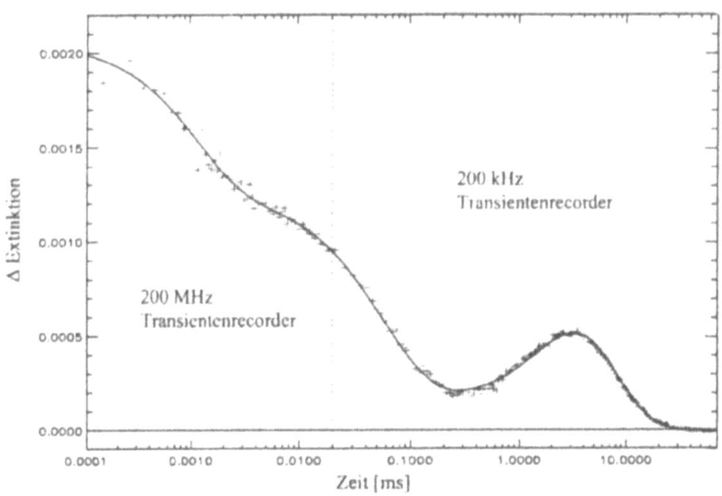


*Fig. 4. An example of a time course at a specific wavenumber. The absorbance change at 1188 $cm^{-1}$ from fig. 3 is shown. Because the data are recorded simultaneously with the 200 kHz and the 200 MHz transient recorder no inconsistency is seen in the overlap region near 20 μs. The band represents a C-C stretching vibration of a 13-cis protonated Schiff base retinal.*

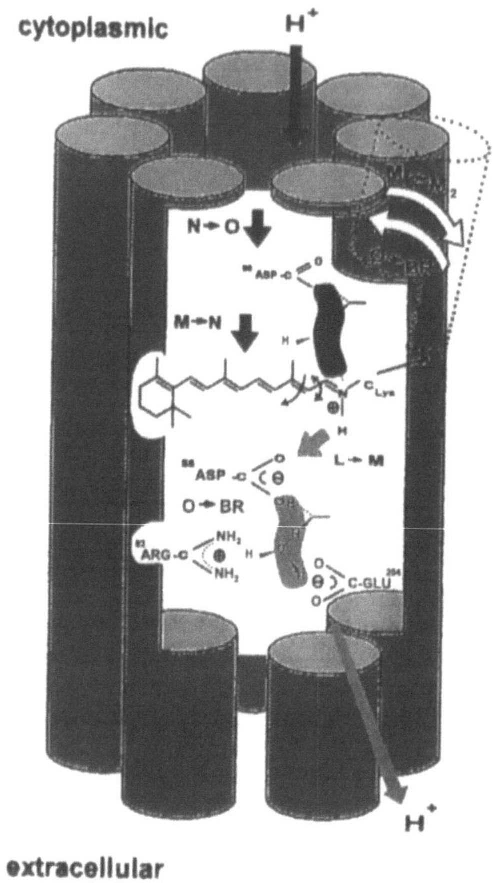


*Fig. 5. The current model of the proton pump mechanism of bacteriorhodopsin, to which many groups have contributed (for references, see text):*
*After the light-induced all-trans to 13-cis retinal isomerization in the BR to K transition, the Schiff base proton is transfered to Asp85 in the L to M transition. Simultaneously a proton is released from an ice-like H-bonded network to the extracellular site. This network is controlled by Glu204 and Arg82. Asp85 reprotonates the network in the O to BR reaction. The Schiff base is oriented in the $M_1$ to $M_2$ transition from the proton release site to the proton uptake site by a small backbone movement and determines thereby the vectoriality of proton pump. A larger backbone movement is observed in the M to N transition as compared to the $M_1$ to $M_2$ transition. Asp96 reprotonates the Schiff base in the M to N transition. Asp96 itself is reprotonated from the cytoplasmic site in the N to O transition. Proton transfer between Asp96 and the Schiff base takes place via a water-like H-bonded network.*

The appearance of this C-C stretching vibration band at 100 ns indicates the all-trans to 13-cis isomerization of retinal, which takes place within 450 fs (compare also the Raman-bands in fig

2a and b). Its disapperance at about 200 μs indicates the deprotonation of the Schiff base. This loss of charge greatly reduces the IR-absorbance of the chromophore. (The Schiff base connects the retinal to Lys216 of the protein (fig. 5).) The deprotonation kinetics of the Schiff base agree nicely with the protonation kinetics of the counterion Asp85, which can be followed at 1761 $cm^{-1}$ [18].The reappearance of the peak at 1188 $cm^{-1}$ in the millisecond time domain indicates the reprotonation of the Schiff base by Asp96[18].Protontransfer between asp 96 and the Schiffbase is performed via a hydrogenbonded network[23]. Recently, also for the protonrelease pathway a hydrogen bonded network proton transfer is identified[24]. Its final disappearance shows the chromophore's relaxation to the all-trans BR-ground state configuration. Based on the FTIR experiments a detailed model of the lightdriven proton pump bacteriorhodopsin is elucidated and presented in fig. 5.

### *H-RAS P21*

The small GTP-binding protein H-ras p21 plays a central role in signalling leading to cell-growth.[11] P21 appears to act as a switch: in the GTP bound form a signal is given to an effector molecule which is part of a cascade leading to cell proliferation and differentiation. The intrinsic GTPase activity is accelerated by GAP (GTPase Activating Protein). The transition from the active GTP bound to the inactive GDP bound form is accompanied by a conformational change. In order to specify in more detail the structural changes occurring in the GTP to GDP reaction, we have started complementary to the x-ray studies investigations by time-resolved FTIR difference spectroscopy.

The reaction is studied using caged GTP complexed to p21 instead of GTP to prevent hydrolysis. Using an intensive UV flash (308 nm, 100 mJ) the caged group is photolysed, GTP released and GTPase activity is precisely started. In Fig. 6 difference spectra between p21-GTP and finally p21-GDP is shown. The region between 1300 and 900 $cm^{-1}$ is dominated by phosphate bands. Using isotopically labelled caged GTP the band at 1237 $cm^{-1}$ is assigned to the released $P_i$ group, the band at 1144 $cm^{-1}$ is assigned to the γ-Phosphate vibration and the band at 1101 $cm^{-1}$ to the β–GDP vibration. The vibration of $P_i$ at 1237 $cm^{-1}$ indicates an $H_2PO_4^-$ species which seem to favour an assoziative mechanism instead of a dissoziative mechanism (Cepus, Goody, Gerwert, submitted)

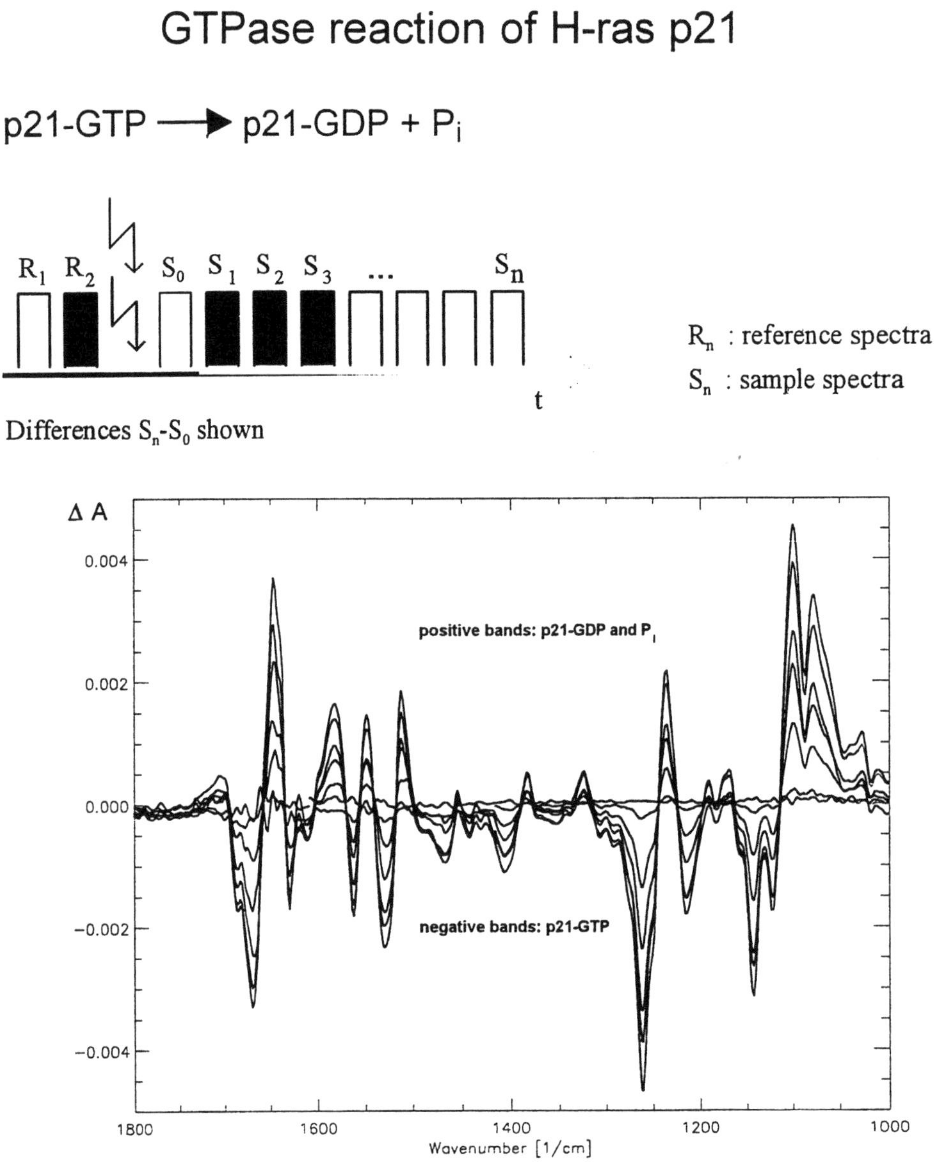


*Fig. 6. Principle of the measurement of the intrinsic p21 GTPase reaction. GTP is released from caged GTP using several UV flashes and GTPase activity starts. Difference spectra are taken between the first spectrum taken directly after GTP ($S_0$) is released and spectra taken during the following GTPase reaction ($S_i$) (upper part of the figure). The IR absorbance changes are represented in the lower part of the figure.*

The results on p21 show that molecular reaction mechanisms of proteins without intrinsic chromophore can now be investigated with time-resolved FTIR. In future studies beside the intrinsic mechanism of p21 the GAP catalysed GTPase mechanism and the interaction with the downstream effector raf will be studied.

*Acknowledgement:*

I would like to thank my Coworkers Dr. Benedik Heßling, Dr. Johannes le Coutre,Valentin Cepus, Ronald Brudler and Robin Rammelsberg for contributions to this work, and for collaboration on p21 Dr. R. Goody and Dr. F. Wittinghofer, Max Plank Institut, Dortmund.
The financial support of the DFG in the SFB394,Teilprojekt B1 and C2, is gratefully acknowledged

***References***

1. K. Gerwert, Current Opinion Structural Biology **3**, 769 (1993)
2. N.B. Colthup, L.H. Daly and S.E.Wiberley, in Introduction to Infrared and Raman Spectroscopy, 3rd edn. Academic Press,Boston (1990)
3. N. Grigorieff, T.A. Ceska, K.H. Downing, J.M. Baldwin and R. Henderson, J.Mol.Biol. **259**, 393 (1996)
4. M.S. Braiman and K. Rothschild, Ann.Rev.Biophys.Chem. **17**, 541 (1988)
5. K. Gerwert, Biochimica et Biophysica Acta **1101**, 147 (1992)
6. F. Siebert in *Biomolecular Spectroscopy*, part A (R.J.H. Clark, R.E. Hester eds.), 1, Wiley, Chichester, U.K. (1993)
7. B. Robert, E. Nabedryk, M. Lutz in *Time Resolved Spectroscopy* (R.J.H. Clark, R.E. Hester, eds.) Wiley, New York (1989)
8. W. Mäntele, in *The Photosynthetic Reaction Center* Vol. II (J. Deisenhofer, J. Norris, eds.) Academic Press, New York, 239 (1993)
9. R. Brudler, H.J.M. de Groot, W.B.S. van Liemt, W.F. Steggerda, R. Esmeijer, P. Gast, A.J. Hoff, J. Lugtenburg, K. Gerwert, EMBO J. **13**, 5523 (1994)
10. J.A. McGray and D. Trentham, Annu.Rev.Biophys.Biophys.Chem. **18**, 239 (1989)

11. A. Wittinghofer, E.F. Pai, Trends Biochem. Sci. **16**, 382 (1991)
12. K. Gerwert, V. Cepus, A. Scheidig, R.S. Goody in *Time Resolved Vibrational Spectroscopy* (A. Lau, F. Siebert, W. Werncke, eds.) Springer-Verlag, Berlin, 256 (1994)
13. A. Barth, W. Kreutz and W. Mäntele, Biochim.Biophys.Acta **1057**, 115 (1991)
14. A. Troullier, K. Gerwert, Y. Dupont, Biophy. J. **71,** 2970 (1996)
15. M. Lübben and K. Gerwert, FEBS Letters **397** 303 (1996)
16. M.S. Braiman, P.L. Ahl and K.J. Rothschild, Proc. Natl. Acad. Sci. USA **84**, 5221 (1987)
17. K. Gerwert, Ber. Bunsenges. Phys. Chem. **92**, 978 (1988)
18. K. Gerwert, G. Souvignier and B. Hess, Proc. Natl. Acad. Sci. USA, **87**, 9774-9778 (1990)
19. M.S. Braiman, O. Bousche and K.J. Rothschild, Proc. Natl. Acad. Sci. USA **88**, 2388 (1991)
20. G. Souvignier, K. Gerwert, Biophy. J. **63**, 1393 (1992)
21. W. Uhmann, A. Becker, C. Taran, F. Siebert, Appl. Spectrosc. **45**, 390 (1991)
22 R. Rammelsberg, B. Heßling, H. Chorongiewski and K. Gerwert, Applied Spectroscopy **51** (1997)

23. J. le Coutre, J. Tittor, D. Oesterhelt and K. Gerwert, Proc. Natl. Acad. Sci. USA **92**, 4962 (1995)
24. J. le Coutre, K. Gerwert, FEBS Letters **398,** 333 (1996)

# Part III

# Phospholipid Signaling

# Downstream Signaling from Phosphoinositide 3-Kinase

Lewis Cantley

Department of Cell Biology, Harvard Medical School and Division of Signal Transduction, Beth Israel Deaconess Medical Center, Boston, MA USA.

## 1 Introduction

Phosphoinositide 3-kinase was discovered because of its co-purification with the activated protein-Tyr kinases v-src and polyoma middle/c-src and its correlation with cell transformation by these oncogenes (Cantley et al., 1991). Interest in this enzyme has intensified with recent discoveries that phosphoinositide 3-kinase is involved in chemotaxis, insulin-dependent recruitment of Glut4, activation of MAP kinases, preventing apoptosis, actin rearrangement and membrane trafficking (Carpenter and Cantley, 1996). In addition, knocking out a PI 3-kinase gene in C. elegans (AGE1 gene) results in a three fold increase in longevity (Morris et al., 1996).

How does a single enzyme become involved in so many apparently unrelated cellular responses? A partial answer to this question is that phosphoinositide 3-kinase makes three different products that mediate separate cellular responses. Originally, this enzyme (previously called type I phosphatidylinositol kinase) was thought to be a phosphatidylinositol 4-kinase (PtdIns 4-K) since it phosphorylated PtdIns to make a product that co-purified with PtdIns-4-P. At that time no other monophosphorylated PtdIns was known to exist. Further investigation of the product of the enzyme revealed it to be PtdIns-3-P, indicating a new signaling pathway in mammalian cells (Whitman et al., 1988). Surprisingly, the purified PI 3-K was found not only to phosphorylate PtdIns at the D-3 position but to also phosphorylate PtdIns-4-P and PtdIns-4,5-$P_2$ at the D-3 position to form two additional novel lipids, PtdIns-3,4-$P_2$ and PtdIns-3,4,5-$P_3$ (Carpenter et al., 1990). Thus, this enzyme is properly called a phosphoinositide 3-kinase (PI 3-K) to indicate its ability to phosphorylate all three lipids. (PtdIns 3-K will be used to refer to enzymes that can only use PtdIns as a substrate; see below). None of the lipids with phosphate at the D-3 position are substrates of the phospholipases type C, suggesting that the lipids themselves act as signaling molecules rather than soluble inositol phosphates derived from the lipids (Serunian et al., 1989). In addition, the lipids with phosphate at the D-3 position are far less abundant than PtdIns-4-P or PtdIns-4,5-$P_2$ suggesting that they are acting as cellular regulators rather than as participants in membrane structure or cellular architecture (Auger et al., 1989b). These discoveries led to the rather complex PtdIns phosphorylation pathways that are indicated in Fig. 1.

NATO ASI Series, Vol. H 102
Interacting Protein Domains
Their Role in Signal and Energy Transduction
Edited by Ludwig Heilmeyer

A

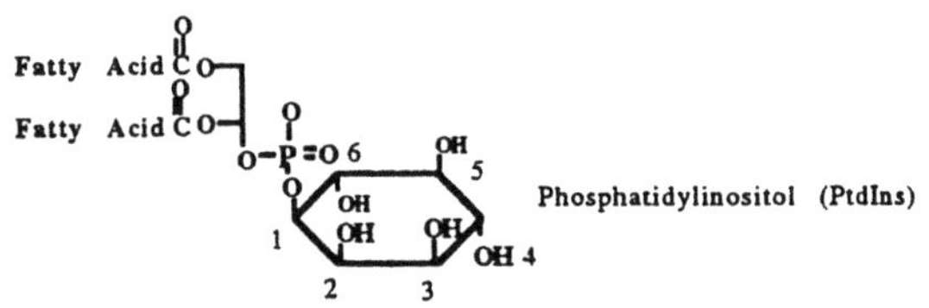


B

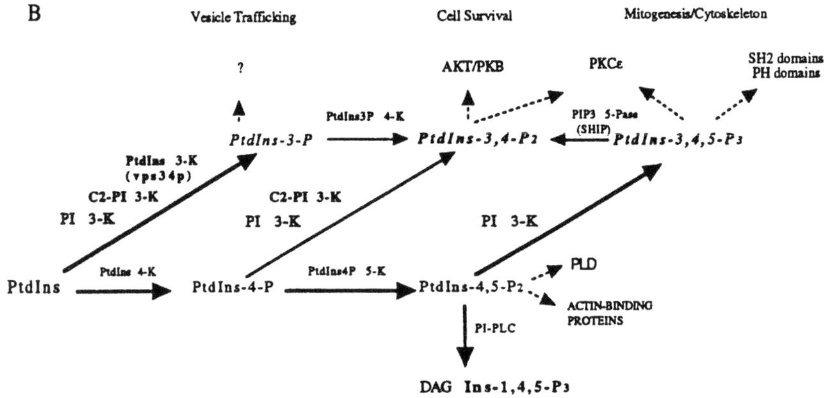


**Figure 1**: A. The structure of Phosphatidyinositol
B. Pathways for synthesis of phosphoinositides. The various enzymes that are capable of carrying out the phosphorylation steps are indicated. Some targets of phosphoinositides are also indicated. See text for details. Multiple phosphatases exist for dephosphorylating the various phosphoinositides. One of these enzymes (SHIP) is discussed in the text.

PI 3-kinase signaling is further complicated by the existence of multiple isoforms of this enzyme with different substrate specificities and different mechanisms of regulation. The original PI 3-kinase that was found to bind to protein-Tyr kinases consists of an 85 kd regulatory subunit and a 110 kd catalytic subunit (Carpenter et al., 1990)(Zvelebil et al., 1996). The 85 kd subunit has two SH2 domains that mediate binding to phosphotyrosine sites on tyrosine kinases and substrates of Tyr kinases (Fig. 2). It also has two Pro-rich sequences that bind to SH3 domains of src-family Tyr kinases (Kapeller et al., 1994). In addition, it has an SH3 domain that is capable of binding to Pro-rich sequences in other signaling proteins. Between the two Pro-rich sequences in p85 is a domain with homology to BCR (BH domain) that mediates binding to the low molecular weight GTP-binding proteins RAC and CDC42 (Zeng et al., 1994)(Tolias et al., 1995). This domain resembles rho/rac-GTPases but does not appear to catalyze GTP hydrolysis on these proteins. The binding of p110 to p85 occurs via a region of p85 between the two SH2 domains.

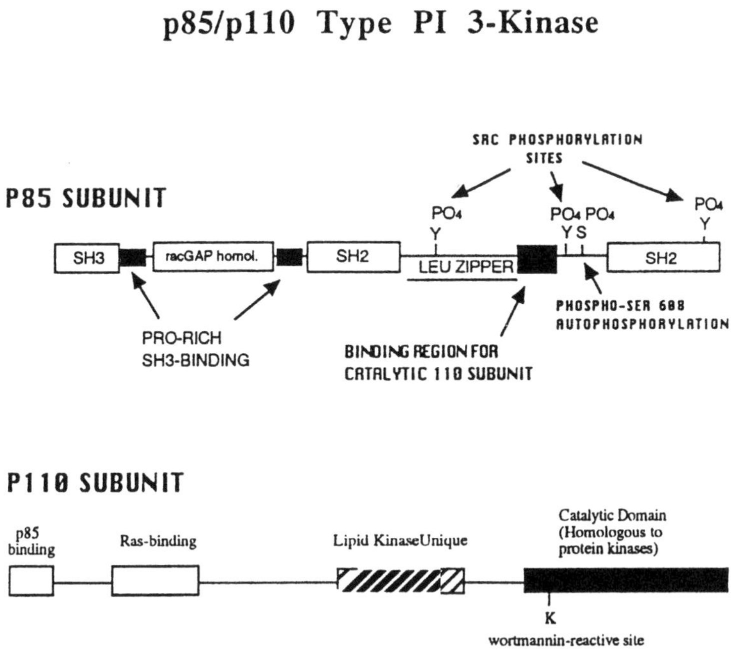


**Figure 2:** The structure of the p85/p110-type PI 3-K. The various domains are discussed in the text.

The p110 catalytic subunit also has domains of homology to other proteins (Zvelebil et al., 1996). The C-terminus of this subunit has distant but significant homology to the catalytic domains of protein-Ser/Thr kinases. In fact, this subunit is capable of phosphorylating a Ser residue on the regulatory subunit at a low rate and thereby turning down the catalytic activity (Carpenter et al., 1993)(Dhand et al., 1994). Thus, PI 3-kinase is a dual-specificity kinase. There is evidence that this subunit can phosphorylate other proteins in vitro at low rates but no other *in vivo* targets have yet been identified. Upstream of the domain with homology to protein kinases is a domain that is not found in protein kinases but is found in all the PI 3-kinases as well as PtdIns 4-kinases. This domain has been called the lipid kinase unique domain (LKU domain). Binding of p85 to p110 occurs via a region of p110 at the N-terminus. Finally, between the p85-binding domain and the LKU domain is a region of p110 that bind to activated RAS (Rodriguez-Viciana et al., 1996). There is evidence that optimal activation of p85/p110 type PI 3-kinases requires simultaneous binding to Tyr-phosphorylated proteins and binding to activated RAS.

As indicated above, the p85/p110 PI 3-kinase is capable of phosphorylating the D-3 position of PtdIns, PtdIns-4-P or PtdIns-4,5-P2 *in vitro*. In most quiescent mammalian cells, PtdIns-3,4-$P_2$ and PtdIns-3,4,5-$P_3$ are nominally absent but PtdIns-3-P, PtdIns-4-P and PtdIns-4,5-$P_2$ are constitutively produced. Agents that activate the p85/p110 type PI 3-kinase cause an acute production of PtdIns-3,4,5-$P_3$ and a delayed production of PtdIns-3,4-$P_2$ suggesting that PtdIns-4,5-$P_2$ is the major *in vivo* target of the p85/p110 type PI 3-kinase and that PtdIns-3,4-$P_2$ is produced by a phosphatase that removes the 5 phosphate from PtdIns-3,4,5-$P_3$ (Stephens et al., 1993)(Damen et al., 1996). Many G-protein coupled receptors (e.g. the thrombin receptor in platelets and formyl-peptide receptor in neutrophils) can also stimulate the production of PtdIns-3,4,5-$P_3$ and PtdIns-3,4-$P_2$. Rather than using the p85/p110 type PI 3-kinase, these receptors appear to activate a related enzyme that is activated by the $\beta\gamma$ subunits of G-proteins (Stephens et al., 1996). The $\beta\gamma$-activated PI 3-kinase has similar substrate specificity to the p85/p110 type PI 3-kinase.

There are also homologues of PI 3-K that are not capable of phosphorylating PtdIns-4,5-$P_2$. For example, the bulk of PtdIns-3-P in cells is not produced by the p85/p110 type PI 3-kinase but rather by a related enzyme called VPS34p (see below) that can only utilize PtdIns as a substrate and is properly called a PtdIns 3-kinase (Volinia et al., 1995). Thus far, there is no evidence for regulation of this enzyme by extracellular ligands. Finally, a PI 3-kinase has been recently cloned that is able to phosphorylate PtdIns and PtdIns-4-P but not PtdIns-4,5-$P_2$ (Virbasius et al., 1996). Thus, this enzyme provides an additional pathway for making PtdIns-3,4-$P_2$ that does not involve the synthesis of PtdIns-3,4,5-$P_3$. The existence of distinct pathways for making the three different phosphoinositides with phosphate at the D-3 position supports the idea discussed below that these three lipids have distinct functions in the cell.

## 2 Cellular Function of PtdIns-3-P

*Saccharomyces cerevisiae* has only a single member of the PI 3-kinase family. The gene for this enzyme was originally discovered as a vacuolar protein sorting mutant (vps34) (De Camilli et al., 1996). Mutations or deletions of this gene result in defects in trafficking of vesicles from the golgi to the vacuole without major defects in other intracellular trafficking. The VPS34p protein can only phosphorylate PtdIns and consistent with this, PtdIns-3-P is found in yeast but PtdIns-3,4-$P_2$ and PtdIns-3,4,5-$P_3$ are not (Stack and Emr, 1994)(Auger et al., 1989a). Mutations in the VPS34 gene eliminate synthesis of PtdIns-3-P. These results argue that PtdIns-3-P mediates membrane trafficking events. However, the target of this lipid responsible for mediating golgi to vacuole trafficking is not yet known.

As discussed above, PtdIns-3-P is constitutively synthesized in mammalian cells and the mammalian homologue of vps34p is likely to be responsible for constitutive synthesis of this lipid in mammalian cells (Volinia et al., 1995). The possibility that a PI 3-kinase family member is involved in membrane trafficking in mammalian cells is supported by the ability of wortmannin (a relatively specific PI 3-kinase inhibitor at concentrations below 100 nM) to alter trafficking of various proteins from late endosomes to the lysosomes. However, more definitive experiments need to be done since wortmannin (at 100 nM) also inhibits a PtdIns 4-kinase that is located in the cytosol and golgi (Meyers and Cantley., 1997). Evidence for the importance of the p85/p110 type PI 3-kinase in PDGF receptor trafficking to the lysosomes is provided by studies in which mutations at the sites of the PDGF receptor that bind the p85/p110 type PI 3-kinase block lysosomal degradation of this receptor (Joly et al., 1994). Wortmannin also blocks trafficking of this receptor to the lysosomes. Since the p85/p110 type PI 3-kinase makes all three lipids and wortmannin inhibits this enzyme completely at 100 nM concentrations, it is not possible to determine which lipid is critical for PDGF receptor trafficking. Also, in mammalian cells there exists a PtdIns-specific 3-kinase that is completely resistant to concentrations of wortmannin greater than 1 μM. Thus, it cannot be concluded that wortmannin-resistant trafficking steps do not involve PtdIns-3-P. More specific mechanisms for blocking PtdIns-3-P production are needed to understand the role of this lipid in vesicle trafficking in mammalian cells.

## 3 Cellular Targets of PtdIns-3,4-$P_2$

PtdIns-3,4-$P_2$ can in theory be produced by three different pathways. There is evidence that in formyl-peptide-stimulated neutrophils, PDGF-stimulated fibroblasts and thrombin-stimulated platelets the bulk of this lipid is produced by a two step process in which PtdIns-3,4,5-$P_3$ is produced by activated PI 3-kinase and the PtdIns-3,4-$P_2$ is produced by hydrolysis of the D-5 phosphate. An SH2-containing phosphatase (SHIP) that is capable of dephosphorylating the D-5 position of PtdIns-3,4,5-$P_3$ has been identified (Damen et al., 1996). This enzyme is Tyr-phosphorylated and binds to SHC and Grb-2 in response to cell stimulation, suggesting a complex regulation. It also binds to ITIM motifs of B-cell inhibitory receptors. In most cells that have been investigated, PtdIns-3,4,5-$P_3$ appears acutely following cell stimulation, peaks within less than a minute and decays to basal levels (Auger et al., 1989b). The duration of this signal depends on the cell type. In contrast, PtdIns-3,4-$P_2$ appears more slowly (consistent with it being derived from PtdIns-3,4,5-$P_3$) and persists for much longer time periods (Stephens et al., 1993). Insulin stimulates not only the activity of PI 3-kinase but also a novel PtdIns-3,4,5-$P_3$ 5-phosphatase (distinct from SHIP) (Czech, 1996). Thus, the ratio of PtdIns-3,4-$P_2$ to PtdIns-3,4,5-$P_3$ in stimulated cells is under strict control.

A second pathway for production of PtdIns-3,4-$P_2$ is the direct phosphorylation of PtdIns-4-P. The p85/p110 type PI 3-kinase can carry out this reaction in vitro, but PtdIns-4-P is a poor substrate compared to PtdIns and PtdIns-4,5-$P_2$ (Carpenter et al., 1990). In contrast, as discussed above, a novel C2-containing PI 3-kinase prefers PtdIns-4-P over PtdIns-4,5-$P_2$ as a substrate (Virbasius et al., 1996). The regulation of this enzyme is not yet understood.

Finally, a PtdIns-3-P 4-kinase has also been identified in platelets and red cells (Yamamoto et al., 1990)(Graziani et al., 1992). The gene for this enzyme has not been cloned. It could potentially produce PtdIns-3,4-$P_2$ in response to signals not yet identified.

Two potential downstream targets for PtdIns-3,4-$P_2$ have been identified. First, examination of baculovirus-expressed protein kinase C (PKC) isoforms revealed that the calcium-independent PKC family members, most notably PKCε, are activated by PtdIns-3,4-$P_2$ and to a lesser extent by PtdIns-3,4,5-$P_3$

(Toker et al., 1994). PtdIns-4,5-$P_2$, which has the same number of phosphates, did not activate PKCε, indicating a high specificity for the PtdIns-3,4-$P_2$ isomer. Evidence that PKCε is an in vivo target of PtdIns-3,4-$P_2$ and/or PtdIns-3,4,5-$P_3$ is provided by the observation that PDGF causes the recruitment of PKCε to the membrane and mutations that block activation of PI 3-kinase eliminate this recruitment (Moriya et al., 1996). Since PKCε has been implicated in a variety of cellular pathways, including the activation of c-raf-1, this enzyme might mediate certain PI 3-kinase-dependent responses.

A second target of PtdIns-3,4-$P_2$ is the AKT/PKB protein-Ser/Thr kinase. The gene for this enzyme was identified as a retrovirus-encoded oncogene. It has high homology to the kinase domains of PKA and PKC (thus called PKB). However, it has a pleckstrin-homology (PH) domain at the N-terminus that is not found in the related enzymes. The cellular AKT is activated by the PDGF receptor and this activation can be blocked by dominant-negative PI 3-kinase, by wortmannin and by mutations in the PDGF receptor that fail to activate PI 3-kinase (Franke et al., 1995). A single point mutation in the PH domain of AKT blocks its activation by the PDGF receptor. *In vitro* and *in vivo* studies have shown that PtdIns-3,4-$P_2$ activates purified AKT (Franke et al., 1997). PtdIns-4,5-$P_2$ has no effect. Surprisingly, PtdIns-3,4,5-$P_3$ also fails to activate AKT and causes some inhibition of the enzyme. This activation is mediated by the PH domain since the same point mutation that blocks activation of AKT *in vivo* blocks activation by PtdIns-3,4-$P_2$ *in vitro*. Consistent with this idea, the PH domain of AKT binds with high specificity to PtdIns-3,4-$P_2$ and this binding is blocked by the same point mutation (Franke et al., 1997). The mechanism of activation of AKT by PtdIns-3,4-$P_2$ appears to involve dimerization of the enzyme(Datta et al., 1995)(Franke et al., 1997). Full activation of the enzyme *in vivo* also involves phosphorylation of AKT by an unidentified protein-Ser/Thr kinase (Andjelkovic et al., 1996).

The function of AKT in cellular responses has not yet been clearly determined. It may be mediating insulin effects on glyconogenesis since it can phosphorylate and inhibit glycogen synthetase kinase-3 (GSK-3). In addition, there is evidence that AKT may play a role in mediating insulin-dependent recruitment of GLUT4-containing vesicles to the plasma membrane. PI 3-kinase has been implicated in both of these insulin-dependent events.

Another interesting function of AKT is prevention of apoptosis. Both PI 3-kinase and AKT have been shown to block apoptosis in several cell systems (Yao and Cooper, 1995)(unpublished results of Songyang, Franke and Cantley). Thus, it is likely that AKT mediates the anti-apoptotic effects of PI 3-kinase (Yao and Cooper, 1995). Since insulin, IGF-1, NGF and IL3 (agents known to protect various cell types from apoptosis) activate PI 3-kinase and AKT, it is likely that these enzymes are playing critical roles in mediating extracellular factor-dependent cell survival.

## 4 Cellular Targets of PtdIns-3,4,5-$P_3$

PtdIns-3,4,5-$P_3$ appears to be produced uniquely by phosphorylation of the D-3 position of PtdIns-4,5-$P_2$. As discussed above, this lipid can activate PKCε *in vitro*, although it appears to be less potent than PtdIns-3,4-$P_2$. In contrast, we discovered that PtdIns-3,4,5-$P_3$ can bind with high affinity to certain SH2 domains, including the SH2 domains of the p85 regulatory subunit of PI 3-kinase and the SH2 domain of $pp60^{c\text{-}src}$ (Rameh et al., 1995). PtdIns-3,4-$P_2$ and PtdIns-4,5-$P_2$ do not bind significantly to these SH2 domains. Binding of PtdIns-3,4,5-$P_3$ to p85 and src SH2 domains prevents them from binding to Tyr-phosphorylated proteins and peptides. Likewise, phosphotyrosine peptides of the appropriate sequence block binding of PtdIns-3,4,5-$P_3$ to SH2 domains. Although the binding of phosphopeptide and PtdIns-3,4,5-$P_3$ to SH2 domains are mutually exclusively, these two molecules do not use the same binding pocket since mutations in the phosphotyrosine pocket of pp60c-src SH2 that block phosphopeptide binding do not prevent PtdIns-3,4,5-$P_3$ binding (Rameh et al., 1995). These results suggest that local production of PtdIns-3,4,5-$P_3$ by receptor-associated PI 3-kinase can result in the recruitment of SH2-containing proteins to the membrane without the need for direct binding to the receptor. This may be a mechanism for amplifying certain signaling pathways.

A third family of targets for PtdIns-3,4,5-$P_3$ includes a subgroup of PH domains. As discussed above, PtdIns-3,4,5-$P_3$ does not substitute for PtdIns-3,4-$P_2$ in activation of AKT via the PH domain. However, we have found that other PH domains bind PtdIns-3,4,5-$P_3$ with high affinity and specificity (unpublished results of Rameh and Cantley). In fact most PH domains that we have investigated bind PtdIns-3,4,5-$P_3$ with much higher affinity than PtdIns-3,4-$P_2$. The PH domain of the B-cell specific Tyr kinase, BTK, has high affinity and specificity for PtdIns-3,4,5-$P_3$ compared to PtdIns-3,4-$P_2$ or PtdIns-4,5-$P_2$. A point mutation in this domain that causes agammaglobulinemia impairs binding of PtdIns-3,4,5-$P_3$ but does not affect binding of PtdIns-4,5-$P_2$. These results suggest that a collection of PH-domain and SH2-domain-containing signaling proteins may be recruited to the membrane by binding to PtdIns-3,4,5-$P_3$.

In summary, PI 3-kinase is implicated in a host of cellular responses. This enzyme can produce three different phosphoinositides and the levels of these three lipids are acutely controlled by kinases and phosphatases. Each of these lipids appears to mediate different cellular responses and protein targets that have specificity for individual lipids have been identified. It is clear that much work is needed to elucidate the mechanism by which the lipid products of PI 3-kinase mediate cellular responses.

**Acknowledgment:**
This research was supported by NIH grants GM36624 and GM41890

**References:**

Andjelkovic, M., Jakubowicz, T., Cron, P., Ming, X. F., Han, J. W. and Hemmings, B. A. (1996). Activation and phosphorylation of a pleckstrin homology domain containing protein kinase (RAC-PK/PKB) promoted by serum and protein phosphatase inhibitors. Proc. Natl. Acad. Sci. U.S.A. *93*, 5699-5704.

Auger, K. R., Carpenter, C. L., Cantley, L. C. and Varticovski, L. (1989a). Phosphatidylinositol 3-kinase and its novel product, phosphatidylinositol 3-phosphate, are present in Saccharomyces cerevisiae. J Biol Chem *264*, 20181-4.

Auger, K. R., Serunian, L. A., Soltoff, S. P., Libby, P. and Cantley, L. C. (1989b). PDGF-dependent tyrosine phosphorylation stimulates production of novel polyphosphoinositides in intact cells. Cell *57*, 167-75.

Cantley, L. C., Auger, K. R., Carpenter, C., Duckworth, B., Graziani, A., Kapeller, R. and Soltoff, S. (1991). Oncogenes and signal transduction [published erratum appears in Cell 1991 May 31;65(5):following 914]. Cell *64*, 281-302.

Carpenter, C. L., Auger, K. R., Duckworth, B. C., Hou, W. M., Schaffhausen, B. and Cantley, L. C. (1993). A tightly associated serine/threonine protein kinase regulates phosphoinositide 3-kinase activity. Mol Cell Biol *13*, 1657-65.

Carpenter, C. L. and Cantley, L. C. (1996). Phosphoinositide kinases. Curr Opin Cell Biol *8*, 153-8.

Carpenter, C. L., Duckworth, B. C., Auger, K. R., Cohen, B., Schaffhausen, B. S. and Cantley, L. C. (1990). Purification and characterization of phosphoinositide 3-kinase from rat liver. J Biol Chem *265*, 19704-11.

Damen, J. E., Liu, L., Rosten, P., Humphries, R. K., Jefferson, A. B., Majerus, P. W. and Krystal, G. (1996). The 145-kDa protein induced to associate with Shc by multiple cytokines is an inositol tetraphosphate and phosphatidylinositol 3,4,5-trisphosphate 5-phosphatase. Proc. Natl. Acad. Sci. U.S.A. *93*, 1689-1693.

Datta, K., Franke, T. F., Chan, T. O., Makris, A., Yang, S. I., Kaplan, D. R., Morrison, D. K., Golemis, E. A. and Tsichlis, P. N. (1995). AH/PH domain-mediated interaction between Akt molecules and its potential role in Akt regulation. Mol Cell Biol *15*, 2304-10.

De Camilli, P., Emr, S. D., McPherson, P. S. and Novick, P. (1996). Phosphoinositides as regulators in membrane traffic. Science *271*, 1533-1539.

Dhand, R., Hiles, I., Panayotou, G., Roche, S., Fry, M. J., Gout, I., Totty, F., Truong, O., Vicendo, P., Yonezawa, K. and Waterfield, M. (1994). PI 3-kinase is a dual specificity enzyme: autophosphorylation by an intrinsic protein-serine kinase activity. EMBO J. *13*, 522-533.

Franke, T. F., Kaplan, D. R., Cantley, L. C. and Toker, A. (1997). Activation of AKT by PtdIns-3,4-P2. Science *(in press)*,

Franke, T. F., Yang, S. I., Chan, T. O., Datta, K., Kazlauskas, A., Morrison, D. K., Kaplan, D. R. and Tsichlis, P. N. (1995). The protein kinase encoded by the Akt proto-oncogene is a target of the PDGF-activated phosphatidylinositol 3-kinase. Cell *81*, 727-36.

Graziani, A., Ling, L. E., Endemann, G., Carpenter, C. L. and Cantley, L. C. (1992). Purification and characterization of human erythrocyte phosphatidylinositol 4-kinase. Phosphatidylinositol 4-kinase and phosphatidylinositol 3-monophosphate 4-kinase are distinct enzymes. Biochem J *284*, 39-45.

Joly, M., Kazlauskas, A., Fay, F. S. and Corvera, S. (1994). Disruption of PDGF receptor trafficking by mutation of its PI-3 kinase binding sites. Science *263*, 684-7.

Kapeller, R., Prasad, K. V., Janssen, O., Hou, W., Schaffhausen, B. S., Rudd, C. E. and Cantley, L. C. (1994). Identification of two SH3-binding motifs in the regulatory subunit of phosphatidylinositol 3-kinase. J Biol Chem *269*, 1927-33.

Meyers, R. and Cantley, L.C. (1997). Cloning and Characterization of a human wortmannin-sensitive PtdIns 4-kinase. J. Biol. Chem. *(in press)*

Moriya, S., Kazlauskas, A., Akimoto, K., Hirai, S., Mizuno, K., Takenawa, T., Fukui, Y., Watanabe, Y., Ozaki, S. and Ohno, S. (1996). Platelet-derived growth factor activates protein kinase C epsilon through redundant and independent signaling pathways involving phospholipase C gamma or phosphatidylinositol 3-kinase. Proc. Natl. Acad. Sci. U.S.A. *93*, 151-155.

Morris, J. Z., Tissenbaum, H. A. and Ruvkun, G. (1996). A phosphatidylinositol-3-OH kinase family member regulating longevity and diapause in Caenorhabditis elegans. Nature *382*, 536-539.

Rameh, L. E., Chen, C. S. and Cantley, L. C. (1995). Phosphatidylinositol (3,4,5)P3 interacts with SH2 domains and modulates PI 3-kinase association with tyrosine-phosphorylated proteins. Cell *83*, 821-30.

Rodriguez-Viciana, P., Warne, P. H., Vanhaesebroeck, B., Waterfield, M. D. and Downward, J. (1996). Activation of phosphoinositide 3-kinase by interaction with Ras and by point mutation. Embo J *15*, 2442-51.

Serunian, L. A., Haber, M. T., Fukui, T., Kim, J. W., Rhee, S. G., Lowenstein, J. M. and Cantley, L. C. (1989). Polyphosphoinositides produced by phosphatidylinositol 3-kinase are poor substrates for phospholipases C from rat liver and bovine brain. J Biol Chem *264*, 17809-15.

Stack, J. H. and Emr, S. D. (1994). Vps34p required for yeast vacuolar protein sorting is a multiple specificity kinase that exhibits both protein kinase and phosphatidylinositol-specific PI 3-kinase activities. J Biol Chem *269*, 31552-62.

Stephens, L., Hawkins, P. T., Eguinoa, A. and Cooke, F. (1996). A heterotrimeric GTPase-regulated isoform of PI3K and the regulation of its potential effectors. Philos Trans R Soc Lond B Biol Sci *351*, 211-5.

Stephens, L. R., Jackson, T. R. and Hawkins, P. T. (1993). Agonist-stimulated synthesis of phosphatidylinositol(3,4,5)-trisphosphate: a new intracellular signalling system? Biochim. Biophys. Acta *1179*, 27-75.

Toker, A., Meyer, M., Reddy, K. K., Falck, J. R., Aneja, R., Aneja, S., Parra, A., Burns, D. J., Ballas, L. M. and Cantley, L. C. (1994). Activation of protein kinase C family members by the novel polyphosphoinositides PtdIns-3,4-P2 and PtdIns-3,4,5-P3. J Biol Chem *269*, 32358-67.

Tolias, K. F., Cantley, L. C. and Carpenter, C. L. (1995). Rho family GTPases bind to phosphoinositide kinases. J Biol Chem *270*, 17656-9.

Virbasius, J. V., Guilherme, A. and Czech, M. P. (1996). Mouse p170 is a novel phosphatidylinositol 3-kinase containing a C2 domain. J Biol Chem *271*, 13304-7.

Volinia, S., Dhand, R., Vanhaesebroeck, B., MacDougall, L. K., Stein, R., Zvelebil, M. J., Domin, J., Panaretou, C. and Waterfield, M. D. (1995). A human phosphatidylinositol 3-kinase complex related to the yeast Vps34p-Vps15p protein sorting system. Embo J *14*, 3339-48.

Whitman, M., Downes, C. P., Keeler, M., Keller, T. and Cantley, L. (1988). Type I phosphatidylinositol kinase makes a novel inositol phospholipid, phosphatidylinositol-3-phosphate. Nature *332*, 644-6.

Yamamoto, K., Graziani, A., Carpenter, C., Cantley, L. C. and Lapetina, E. G. (1990). A novel pathway for the formation of phosphatidylinositol 3,4- bisphosphate. Phosphorylation of phosphatidylinositol 3-monophosphate by phosphatidylinositol-3-monophosphate 4-kinase. J Biol Chem *265*, 22086-9.

Yao, R. and Cooper, G. M. (1995). Requirement for phosphatidylinositol-3 kinase in the prevention of apoptosis by nerve growth factor. Science *267*, 2003-2006.

Zeng, Y., Bagrodia, S. and Cerione, R. A. (1994). Activation of phosphoinositide 3-kinase activity by cdc42Hs binding to p85. J. Biol. Chem. *269*, 18727-18730.

Zvelebil, M. J., MacDougall, L., Leevers, S., Volinia, S., Vanhaesebroeck, B., Gout, I., Panayotou, G., Domin, J., Stein, R., Pages, F. and et al. (1996). Structural and functional diversity of phosphoinositide 3-kinases. Philos Trans R Soc Lond B Biol Sci *351*, 217-23.

# PHOSPHOINOSITIDE-3-KINASE MEDIATED ACTIVATION OF JNK BY THE βPDGFR

López-Ilasaca, M.A.[1], Gutkind, J.S.[2] and Heidaran, M.A.[3]

[1]Max Planck Research Unit "Molecular Cell Biology", Medical Faculty, University of Jena, D-07747 Jena, Germany
[2] Molecular Signaling Unit, Oral and Pharyngeal Cancer Branch, National Institute of Dental Research, NIH, Bethesda, MD 20892, USA
[3] Laboratory of Cellular and Molecular Biology, National Cancer Institute, NIH, Bethesda, MD 20892, USA.

## 1 INTRODUCTION

Stimulation of a variety of cell surface receptors leads to a rapid elevation of the enzymatic activity of a family of closely related serine-threonine kinases, known as a mitogen-activated protein kinases (MAPKs) (Cano & Mahadevan, 1995). MAPKs have been classified into three subfamilies: extracellular signal regulated kinases (ERKs), also known as $p44^{mapk}$ and $p42^{mapk}$; stress activated protein kinases (SAPKs), also termed as c-jun N-terminus kinases (JNKs); and p38 kinase. Translocation of activated MAPK to the nucleus and subsequent phosphorylation of a variety of transcription factors including c-Myc, Elk-1, and ATF2 support the involvement of MAPK in transducing cytoplasmic signals to nuclear responses. JNKs phosphorylate the amino-terminal transactivating domain of c-Jun and ATF2, thereby increasing their transcriptional activity. p38 function is still unknown, although recent data suggest the involvement of this kinase in the inflamatory response (Post & Brown, 1996).

A series of protein-protein interactions have been implicated in signalling by tyrosine kinase receptors. Cytoplasmic molecules that interact with tyrosine phosphorylated growth factor receptors include the adaptor proteins Shc and Grb2, phospholipase C-γ, PI-3 kinase, and GAP. Whereas, receptor tyrosine kinase activate JNKs poorly in fibroblasts (Kyriakis *et al*, 1994) this class of receptors have been shown to induce activation of JNKs in other cells of epithelial origin (Coso *et al*., 1995). Thus we sought to investigate the mechanism of JNK activation by this class of receptors using a transient expression system in COS-7 cells.

## 2 MATERIAL AND METHODS

**2.1 Cell lines and Transfection.** COS-7 cells were cultured in DMEM media supplemented with 10% of fetal bovine serum. Cells were transfected by the DEAE-dextran technique, adjusting the total amount of DNA to 5-10 μg per plate with vector DNA (pcDNA3; Invitrogen), when necessary (Coso *et al*., 1995a).

NATO ASI Series, Vol. H 102
Interacting Protein Domains
Their Role in Signal and Energy Transduction
Edited by Ludwig Heilmeyer

**2.2 Expression Plasmids.** Expression plasmids for an epitope-tagged JNK (pcDNA3-HA-JNK), for the dominant negative mutants of the small GTP-binding proteins Ras, RhoA, Rac1 and Cdc42 have been described (Coso *et al.*, 1995b). Plasmids expressing βPDGF receptor wild type and the mutants Y740F+Y751F or Y1009F+Y1021F have been previously described (Valius & Kazlauskas, 1993)

**2.3 JNK and MAPK Assays.** JNK activity in lysates from COS-7 cells transfected with an expression vector for an epitope-tagged JNK (pcDNA3-HA-JNK) was determined upon immunoprecipitation with the anti-HA specific monoclonal antibody 12CA5 (Babco) using bacterially expressed GST-ATF2(96) fusion protein as a substrate, as previously described (Coso *et al.*, 1995b). The products of the kinase reactions were fractionated in 12% SDS-PAGE, and radioactivity incorporated into GST-ATF2(96) was determined with the use of a scanning densitometer (PDI, Inc). MAPK activity in COS-7 cells transfected with an epitope-tagged MAPK was determined using myelin basic protein as a substrate (Crespo *et al.*, 1994). In each case, parallel samples were immunoprecipitated with anti-HA antibody and processed by Western blot analysis using the appropriate antibody.

## 3 RESULTS AND DISCUSSION

### 3.1 Ligand-Induced Tyrosine Phosphorylation of The βPDGFR Leads To Enhancement of JNK Activity in COS-7 Cells.

To determine the molecular basis for activation of JNKs by receptor tyrosine kinases (RTKs), we used a transient expression of the wild type βPDGFR (βRWT) with an epitope-tagged JNK (HA-JNK) in COS-7 cells. As shown in figure 1 (upper panel), PDGF-BB potently induced an increase in JNK activity in cells expressing βRWT, as judged by ability of anti-HA-immunoprecipitates to phosphorylate GST-ATF2(96) fusion protein *in vitro.*

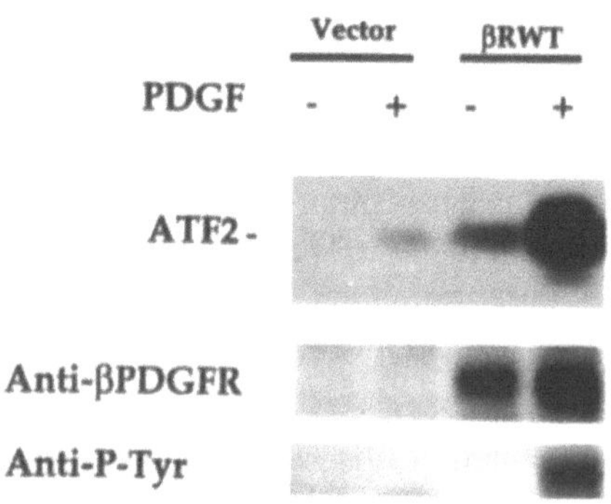


**Fig. 1. PDGFR activation induce JNK activity.** COS-7 cells were transfected with vector alone or the βRWT and treated or untreated with PDGF-BB by 10 min. Cells were lysed and JNK activity was determined in the HA-immunoprecipitates using GST-ATF2(96) as substrate. In the lower panels are shown the Western blot analysis of total cell lysates probed with anti-PDGFR or anti-PY antibodies.

To verify the expression and the activation of the βPDGFRWT, total cell lysates were subjected to immunoblot analysis using anti-βPDGFR antibody or anti-P-Tyr. Results shown in figure 1 (lower panels) indicate that βPDGFR is overexpressed in COS-7 transfectans. Moreover, the tyrosine phosphorylation of the βPDGFR was only detectable in transfected cells treated with PDGF-BB These findings indicate that stimulation of the βPDGFR causes efficient activation of JNK activity in COS-7 cells.

### 3.2 The Ligand-Dependent Association of βPDGFR with PI-3 kinase is Required for JNK Activation.

Accumulating evidence indicates that ligand-dependent PDGFR association with PLCγ and PI-3 kinase is sufficient for efficient biological signaling by this growth factor receptor (Valius M & Kazlauskas A, 1993; Cantley *et al.*, 1991). In order to dissect the role of these two signaling molecules in activation of JNKs, we overexpressed the βPDGFR mutant lacking either binding site for PLCγ/SHP2 or PI3-kinase along with HA-epitope tagged JNK in COS-7 cells. As shown in Figure 2 (upper panel), the βRWT and βPDGFR mutant lacking binding sites for PLCγ and SHP2 (βRY1009F+Y1021F) each induced similar JNK activity upon PDGF stimulation.

In contrast, the βPDGFR mutant lacking the binding site for PI-3 kinase (βRY740F+Y751F) failed to induce J NK activity under the same conditions. The expression and tyrosine phosphorylation level of each mutant and βRWT was shown to be comparable (Figure 2, middle and lower panel). As a control, we also investigated the ability of βPDGFR mutants to enhance the enzymatic activity of MAPKs. The level of PDGF-dependent activation of MAPK by βRWT was comparable to that shown by βRY1009F+Y1021F or βRY740F+Y751F (not shown). Thus, our results suggest that association of PI-3 kinase alpha, but not PLCγ or SHP2, is specifically required for activation of JNK.

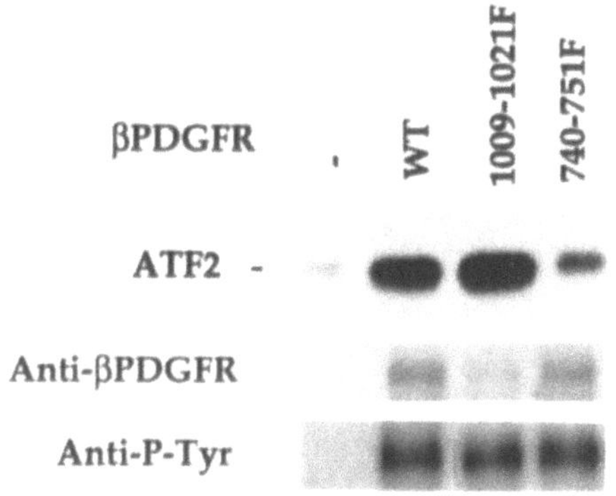


**Fig. 2. Requirement of PDGF-induced receptor association with PI-3Kinase for activation of JNK.** COS-7 cells were transfected with HA-JNK together DNA control or an expression vector coding for βRWT, or the indicated mutants. After 48 h of expression the cells were lysed and the JNK activity was determined as described before. In the lower panels are shown the Western blot analysis of total cell lysates probed with anti-PDGFR or anti-PY antibodies.

We next tested the effect of Wortmannin, a known inhibitor of PI-3 kinase, on JNK activity. Wortmannin caused a significant inhibition of JNK activity at half-maximal concentration of around 20-30 nM in COS-7 transfected with βPDGFR (data not shown). Therefore, our results indicate that PI-3 kinase catalytic domain must regulate the JNK activity.

### 3.3 PI-3Kinase is sufficient to activate JNK in a Ras/Rac dependent pathway.

We next examined the effect of overexpressing p110α, a catalytic domain of PI-3 kinase, and/or p85 the regulatory subunit of this enzyme on JNK activity. As shown in Figure 3 (and data not shown), transfection of p85 alone failed to stimulate JNK activity, while overexpression of p110α is sufficient for activation of JNK.

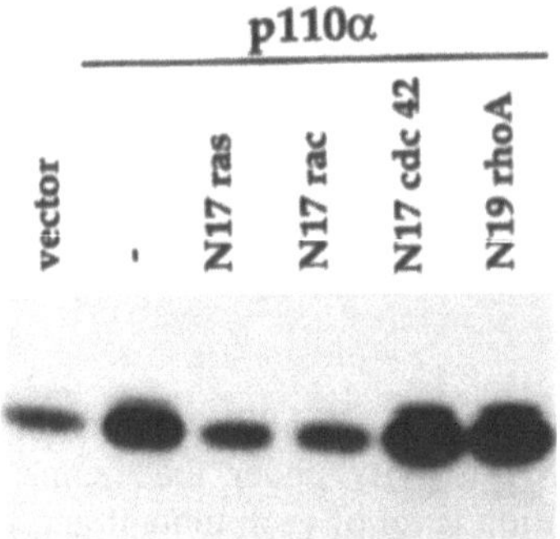


**Fig. 3. Dominant Negative Rac and Ras But not Rho or Cdc42 Inhibit Activation of JNK by p110α.** COS-7 cells were transfected with HA-JNK and p110α together with either pcDNA3 (vector) without insert or carrying cDNAs for N17ras, N19rhoA, N17rac1 or N17cdc42. Cells were lysed and JNK activity was determined in the HA-immunoprecipitates using GST-ATF2(96) as substrate.

Since an activated form of Rac or Cdc42 can stimulate JNK activity (Coso *et al.*, 1995), we sought to determine the role of the small GTP-binding proteins play in JNK activation by PI-3 kinase. As shown in Figure 3, p110α-induced activation of JNK was inhibited by dominant negative Ras (N17ras) or Rac (N17Rac). These results strongly suggest that PI-3 kinase may be upstream of ras and rac activation and that Rho or Cdc42 may not be downstream effectors of p110α.

## 4 REFERENCES

Cano E. and Mahadevan, LC (1995) Trends Biochem. Sci. **20**, 117-22

Cantley LC, Auger KR, Carpenter C. Duckworth B, Graziani A, Kapeller R and Soltoff S (1991) Cell **64**, 281-302

Coso OA, Chiarello M, Kalinec G, Kyriakis JM, Woodgett J. and Gutkind JS (1995a) J. Biol. Chem. **270**, 5620-5624

Coso OA, Chiarello M, Yu JC, Teramoto H, Crespo P, Xu N, Miki T. and Gutkind JS. (1995b) Cell **81**, 1137-1146

Crespo P, Xu N, simonds WF, and Gutkind JS. (1994) Nature **369**, 418-420

Kyriakis J, Banerjee P, Nikolakaki E, Dai T, Rubie E, Ahmad M, Avruch J, and Woodgett J. (1994) Nature **369**, 156-160

Post GR and Brown JH (1996) FASEB J. **10**, 741-749

Valius M. and Kazlauskas A. (1993) Cell **13**, 133-143

# Regulation of Phospholipase C isozymes

Sue Goo Rhee and Yun Soo Bae

Laboratory of Cell Signaling, National Heart, Lung, and Blood Institute,
National Institutes of Health, Bethesda, MD 20892

## 1. Lipid-derived second messengers and phospholipase C

The phospholipases play a crucial role in generating the lipid second messengers implicated in signal transduction processes (1). These enzymes are defined and named by the position they attack on the phospholipid backbone. Phospholipase $A_2$ ($PLA_2$) catalyzes the hydrolysis of the fatty acid group on the middle (sn-2) position while phospholipase C (PLC) catalyzes the hydrolysis of the phosphodiester bond on the third (sn-3) position of the glycerol and phospholipase D (PLD) catalyzes the hydrolysis of the phosphodiester bond between the phosphate and the polar group (which can be choline, ethanolamine, serine, glycerol, or inositol). Figure 1 indicates the sites of the action of these enzymes.

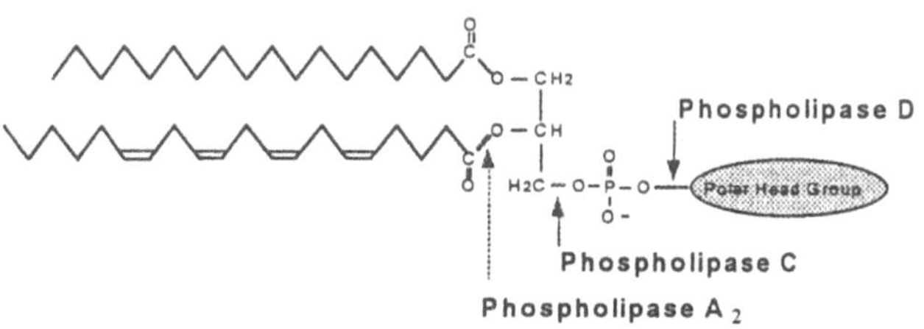


Fig. 1. Signal-activated phospholipases and the sites of their action.

While these and other phospholipases are important in cell signaling, this paper will focus on PLC. The hydrolysis of a minor membrane phospholipid, phosphatidylinositol 4,5-bisphosphate($PIP_2$), by a specific PLC is one of the earliest key events by which more than 100 extracellular signaling molecules are known to regulate functions of their target cells. The hydrolysis produces two intracellular messengers, inositol trisphosphate ($IP_3$) and diacylglycerol (DG). $IP_3$ induces the release of calcium from internal stores and DG activates protein kinase C (PKC) (2-6).

## 2. PLC isozymes

Like many other proteins involved in cell signaling, PLC exists in multiple isoforms. All mammalian PLC enzymes identified up to now are single polypeptides and can be divided into three types (β, γ, and δ) exemplified by the 150-kDa PLC-β1, the 145-kDa PLC-γ1, and the 85-kDa PLC-δ1, on the basis of size and amino acid sequence (2,5,6). Ten mammalian enzymes have been identified, which include four β, two γ, and four δ members. There are only two regions of high homology shared by the three types, designated X and Y, which are ~60% and ~40% identical, respectively among the isozymes (Fig. 2). The X and Y regions comprise ~170 and ~260 amino acids, respectively, and appear to constitute the catalytic domain. All mammalian PLC isoforms contain an amino-terminal region of ~300 amino acids that precedes the X-region. Whereas PLC-β and PLC-δ isozymes contain a short sequence of 50 to 70 amino acids that separate the X and Y regions, PLC-γ isozymes have a long sequence of ~400 amino acids that contains Src homology (two SH2 and one SH3) domains. The SH2 and SH3 domains are small modules of protein

NATO ASI Series, Vol. H 102
Interacting Protein Domains
Their Role in Signal and Energy Transduction
Edited by Ludwig Heilmeyer

structures that comprise ~100 and ~50 amino acids, respectively, and govern protein-protein interactions; the SH2 domain targets the protein molecule to tyrosine phosphorylated sequences present in other proteins, and the SH3 domain targets it to proline-rich sequences present in cytoskeletal components. All mammalian PLC isozymes also contain one or two domains known as pleckstrin homology (PH) domains, ~100-residue protein modules that have recently been shown to be present in many signaling proteins (7). A PH domain is found in the amino-terminal region, preceding the X domain, in all three types of PLC. PLC-γ isozymes contain an additional PH domain that is split by the SH domain. PH domains mediate interaction with the membrane surface by binding to $PIP_2$ (8). The three dimensional structures of PH (PLC-δ1) (8), SH2 (PLC-γ1) (9), and SH3 (PLC-γ1) (10) domains have been determined.

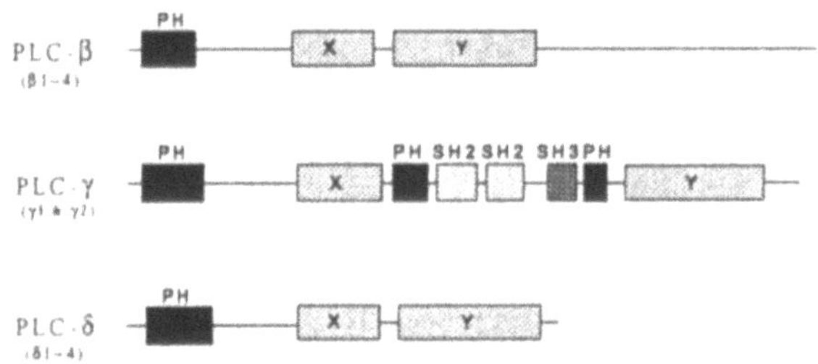


Fig. 2. Domain structure of the PLC family. All membersm contain highly homolous X and Y regions, whichconstitute the catalytic domain. Src homology domains, SH2 and SH3, and pleckstrin homology (PH) domain are indicated.

## 3. Activation of PLC-β isozymes

It is generally accepted that there are at least two distinct mechanisms to link receptor occupancy to the activation of PLC isozymes as shown in Fig.3; PLC-β isozymes by heterotrimeric G proteins and PLC-γ isozymes by tyrosine phosphorylation (2-6). The α subunits ($\alpha_q$, $\alpha_{11}$, $\alpha_{14}$, and $\alpha_{16}$) of all four members of the $G_q$ subfamily of heterotrimeric G proteins activate PLC-β isozymes but not PLC-γ and PLC-δ isozymes (2-6,11). The receptors that utilize this $G_q\alpha$/PLC-β pathway include those for thromboxane $A_2$, bradykinin, bombesin, angiotensin II, histamine, vasopressin, muscarinic acetylcholine (m1, m2, and m3), $\alpha_1$-adrenergic agonist, thyroid stimulating hormone, C-C and C-X-C chemokines, and endothelin-1 (5,11,12). PLC-β isozymes are also activated by the βγ subunits of G proteins (2-5,11,13-15). The m2 and m4 muscarinic acetylcholine receptors, lutenizing hormone receptor, V2 vasopressin receptor, β1- and β2-adrenergic receptors, and the receptors for the chemoattractants interleukin-8, formyl-Met-Leu-Phe, and complementation factor 5a activate this Gβγ/PLC-β pathway (16,17).

The region of PLC-β isozymes that interacts with $G_q\alpha$ subunits differs from that responsible for interaction with Gβγ dimers (6). Whereas the carboxyl-terminal region following the Y-region is important for activation by $G_q\alpha$ subunits (18,19), the site of interaction of PLC-β2 with Gβγ dimers was be localized to the region spanning Glu 435 to Val 641(20). Thus, $G_q\alpha$ and Gβγ subunits may independently modulate a single PLC-β molecule concurrently. Several positively charged residues important for interaction with $G_q\alpha$ have been identified in the COOH-terminal region of PLC-β1 (21). The COOH-terminal 14 residues of the Gβ subunit were also shown to be important for PLC-β activation (22).

The sensitivity of PLC-β isozymes to $G_q\alpha$ or Gβγ subunits is different: Whereas $G_q\alpha$ subunits activate PLC-β isozymes according to the hierarchy of PLC-β1 ≥ PLC-β3 〉 PLC-β4 〉 PLC-β2 (23,24), the sensitivity to Gβγ subunits decreases in the order of PLC-β3 〉 PLC-β2 〉 PLC-β1 (15). The Gβγ subunits does not activate PLC-β4 (25). These results suggest that cell-specific expression of PLC-β isozymes and G protein subunits contributes to diversity in the type and magnitude of enzyme responses observed.

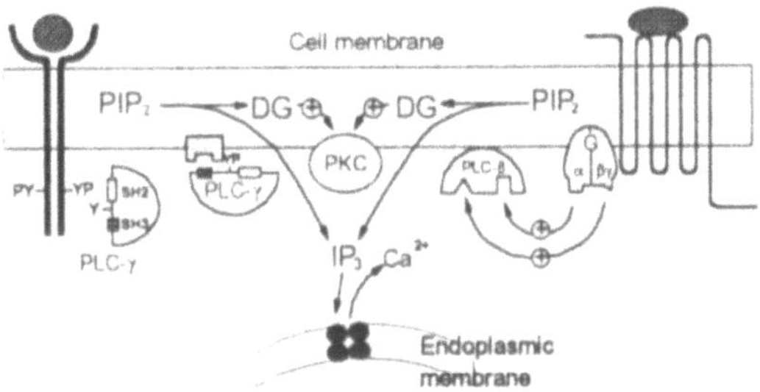


Fig. 3. Activation mechanisms for PLC-γ and PLC-β isozymes. Abbreviations are: YP, phosphotyrosine; M, a putative membrane-associated protein with proline-rich sequence that might serve as the site of interaction with the SH3 domain of PLC-γ; $PIP_2$, phosphatidylinositol 4,5-bisphosphate; DG, diglyceride; $IP_3$, inositol 1,4,5-trisphosphate; G, heterotrimeric GTP-binding protein consisting of α and βγ subunits; and PKC, protein kinase C.

Maximal activation of PLC-β isozymes in permeabilized HL-60 cells was found to require phosphatidylinositol transfer protein (PI-TP) in addition to the receptor and the G protein (26). As the supply of substrate $PIP_2$, which is synthesized from PI is likely to be limited in the vicinity of activated PLC, this additional requirement of PI-TP is believed to serve to ensure a steady substrate supply. PI-TP is likely to play its substrate supply role also for other types of PLC.

## 4. Activation of PLC-γ isozymes

Polypeptide growth factors, such as platelet-derived growth factor (PDGF), epidermal growth factor (EGF), fibroblast growth factor (FGF), and nerve growth factor (NGF) are known to induce stimulation of $PIP_2$ turnover by activating PLC-γ1 in a wide variety of cells. Binding of a growth factor to its receptor results in dimerization of receptor subunits and activation of the intrinsic tyrosine kinase activity of the receptor. This leads to tyrosine phosphorylation of numerous proteins, including the receptor itself and PLC-γ1 (2-5). The receptor autophosphorylation creates high affinity binding sites for several SH2-containing proteins including PLC-γ1, the 85-kDa subunit of PI 3-kinase, and Ras GTPase activating protein (Ras Gap). In the PDGF receptor (β chain), for example, eight autophosphorylation sites have been identified for the specific binding of Src family tyrosine kinases (Tyr-579 and Tyr-581), growth factor receptor bound 2 (Grb-2) (Tyr-716), PI 3-kinase subunit (Tyr-740 and Tyr-751), Ras Gap (Tyr-771), protein tyrosine phosphatase 1D (Tyr-1009), and PLC-γ1 (Tyr-1021) (27,28). Also, in FGF receptor and NGF receptor (Trk), Tyr-766 and Tyr-785, respectively, are known to specifically bind PLC-γ1 (29,30). However, in the case of EGF receptor, which contains five autophosphorylation sites located in the carboxyl-terminal region, individual autophosphorylation sites are not strictly required for the recognition and association of different SH2-containing proteins (31,32). Thus, PLC-γ1 can bind any of these five phosphorylated tyrosine residues.

Phosphorylation of PLC-γ1 by PDGF, EGF, FGF, and NGF receptors occurs at identical sites, namely tyrosine residues 771, 783, and 1254 (33). The role of tyrosine phosphorylation of PLC-γ1 was investigated by substituting phenylalanine for tyrosine at these three sites and expressing the mutant enzymes in NIH 3T3 cells (34). Phenylalanine substitution at Tyr 783 completely blocked the activation of PLC by PDGF. However, like the wild-type enzyme, PLC-γ1 with a Tyr-783→Phe point mutation associated with the PDGF receptor. In another experiments, the binding of PLC-γ1 to PDGF and FGF receptors was disrupted by replacing Tyr-1021 ( PDGF receptor) and Tyr-766 (FGF receptor) with phenylalanine (35-37). As expected, the mutant receptors failed to associate with PLC-γ1 and were unable to mediate the growth factor-dependent production of Ins(1,4,5)$P_3$. However, the mutant receptors mediated readily detectable levels of growth factor-dependent PLC-γ1 tyrosine phosphorylation. Thus, the growth factor-induced activation of PLC-γ1 requires not only PLC-γ1 tyrosine phosphorylation but also the association of the enzyme with the growth factor receptor.

Tyrosine phosphorylation of PLC-γ1 appears to promote its association with actin components of cytoskeleton (38,39). The SH3 domain of PLC-γ1 has been shown to be responsible for targeting it to actin microfilament network (40). Whether this cytoskeletal association serves to bring the enzyme to its substrate is or whether it promotes interaction with another protein component essential for its activation is unknown.

Nonreceptor protein tyrosine kinases also phosphorylate and activate PLC-γ isozymes in response to the ligation of certain cell surface receptors. Such receptors include the T cell antigen receptor (TCR), the membrane immunoglobulin M (mIgM), the high-affinity IgE receptor (FcεRI), the IgG receptors (FcγRs), the IgA receptor, CD20, CD38, the α2-macroglobulin receptor, integrins, and several receptors for cytokines such as ciliary neurotrophic factor, leukemia inhibitory factor, oncostatin M, IL-1, IL-4, IL-6, and IL-7 (41-45). These receptors, which comprise multiple polypeptide chains, are not protein tyrosine kinases themselves, but activate a wide variety of nonreceptor protein tyrosine kinases like the members of Src, Syk, and Jak/Tyk families. Some of these kinases have been shown to associate with PLC-γ1.

PLC-γ2, which is mainly expressed in hematopoietic cells, is also phosphorylated at two residues, Tyr-753 and Tyr-759, by the growth factor receptors as well as nonreceptor protein tyrosine kinases.

Tyrosine phosphorylation of PLC-γ has also been observed in response to the ligation of several heptahelical, G protein-coupled receptors, including m5AChR in CHO cells, the angiotensin II and thrombin receptors in vascular smooth muscle cells, the platelet-activating factor (46-49). Src appears to be responsible for the phosphorylation of PLC-γ1 in vascular smooth muscle cells and platelets; electroporation of antibodies to Src inhibited the tyrosine phosphorylation of PLC-γ1 elicited by angiotensin II or platelet-activating factor. Although activation of Src family PTKs in response to stimulation of a variety of G protein-coupled receptors has been demonstrated (50), the mechanism by which the enzymes are coupled to the receptors is not clear. One possible mechanism is through a member of the recently identified proline-rich PTK (pyk) family: Stimulation of receptors coupled to the G proteins Gi or Gq in neuronal cells resulted in tyrosine phosphorylation of pyk-2, binding of the SH2 domain of Src to the phosphorylated pyk-2, and activation of Src (51).

## 5. PTK-independent activation of PLC-γ

PLC-γ isozymes can be activated directly by several lipid-derived second messengers in the absence of tyrosine phosphorylation. Phosphatidic acid produced by the action of PLD, activates purified PLC-γ1 by acting as an allosteric modifier (52). PLC-γ isozymes are also stimulated by arachidonic acid (AA) in the presence of the microtubule-associated protein tau (in neuronal cells) or taulike proteins (in nonneuronal cells) (53). The effect of tau and AA was specific to PLC-γ isozymes in the presence of submicromolar concentrations of $Ca^{2+}$ and was markedly inhibited by phosphatidylcholine (PC). These observations suggest that the activation of PLC-γ1 by tau or taulike proteins might be facilitated by a concomitant decrease in PC concentration and increase in AA concentration, both of which occur in cells on activation of an 85-kDa cytosolic phospholipase $A_2$ ($cPLA_2$). This enzyme is coupled to various receptors and preferentially hydrolyzes PC containing AA. Therefore, activation of PLC-γ isozymes may occur secondarily to receptor-mediated activation of $cPLA_2$. Several studies are consistent with the notion that stimulation of PLC by endogenously released AA occurs in cells (53).

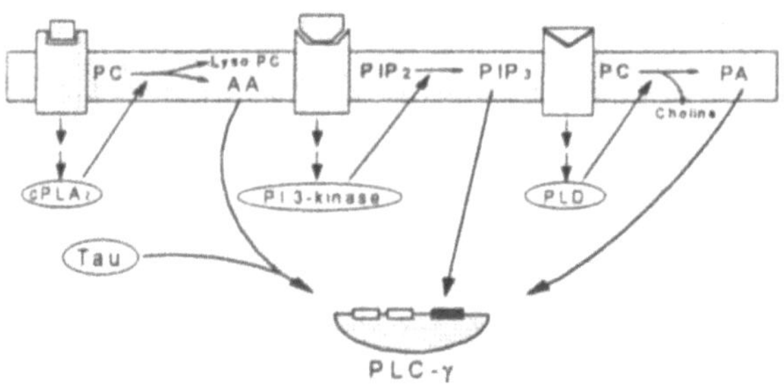


Fig. 4. Receptor-induced activation of PLC-γ isozymes by tau and AA generated by $cPLA_2$ (left), $PIP_3$ generated by PI 3-kinase (middle), and phosphatidic acid (PA) generated by PLD (right).

Ligation of a variety of receptors results in the activation of PI 3-kinase, which phosphorylates the D3 position of $PIP_2$ to produce phosphatidylinositol 3,4,5-trisphosphate ($PIP_3$). $PIP_3$ activates purified PLC-γ isozymes specifically by interacting with their SH2 domains (54). In addition, incubation of NIH 3T3 cells with $PIP_3$ resulted in a transient increase in the intracellular $Ca^{2+}$ concentration, an effect that was blocked in the presence of a PLC inhibitor. Thus, receptors coupled to PLD, $cPLA_2$, or PI 3-kinase may activate PLC-γ isozymes indirectly, in the absence of tyrosine phosphorylation, through the generation of lipid-derived second messengers (Fig. 4).

## 6. Activation of PLC-δ

Although four distinct PLC-δ isoforms are known, the mechanism by which these isozymes are coupled to membrane receptors remains unclear. A new class of GTP-binding protein, termed $G_h$ and containing 75- to 80-kDa α and ~50-kDa β subunits, has been shown to be associated with agonist-bound $\alpha_1$-adrenergic receptors ($\alpha_1$ AR). The $G_h\alpha$ subunit, a multifunctional protein that also possesses tissue transglutaminase activity (55), activates purified PLC-δ1 and forms a complex with PLC-δ1 in cells stimulated via $\alpha_1$ AR (56). Furthermore, overexpression of $G_h\alpha$ in COS cells enhanced the activation of PLC induced by ligation of the $\alpha_1$ AR. These results suggest that $G_h\alpha$ directly couples α1 AR to PLC-δ1. It is not yet known whether other PLC-δ isozymes are also activated by $G_h\alpha$, what other receptors couple to $G_h\alpha$, and how the tissue transglutaminase activity of $G_h\alpha$ is related to its PLC-δ1-activating function. The GTPase-activating protein for the small GTP-binding protein RhoA (RhoGAP) also activates purified PLC-δ1 (57); PLC-δ1 activation was thus suggested to occur downstream of RhoA activation.

## References


1. Dennis, E.A., Rhee, S.G., Billah, M.M., and Hunnun, Y.A. (1991) *FASEB* **5**, 2068-2077
2. Rhee, S.G., and Choi, K.D. (1992) *J. Biol. Chem.* **267**, 12393-12396
3. Cockcroft, S. and Thomas, G. M. H. (1992) *Biochem. J.* **288**, 1-14
4. Berridge M. J. (1993) *Nature* **361**, 315-325
5. Noh, D.-Y., Shin, S.H., Rhee, S.G. (1995) *Biochim. Biophys. Acta* **1242**, 99-114
6. Lee, S.B., and Rhee, S.G. (1995) *Curr. Opin. Cell Biol.* **7**, 183-189.
7. Parker, P. J., Hemmings, B. A., and Gierschik, P. (1994) *Trends. Biochem. Sci.* **19**, 54-55
8. Ferguson, K.M., Lemmon, M.A., Schlessinger, J. and Sigler, P.B. (1995) *Cell* **83**,1037-1046
9. Pascal, S.M., Singer, A.U., Gish, G., Yamazaki, T., Shoelson, S.E., Pawson, T., Kay, L.E., and Forman-Kay, J.D. (1994) *Cell* **77**, 461-472
10. Kohda, D., Hatanaka, H., Odaka, M., Mandiyan, V., Ullrich, A., Schlessinger, J., and Inagaki, F. (1993) *Cell* **72**, 953-960
11. Sternweis, P. C., and Smrcka, A. V. (1992) *Trends Biochem. Sci.* **17**, 502-506
12. Kuang, Y., Wu, Y., Jiang, H., and Wu, D. (1996) *J. Biol. Chem.* **271**, 3975-3978
13. Camps, M., Carozzi, A., Schnabel, P., Scheer, A., Parker, P.J. and Gierschik, P. (1992) *Nature* **360**, 684-686.
14. Katz, A., Wu, D. and Simon, M.I. (1992) *Nature* **360**, 686-689.
15. Park, D., Jhon, D. Y., Lee, C. W., Lee. K. H., and Rhee, S. G. (1993) *J. Biol. Chem.* **268**, 4573-4576
16. Zhu, X., and Birnbaumer, L. (1996) *Proc. Natl. Acad. Sci. USA* **93**, 2827-2831
17. Jiang, H., Kuang, Y., Wu, Y., Smrcka, A., Simon, M.I., and Wu, D. (1996) *J. Biol. Chem.* **271**, 13430-13434
18. Park, D., Jhon, D.Y., Lee, C.W., Ryu, S.H. and Rhee, S.G. (1993) *J. Biol. Chem.* **268**, 3710-3714
19. Lee, S.B., Shin, S.H., Hepler, J.R., Gilman, A.G., and Rhee, S.G. (1993) *J. Biol. Chem.* **268**, 25952-25957
20. Kuang, Y., Wu, Y., Smrcka, A., Jiang, H., Wu, D. (1996) *Proc. Natl. Acad. Sci.USA* **93**, 2964-2968
21. Kim, C.G., Park, D., and Rhee, S.G. (1996) *J. Biol. Chem.* **271**, 21187-21192
22. Zhang, S., Coso, O.A., Collins, R., Gutkind, J.S., and Simonds, W.F. (1996) *J. Biol. Chem.* 271, 20208-20212
23. Jhon, D. Y., Lee, H. H., Park, D., Lee, C. W., Lee, K. H., Yoo, O. J., and Rhee, S. G. (1993) *J. Biol. Chem.* **268**, 6654-6661
24. Smrcka, A. V. and Sternweis, P. C. (1993) *J. Biol. Chem* **268**, 9667-9674
25. Lee, C.W., Lee, K.H., Lee, S.B., Park, D., and Rhee, S.G. (1994) *J. Biol. Chem.* **269**, 25335-25338
26. Thomas G.M.H., Cunningham E., Fensome A., Ball A., Totty N.F., Truong O., Hsuan J.J., and Cockcroft S. (1993) *Cell* **74**, 919-928

27. Claesson-Welsh L. (1994) *J. Biol. Chem.* **51**, 32023-32026
28. Kashishian, A. and Cooper. J. A. (1993) *Molecular Biology of the Cell* **4**, 49-57.
29. Mohammadi, M., Honegger, A. M., Rotin, D., Fischer, R., Bellot, F., Li, W., Dionne, C. A., Jaye, M., Rubinstein, M., and Schlessinger, J. (1991) *Molecular and Cellular Biology* **11**, 5068-5078.
30. Obermeier, A., Halfter, H., Wiesmüller, K.-H., Jung, G., Schlessinger, J., and Ullrich, A., (1993) *EMBO J.* **12**, 933-941.
31. Vega, Q. C., Cochet, C., Filhol, O., Chang, C.-P., Rhee, S. G., and Gill, G. N. (1992) *Molecular and Cellular Biology* **12**, 128-135.
32. Soler, C., Beguinot, L., and Carpenter, G. (1994) *J. Biol. Chem.* **269**, 12320-12324.
33. Kim, J. W., Sim, S. S., Kim, U.-H., Nishibe, S., Wahl, M., Carpenter, G., and Rhee, S.G. (1990) *J. Biol. Chem.* **265**, 3940-3943.
34. Kim, H. K., Kim, J. W., Zilberstein, A., Margolis, B., Kim, J. G., Schlessinger, J., and Rhee, S. G., *Cell* **65**, 435-441.
35. Valius, M. and Kazlauskas, A. (1993) *Cell* **73**, 321-334.
36. Mohammadi, M., Dionne, C. A., Li, W., Spivak, T,. Honegger, A. M., Jaye, M., Schlessinger, J. (1992) *Nature* **358**, 681-684.
37. Peters, K. G., Marie, J., Wilson, E., Ives, H. E., Escobedo, J., Del Rosario, M., Mirda, D., and Williams, L. T. (1992) *Nature* **358**, 678-681.
38. McBride, K., Rhee, S. G., and Jaken, S. (1991) *Proc. Natl. Acad. Sci. USA* **88**, 7111-7115.
39. Yang, L. J., Rhee, S. G., and Williamson, J. R. (1994) *J. Biol. Chem.* **269**, 7156-7162.
40. Bar-Sagi, D., Rotin, D., Batzer, A., Mandiyan, V., and Schlessinger, J. (1993) *Cell* **74**, 83-91.
41. Boulton, T. G., Stahl, N., and Yancopoulos, G.D. (1994) *J. Biol. Chem.* **269**,11648-11655.
42. Keely, P. J., and Parise, L. V. (1996) *J. Biol.Chem.* **271**, 26668-26676
43. Gómez-Guerrero, C., Duque, N., and Egido, J. (1996) *J. Immunol.* **156**, 4369-4376
44. Misra, U.K., Gawdi, G., and Pizzo, S.V. (1995) *Biochem. J.* **309**, 151-158
45. Deans, J.P., Schieven, G.L., Shu, G.L., Valentine, M.A., Gilliland, L.A., Aruffo, A., Clark, E.A., and Ledbetter, J.A. (1993) *J. Immunol.* **151**, 4494-4504
46. Gusovsky, F., Lueders, J.E., Kohn, E.C., and Felder, C.C. (1993) *J. Biol. Chem.* **268**, 7768-7772
47. Marrero, M.B., Schieffer, B., Paxton, W.G., Schieffer, E., and Bernstein, K.E., (1995) *J. Biol. Chem.* **270**, 15734-15738
48. Rao, G.N., Delafontaine, P., and Runge, M.S. (1995) *J. Biol. Chem.* **270**, 27871-27875
49. Dhar, A., and Shukla, S.D. (1994) *J. Biol. Chem.* **269**, 9123-9127
50. Sadoshima, J. and Izumo, S. (1996) *EMBO J.* **15**, 775-787
51. Dikic, I., Tokiwa, G., Lev, S., Courtneidge, S.A., and Schlessinger, J. (1996) *Nature* **383**, 547-550
52. Jones, G.A., and Carpenter, G. (1993) *J. Biol. Chem.* **268**, 20845-20850
53. Hwang, S.C., Jhon, D.-Y., Bae, Y.S., Kim, J.H., and Rhee, S.G. (1996) *J. Biol. Chem.* **271**, 18342-18349
54. Bae, Y.S., Cantley, L.G., Chen, C.-S., Kim, S.-R., Kwon, K.-S., and Rhee, S.G. (1996) submitted
55. Nakaoka, H., Perez, D. M., Baek, K. J., Das, T., Husai, A., Misono, K., Im, M.-J., and Graham, R. M. (1994) *Science* **264**, 1593-1596
56. Feng, J.-F., Rhee, S.G., and Im, M.-J. (1996) *J. Biol. Chem.* **271**, 16451-16454
57. Homma, Y., and Emori, Y. (1995) *EMBO J.* **14**, 286-291

# Substrate Binding and Catalytic Mechanism in Phospholipase C from *Bacillus cereus*

Dorthe da Graça Thrige, Jette Raun Byberg Buur and Flemming Steen Jørgensen

Department of Medicinal Chemistry, Royal Danish School of Pharmacy, Universitetsparken 2, DK-2100 Copenhagen, Denmark

## 1. Introduction

Phospholipase C from *Bacillus cereus* ($PLC_{Bc}$) is a monomeric extracellular zinc containing enzyme consisting of 245 amino acids.[1] Like other PLCs it cleaves membrane phospholipids at the phospho moiety liberating the polar head group and *sn*-1,2-diacylglycerol (DAG) or ceramide. Although $PLC_{Bc}$ is a non-specific enzyme, it prefers phosphatidylcholine (PC) as a substrate. In bacteria PLC is part of a phosphate retrieval system, and in mammals PLC plays an important role in generation of second messengers, which are involved in control of cell metabolism, differentiation and growth.[2,3] Presently, no structural information on mammalian PC-PLCs is available, but $PLC_{Bc}$ is regarded as a useful model of mammalian PC-PLCs, since antibodies against $PLC_{Bc}$ cross react with a PC-PLC in mammalian cells[4] and since they neutralize the PC-PLC activity in *Xenopus Oocytes*.[5,6]

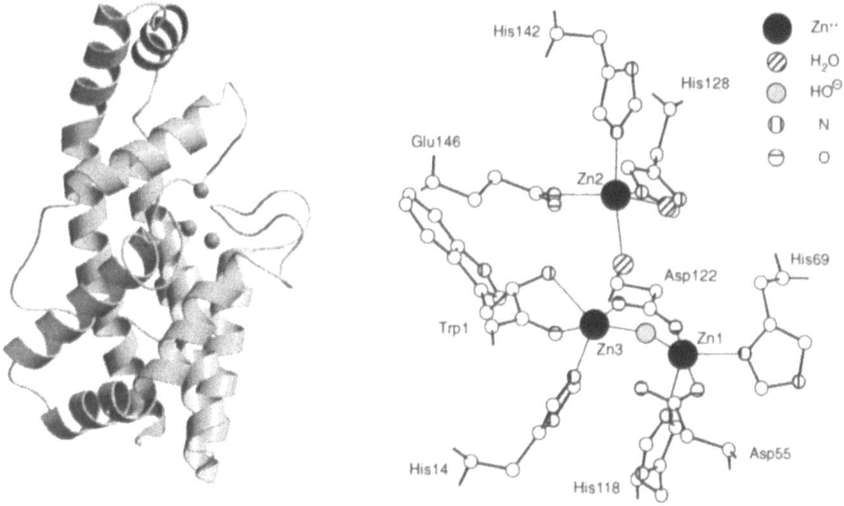


Figure 1. Overall structure (left) and zinc coordination (right) in $PLC_{Bc}$

The crystal structure of $PLC_{Bc}$ revealed an all-helix enzyme comprising 10 α-helices, which are folded into a single domain with a deep cleft on the surface,[7] cf. Figure 1. Three closely placed zinc ions are located at the bottom of this cleft,

NATO ASI Series, Vol. H 102
Interacting Protein Domains
Their Role in Signal and Energy Transduction
Edited by Ludwig Heilmeyer

which numerous biochemical and X-ray crystallographic experiments have proven to be the active site of the enzyme.[8,9] Although much structural information concerning $PLC_{Bc}$ is known today, the catalytic mechanism still has to be proven.

Here we report a consistent mechanism for the hydrolysis of phospholipids in $PLC_{Bc}$. Using the GRID program,[10] which determines favorable binding sites on macromolecules for probes by calculating intermolecular interaction energies, two catalytically important water molecules could be identified in the active site. One water molecule is believed to initiate the catalysis by attacking the phosphate group of the substrate, and the second water molecule is necessary for completing the hydrolysis. The GRID calculations were based on a completely energy minimized structure of $PLC_{Bc}$, which was obtained by a stepwise energy minimization procedure.

## 2. Computational Details

The high resolution X-ray structure of $PLC_{Bc}$ was prepared for energy minimization by addition of hydrogen atoms considering a neutral pH for the acidic and basic groups. Crystallographically determined water molecules were ignored, except the three involved in zinc coordination. One of these was considered as a hydroxide ion, since it bridges two closely located zinc ions, Zn1 and Zn3, cf. Figure 1.

Kollman´s all-atom partial charges[11] were used for the molecular mechanics calculations on $PLC_{Bc}$ except the zinc ions and coordinating residues and water molecules. For these atoms partial charges were calculated using the semi-empirical MNDO method.[12] Furthermore, MNDO derived charges were applied on the substrate as well as on the water molecules identified by GRID prior to the molecular mechanics calculations on the enzyme-substrate complex.

All energy minimizations were performed using the Tripos force field[13] in SYBYL.[14] A stepwise energy minimization, where the protein was gradually relaxed, was performed, and the final model of $PLC_{Bc}$ was completely energy minimized. A dielectric constant on 20 was applied for the calculations.

## 3. Results and Discussion

Due to strong electrostatic repulsion between the three zinc ions in $PLC_{Bc}$ it was necessary to take delocalization of the charges in the active site into account when energy minimizing the protein. This was achieved by a semi-empirical charge calculation of the zinc ions and their coordinating residues. By this procedure the major part of the positive charge on a zinc ion was transferred to the zinc coordinating histidines.

The crystal structure and the energy minimized structure of $PLC_{Bc}$ turned out very similar. The majority of the Cα-atoms have RMS-deviations less than 1 Å. Only two areas within the energy minimized structure display significant movements relative to the X-ray structure, the surface loop between the first and the second α-helix and the C-terminal end, respectively. None of these areas are close to the active site cleft. Although keeping almost the same metalcoordination pattern, all zinc-zinc distances and zinc-coordination distances increased during the energy minimization. RMS-deviations are smaller in the active site relative to the whole enzyme, indicating that the active site is less flexible or more stable than the remaining part of the protein.

As described in our previous work,[15] the substrate (PC) was modelled into the active site of $PLC_{Bc}$, where it replaced the three water molecules. It is bound to the enzyme via coordination of the two phosphate oxygen atoms to the three metal ions and coordination of the ester carbonyl oxygen atom to the C(2) fatty acid chain to Zn2. The choline head group protrudes from the active site with the positively charged nitrogen atom close to the phenolic oxygen atom of Tyr56 and to the carboxylate groups of Glu4 and Asp55. In many aspects, the position and orientation of the substrate is similar to that found in a crystallographic study of $PLC_{Bc}$ complexed with a substrate analog.[9]

Generally, the cleavage of phospholipids is assumed to occur by an in-line associative mechanism where a nucleophile attacks the phosphorus atom in the substrate leading to a penta-coordinated trigonal bipyramidal intermediate.[16] In proteins the nucleophile is often assumed to be a hydroxide ion, derived from a water molecule, which has been attacked by an acidic residue in the vicinity of the scissile bond in the substrate.

Presently, three different catalytic mechanisms have been proposed for $PLC_{Bc}$. None of these are complete or have been proven. In a previous study we suggested that the catalysis could be initiated by Glu4, a non-metal binding residue located on the entrance to the active site.[15] In a crystallographic study of $PLC_{Bc}$ complexed with a phosphonate inhibitor (substrate analog), Glu146 is assumed to initiate the cleavage of the substrate.[9] The catalytic mechanism in our former work seems unlikely since it involves charge transfer through more than a single water molecule. In the latter work a water molecule is hydrogen bonded to Glu146, but the distance for a nucleophilic attack on phosphorus is too long. Furthermore, neither of the models could explain how the protonization of the DAG leaving group takes place. In another theoretical study it has been proposed that the hydrolysis occurs by attack of the bridging hydroxide ion on the phosphorus atom in the substrate.[17] According to this mechanism the substrate never reaches the bottom of the active site, and it never binds to the enzyme as inhibitors do in X-ray crystallographic studies of $PLC_{Bc}$-inhibitor complexes.

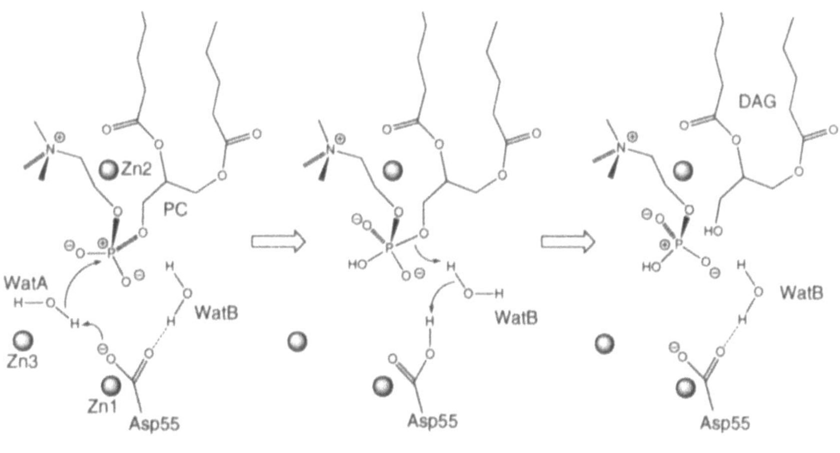


Figure 2. Schematic drawing of the proposed catalytic mechanism in $PLC_{Bc}$.

In this study it was possible to determine the position of two water molecules close to the phosphate group in the substrate by GRID calculations. The nucleophilic water molecule, WatA, is hydrogen bonded to Oδ1 in Asp55 and to N in Trp1, and the second water molecule, WatB, is hydrogen bonded to Oδ2 in Asp55. Based on these observations a new catalytic mechanism is proposed, cf. Figure 2.

## 4. Conclusion

A consistent mechanism for the catalytic cleavage of phospholipid substrates in $PLC_{Bc}$, based on a fully energy minimized structure, has been proposed. The mechanism involves two water molecules. Further investigations of the catalytic mechanism of the enzyme require site directed mutagenesis experiments to be carried out. X-ray crystallographic studies of $PLC_{Bc}$ complexed with other inhibitors, substrate analogs and transition state analogs as well as stereochemical studies would also be of considerable importance for further understanding of the catalytic mechanism in $PLC_{Bc}$.

## References

1. Johansen, T., Holm, T., Guddal, P. H., Sletten, K., Haugli, F. B. and Little, C. (1988) *Gene* **65**, 303-314.
2. Berridge, M. J. and Irvine, R. F. (1989) *Nature* **314**, 197-205.
3. Exton, J. H. (1990) *J. Biol. Chem.* **265**, 1-4.
4. Clark, M. A., Shorr, R. G. L. and Bomalaski, J. S. (1896) *Biochem. Biophys. Res. Com.* **140**, 114-119.
5. Garcia de Herreros, A., Domingues, I., Diaz-Meco, M. T., Cornet, M. E., Guddal, P. H., Johansen, T. and Moscat, J. (1990) *J. Biol. Chem.* **266**, 6825-6829.
6. Dominguez, I., Marshall, M. S., Gibbs, J. B., Garcia de Herreros, A., Cornet, M. E., Graziani, G., Diaz-Meco, M. T., Johansen, T., McCormick, S. and Moscat, J. (1991) *EMBO J.* **10**, 3215-3220.
7. Hough, E., Hansen, L. K., Birknes, B., Jynge, K., Hansen, S., Hordvik, A., Little, C., Dodson, E. and Derewenda, Z. (1989) *Nature* **338**, 357-360.
8. Hansen, S., Hansen, L. K., and Hough, E. (1992) *J. Mol. Biol.* **225**, 543-549.
9. Hansen, S., Hough, E., Svensson, L. A., Wong, Y.-L. and Martin, S. F. (1993) *J. Mol. Biol.* **234**, 179-187.
10. Goodford, P. J. (1985) *J. Med. Chem.* **28**, 849-857.
11. Weiner, S. J., Kollman, P. A., Nguyen, D. T. and Case, D. A. (1986) *J. Comp. Chem.* **7**, 230-252.
12. Clark, M., Cramer III, R. D. and Van Opdenbosch, N. (1989) J. Comput. Chem. **10**, 982-1012.
13. Steward, J. J. P. (1990) *J. Comput.-Aided Mol. Des.* **4**, 1-105
14. Tripos Associates Inc., St. Louis, USA.
15. Byberg, J. R. , Jørgensen, F. S., Hansen, S. and Hough, E. (1992) *Proteins Struct. Funct. Genet.* **12**, 321-328.
16. Fersht, A. (1985) *Enzyme Structure and Mechanism*, 2nd ed. W. H. Freeman and company, New York, pp. 236-243.
17. Sundell, S., Hansen, S. and Hough, E. (1994) *Protein Engineering* **7**, 571-577.

# Identification of Three Active Site Residues Involved in Substrate Binding by Human 43 kDa-D-*myo*-Inositol 1,4,5-*Tris*phosphate 5-Phosphatase

David Communi and Christophe Erneux

Institute of Interdisciplinary Research, University of Brussels, Campus Erasme, Bldg C, Route de Lennik 808, B-1070 Brussels, Belgium.

## 1 Introduction

In a wide variety of cell types, Ins(1,4,5)$P_3$ and 1,2-diacylglycerol are generated from phosphatidylinositol 4,5-*bis*phosphate (PtdIns(4,5)$P_2$) by receptor-mediated activation of phospholipase C (1). Alternatively, PtdIns(4,5)$P_2$ is the substrate for a specific receptor-activated 3-kinase to generate another signal molecule, i.e. phosphatidylinositol 3,4,5-*tris*phosphate. Ins(1,4,5)$P_3$ is the substrate of a 5-phosphatase to produce Ins(1,4)$P_2$ and of a 3-kinase to produce Ins(1,3,4,5)$P_4$. Some evidence supports a role for Ins(1,3,4,5)$P_4$ in the regulation of intracellular free calcium concentration in concert with Ins(1,4,5)$P_3$. Recently, a specific Ins(1,3,4,5)$P_4$-binding protein has been isolated and identified as a member of the GAP1 family, suggesting a connection between phospholipase C-derived signals and a proliferative cascade involving Ras (2).

At least six distinct isoforms of phosphatidylinositol and inositol polyphosphates 5-phosphatases have been isolated and characterized: 43 kDa- and 75 kDa-5-phosphatases, synaptojanin, OCRL, SHIP and SIP proteins (3). In brain, 43 kDa-Ins(1,4,5)$P_3$/Ins(1,3,4,5)$P_4$ 5-phosphatase is the major enzyme hydrolysing the calcium-mobilizing second messenger Ins(1,4,5)$P_3$.

NATO ASI Series, Vol. H 102
Interacting Protein Domains
Their Role in Signal and Energy Transduction
Edited by Ludwig Heilmeyer

Arginyl residues are known to act as anionic binding sites in proteins and may thus assist in the binding of substrates or enzyme catalysis. Previous studies in crude rat brain have shown that thiol blocking agents have a marked inhibitory effect on Ins(1,4,5)$P_3$ 5-phosphatase activity. We therefore investigated the possibility that active site arginines and cysteines may play a critical role in enzyme function for 43 kDa-Ins(1,4,5)$P_3$ 5-phosphatase.

## 2 Results and discussion

**Table 2.1.** Effect of various modifying agents on 43 kDa-Ins(1,4,5)$P_3$/Ins(1,3,4,5)$P_4$ 5-phosphatase activity and protection by substrates. The enzyme was preincubated at 23°C in the presence of each modifying agent at 20 mM for 10 min in the presence or absence of each substrate (at 50 μM), before assaying for residual activity at 10 μM Ins(1,4,5)$P_3$. Phglx is for phenylglyoxal, NEM for *N*-ethylmaleimide, DTNB for dithionitrobenzoic acid and DCCD for dicyclohexylcarbodiimide. Residual activity values are means of triplicates ± S.D.

| Reagent | Residual enzymic activity (%) | | |
|---|---|---|---|
| | without substrate | with Ins(1,4,5)$P_3$ | with Ins(1,3,4,5)$P_4$ |
| None | 100 ± 3 | 102 ± 5 | 97 ± 8 |
| Phglx | 10 ± 2 | 86 ± 5 | 93 ± 4 |
| NEM | 0 | 95 ± 6 | 93 ± 3 |
| DTNB | 4 ± 3 | 16 ± 4 | 9 ± 2 |
| Iodoacetate | 0 | 0 | 0 |
| DCCD | 0 | 3 ± 2 | 5 ± 3 |

Several agents are able to inactivate the enzyme, e.g. phenylglyoxal (Phglx) and *N*-ethylmaleimide (NEM) which covalently modify arginines and cysteines, respectively (see Table 2.1). This covalent modification is prevented in the presence of both substrates (see Table 2.1) or 2,3-*bis*phosphoglycerate. Our results indicate that the amount of Phglx and NEM labelling paralleles the loss in enzymic activity. The covalent modification by the two agents results also in a drastic decrease in Ins(1,4,5)$P_3$ binding. The peptide mapping of the protein, which had been labelled with radioactive Phglx or NEM, enabled us to isolate in each case a peptide which was preferentially labelled in the absence of substrate. We have identified two reactive arginyl residues (i.e. Arg-343 and Arg-350) and a unique cysteinyl residue (i.e. Cys-348) within the active site of 43 kDa-Ins(1,4,5)$P_3$/Ins(1,3,4,5)$P_4$ 5-phosphatase. The similarity in circular dichroism spectra between the Arg343Ala, Arg350Ala, Cys348Ala and Cys348Ser mutants and the wild-type enzyme indicate that these mutations did not affect the gross secondary structure of the enzyme. However, the four mutants display a drastically decreased enzymic activity (4,5). Arg343Ala and Arg350Ala mutants show increased $K_m$ values for substrate. Cys348Ala and Cys348Ser mutants are nearly inactive.

Altough primary structures corresponding to inositol and phosphatidylinositol polyphosphates 5-phosphatases present little amino acid identity, it is intriguing to note that reactive cysteinyl and arginyl residues identified in this study take part of a C-terminal sequence segment, i.e. $^{343}$R-C-P-A-W-C-D-R-I-L$^{352}$, which is well conserved between sequences corresponding to inositol and phosphatidylinositol polyphosphates 5-phosphatases. This 10 amino acids-long peptide, which is critical for substrate binding in 43 kDa-Ins(1,4,5)$P_3$/Ins(1,3,4,5)$P_4$ 5-phosphatase, could represent a diagnostic motif for this growing family of 5-phosphatases.

## 3 References

1. Berridge, M.J. (1993) *Nature* 361, 315-325

2. Cullen, P.J., Hsuan, J.J., Truong, O., Letcher, A.J., Jackson, T.R., Dawson, A.P. and Irvine, R.F. (1995) *Nature* 376, 527-530

3. Majerus, P.W. (1996) Genes & Devel. 10, 1051-1053

4. Communi, D., Lecocq, R. and Erneux, C. (1996) *J. Biol. Chem.* 271, 11676-11683

5. Communi, D. and Erneux, C. (1996) *Biochem. J.* 320, 181-186

# Structural basis of protein-ligand interactions in the Pleckstrin Homology domain

M. Wilmanns, M. Hyvönen & M. Saraste, EMBL Heidelberg, Postfach 10.2209, D-69117 Heidelberg, Germany.

The Pleckstrin Homology (PH) domain has been identified as a common modular domain in a variety of signaling proteins and cytoskeletal proteins [1,2]. More than one hundred known PH sequences show very strong diversity and are lacking any invariant residues. Therefore, it has not been possible to assign a unique site that could lead to the identification of a common function in all PH domains.

If there is a general role of PH domains then it is most likely tethering other protein domains to the membrane. Most of the PH domains are identified in proteins that are membrane associated [3]. Strong evidence for membrane association through PH domains originates from the comparison of the related rhodopsin kinase and β-adrenergic receptor kinase. Apart from very few exceptions, namely the Btk PH domain [4] and the Akt/PKB PH domain [5,6], very little is yet known about the *in vivo* function of PH domains.

Phosphatidylinositol lipids, in particular phosphatidylinositol-4,5-bisphosphate (PIP2), have been identified as specific ligands for several PH domains [7-9]. Inositol-1,4,5-trisphosphate (IP3) inhibits binding of PIP2 to PH domains suggesting binding of PIP2 through the inositol-4,5-bisphosphate head group [8]. An alternative way of membrane association of some PH domains may be achieved by binding to βγ-subunits of heterotrimeric G proteins [10]. Binding to βγ subunits might, however, require flanking sequences C-terminal to the PH domain.

The 3D structure of a PH domain was first determined by NMR spectroscopy [11], succeeded by a number of high resolution X-ray structures of the dynamin PH domain [12], the β-spectrin PH domain in complex with IP3 [13] and the PLCδ PH domain in complex with IP3 [14]. The two complex structures are shown in Figure 1. Despite of very low sequence similarity all these three PH domains have the same core fold that consists of an orthogonal seven-stranded β-sheet followed by a C-terminal α-helix. The loop regions are, however, highly variable in length and conformation. The spectrin PH domain contains an additional α-helix in the loop that connects strands 3 and 4 (β3-β4). The PLCδ PH domain, in contrast, contains two α-helices at the N-terminus and within the loop β5-β6. The surfaces of all known PH domain structures display strong elostrostatic polarity.

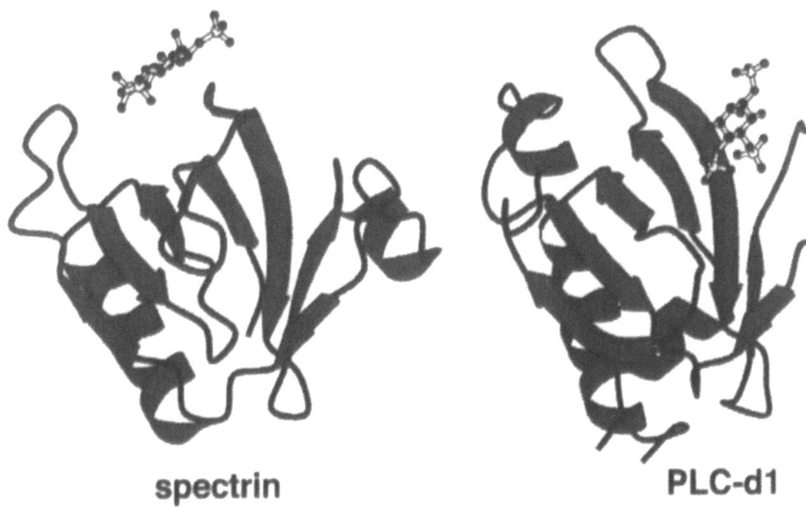


Figure 1: Ribbon representations of (A) spectrin PH : IP3 complex (PDB code: 1BTN) and (B) the PLCδ PH domain : IP3 complex (PDB code: 1MAI). The locations of the IP3 ligands are shown in ball-and-stick.

NATO ASI Series, Vol. H 102
Interacting Protein Domains
Their Role in Signal and Energy Transduction
Edited by Ludwig Heilmeyer

The phosphatidylinositol binding site of the spectrin PH domain was identified by our X-ray structure of this PH domain in complex with IP3 [13]. IP3 is bound in a cavity formed between the two loops β1-β2 and β5-β6 (Figure 1A). Most of the specific interactions are formed between the 4- and 5-phosphate groups of IP3 and polar residues of this PH domain. A key residue is K8 that forms salt bridges with both phosphate groups and becomes entirely buried upon IP3 binding. The three free hydroxyl groups of the inositol ring (in the 2, 3, and 6 position) are not involved in any specific interactions with this PH domain. The 1-phosphate group of the inositol ring is highly solvent exposed allowing modeling of a PH domain : PIP2 complex without steric clashes. Thus, we assume that the PIP2 binding site is identical with the IP3 binding site. Near-UV CD measurements did not indicate preferential binding of inositol-1,3,4,5-tetraphosphate (IP4) compared to IP3. These data are supported by a model of the PH domain : IP4 complex where the binding cavity can accommodate an additional 3-phosphate group at the inositol ring but does not provide any new specific interactions.

The IP3 binding site of this complex structure is surrounded by an excess of positively charged residues on a flat triangular surface [13] which represents one pole of the strong electrostatic sideness of this PH domain [11]. From these data we have created a model in which the PH domain binds specifically to PIP2 that is anchored in the membrane by its fatty acid tails. The positively charged surface around the IP3 binding site enhances membrane association by unspecific interactions with the polar, predominantly anionic, head groups of the membrane (Figure 2).

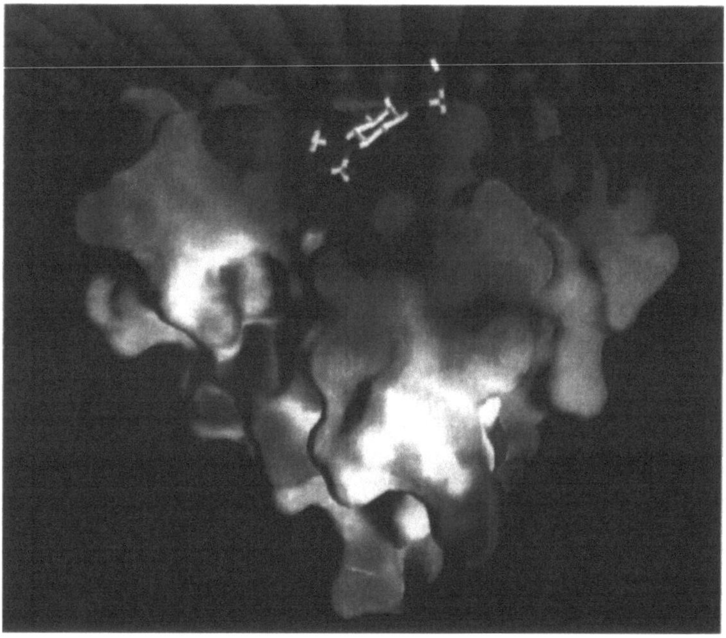

Figure 2: Model for lipid association through PH domains. The electrostoatic potential of the surface of the spectrin PH domain is shown by a colour range from blue (positive) to red (negative).

The localization of the IP3/PIP2 binding site in the spectrin PH domain is consistent with the mapping of the same site by difference NMR spectroscopy [13]. Prior this structure determination, mapping of the IP3/PIP2 binding site in the N-terminal PH domain of pleckstrin, however, indicated a different binding site [8]. In this PH domain the binding site is apparently located in a cavity formed by loops β1-β2 and β3-β4 (Figure 3) adjacent to the IP3/PIP2 binding site in the spectrin PH domain.

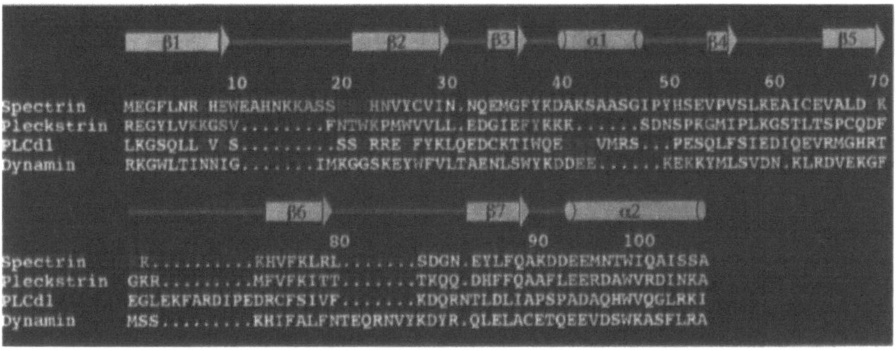


Figure 3: Binding of IP3/PIP2 to the PH domains of spectrin [13], pleckstrin [8], PLCδ [14] and dynamin [15]. Residues affected upon IP3/PIP2 binding are shown in yellow (NMR) and red (X-ray). Red residue positions in the spectrin PH domain have been identified by the X-ray complex structure and difference NMR spectroscopy. The positions of the secondary structure elements found in the spectrin PH domain : IP3 complex [13] are indicated by cylinders and arrows for α-helices and β-strands, respectively.

The subsequently solved X-ray structure of the PLCδ PH : IP3 complex shows that the IP3/PIP2 binding site of the PLCδ PH domain is located in the cavity between loops β1-β2 and β3-β4 [14], which is the predicted binding site of the N-terminal PH domain of pleckstrin but not of the spectrin PH domain. Despite this striking difference IP3 is bound to both PH domains in very similar ways, summarized in the following:

1. in both complexes most of the specific interactions are between the 4- and 5-phosphate groups of the inositol ring and polar residues of the PH domain;
2. despite different binding sites a lysine from loop β1-β2 (K8 in the spectrin PH domain; K30 in the PLCδ PH domain; see Figure 3) provides two salt bridges to the 4-phosphate and 5-phosphate groups of IP3 in each complex and plays a pivotal role in IP3 binding;
3. in both complex structures the free hydroxyl groups of the inositol ring are not involved in any specific interactions;
4. the 1-phosphate groups of the IP3 ligand are hydrogen bonded by one residue of each PH domain (S22 in the spectrin PH domain; W36 in the PLCδ PH domain). In both structures the 1-phosphate group is highly solvent exposed;
5. in both structures the IP3 binding site is surrounded by a positively charged surface. Virtually identical models for membrane association of PH domains by specific interactions with PIP2 and unspecific interactions with polar head groups of the membrane have been proposed [13, Figure 7; 14, Figure 6].

It remains unclear why different IP3/PIP2 binding sites have been found in these two complex structures. It is interesting to note that in the spectrin PH domain the loop β3-β4 contains an α-helix but does not contribute to the IP3 binding site (Figure 1A). The converse situation is observed in the PLCδ PH domain complex where loop β5-β6 contains a helix but the IP3/PIP2 binding site is located in the cavity between β1-β2 and β3-β4 (Figure 1B). Does the presence of these α-helices prevent IP3/PIP2 binding to the pockets they frame? Another possible extrapolation could lead to the suggestion of the presence of multiple phosphatidylinositol binding sites, may be with preferential binding affinities and specificities.

Can the comparison of the two complex structures explain why the PLCδ PH domain binds IP3 about 200 times stronger than the spectrin PH domain [13, 14]? Counting the number of specific interactions and snug contacts obviously cannot provide a satisfactory answer. Ferguson et al. [14] speculate that conformational rearrangements upon IP3/PIP2 binding might contribute to binding affinity and specificity in the case of PLCδ PH domain. A 3D-structure of the apo-PLCδ PH domain would be required to answer this hypothesis. Comparison of the apo-spectrin PH domain [11] with the spectrin PH domain : IP3 complex [13], at least, does not indicate significant conformational changes upon IP3 binding.

We are left with a detailed knowledge of the 3D structures and IP3/PIP2 binding sites of divergent PH domains, supported by an accumulation of strong *in vitro* data. The spectrin PH domain has been identified as a module that associates spectrin with the membrane through binding to PIP2 [16]. In PLCδ the N-terminal PH domain is a high affinity binding site for PIP2/IP3, and excess of IP3 affects the catalytic activity of PLCδ [17]. However, *in vivo* evidence for a common and general function of PH domains is still missing. Despite PH domains have been quite accessible for their structural analysis the discovery of their function or functions still remains a formidable task.

**References:**


1. Haslam, R.J., Kolde, H.B. & Hemmings, B.A. (1993) *Nature* **363**, 309-310.
2. Mayer, B.J., Ren, R., Clark, K. & Baltimore, D. (1993) *Cell* **73**, 629-630.
3. Gibson, T.J., Hyvönen, M., Musacchio, A., Saraste, M. & Birney, E. (1994) *Trends Biol. Sci.* **19**, 349-353.
4. Thomas, J.D., Sideras, P., Smith, C.I.E., Vorechovsky, I., Chapman, V. & Paul, W.E. *Science* **261**, 355-358 (1993).
5. Burgering, B.M.T. & Coffer, P.J. (1995) *Nature* **376**, 599-602.
6. Andjelkovic, M., Jakubowicz, T., Cron, P., Ming, X.-F., Han, J.-W. & Hemmings, B.A. (1996) *Proc. Natl. Acad. Sci. USA* **93**, 5699-5704.
7. Cifuentes, M.E., Honkanen, L. & Rebecchi, M.J. (1993) *J. Biol. Chem.* **268**, 11586-11593
8. Harlan, J.E., Hajduk, P.J., Yoon, H.S. & Fesik, S.W. (1994) *Nature* **371**, 168-170.
9. Lemmon, M.A., Ferguson, K.M., Sigler, P.B. & Schlessinger, J. (1995) *Proc. Natl. Acad. Sci. USA* **92**, 10472-10476.
10. Inglese, J., Koch, W.J., Touhara, K. & Lefkowitz, R.J. (1995)*Trends Biol. Sci.* **20**, 151-156.
11. Macias, M.J., Musacchio, A., Ponstingl, H., Nilges, M., Saraste, M. & Oschkinat, H. (1994). *Nature* **369**, 675-677.
12. Ferguson, K.M., Lemmon, M.A., Schlessinger, J. & Sigler, P.B. (1994) *Cell* **79**, 199-209.
13. Hyvönen, M., Macias, M.J., Nilges, M., Oschkinat, H., Saraste, M. & Wilmanns, M. (1995) *EMBO J.* **14**, 4676-4685.
14. Ferguson, K.M., Lemmon, M.A., Schlessinger, J. & Sigler, P. (1995)*Cell* **83**, 1037-1046.
15. Zheng, J., Cahill, S.M., Lemmon, M., Fushman, D., Schlessinger, J. & Cowburn, D. (1996) *J. Mol. Biol.* **255**, 14-21.
16. Wang, D.-S. & Shaw, G. (1995) *Biochem. Biophys. Res. Comm.* **217**, 608-615.
17. Garcia, P., Gupta, R., Shah, S., Morris, A.J., Rudge, S.A., Scarlata, S., Petrova, V., McLaughlin, S. & Rebecchi, M.J. (1995) *Biochemistry* **34**, 16228-16234.

# Part IV

# Tyrosine Phosphorylation and Downstream Signaling

# Cell Signaling by Tyrosine Phosphorylation: The Other Side of the Coin

Edmond H. Fischer
Department of Biochemistry, University of Washington, Seattle, WA 98195

The study of protein tyrosine phosphatases was undertaken with the assumption that if overexpression or mutations of tyrosine kinases could bring about oncogenicity then, necessarily, overexpression of tyrosine phosphatases would block or reverse transformation. This assumption turned out to be incorrect.

A tyrosine phosphatase was isolated in homogeneous form from human placenta (1,2). Surprisingly the amino acid sequence of the enzyme showed no homology with any of the other Ser/Thr phosphatase but a structural relationship to an already well-known surface antigen, namely, the leukocyte common antigen CD45 (3).

CD45 comprises a broad family of molecules differentially expressed on leukocyte subsets having a single transmembrane segment and two, internally homologous domains of ca. 30 kDa each structurally related to the placenta PTPase. CD45 has been implicated in the regulation of lymphocyte function, including signaling through the antigen receptor, cytotoxicity, proliferation and differentiation (4).

But, of course, the important question is the role CD45 might play in lymphocyte activation. Reaction of a T-cell with an antigen presenting cells through its T-cell receptor/CD3 complex, immediately triggers the activation of various intracellular src family tyrosine kinases (particularly lck and fyn). This results in the tyrosine phosphorylation of the TCR/CD3 complex on specific recognition motifs (designated as ARAMs or ITAMs) and many other proteins thereafter. But for these reactions to occur, CD45 is required. Mutant forms of human leukemia T-cell lines lacking CD45 fail to signal through the T-cell receptor and the overall level of tyrosine phosphorylation is almost abolished (5,6). Therefore, the transmembrane phosphatase is needed for tyrosine phosphorylation to occur (7).

Since then, a great variety of PTP receptor forms have been identified. All but a few have two catalytic domains in their cytoplasmic portion, as found in CD45, but show a considerable diversity in the structure of their external segments. Strikingly, most display all the hallmarks of cell adhesion molecules. Some, such as the LARs have Ig-like and FNIII repeats, and present the characteristics of cell adhesion molecules belonging to the immunoglobulin super family, that include the N-CAMs, Ng-CAMs, neurofascin, contactin, fasciclin 2 and 3, etc. These molecules promote neurite outgrowth and are responsible for the generation of tissue pattern and form during embryonic development (8,9).

NATO ASI Series, Vol. H 102
Interacting Protein Domains
Their Role in Signal and Energy Transduction
Edited by Ludwig Heilmeyer

R-PTP κ & μ contain on the outside 4 FNIII repeats, an Ig-like domain, and, at their N-terminus, a 170-residue globular, so-called MAM motif - for Meprin, (an enterokinase), the A5 neuronal antigen of Xenopus and receptor μ. They undergo homophilic interaction (10,11).

The most unexpected receptors, RTP β or ζ (or γ), contain at the end of either a short or very long arm, one FNIII motif linked to a globular molecule almost identical to carbonic anhydrase (CAH) (12.13). These are widely expressed on, and restricted to, glial cells in the embryonic nervous system; a third soluble form lacking the transmembrane and catalytic domains exists in large amounts in adult brain. It was already known as a chondroitin sulfate proteoglycan termed "phosphacan" and shown to interact with the N-CAM, Ng-CAM and the extracellular matrix protein tenascin (14). What is exciting is that good evidence has recently been presented that the ligand for these receptors is contactin (15). Contactins (F3/F11), represent a subgroup of molecules containing 6 Ig and 4 FNIII domains that are expressed on the surface of specific neuronal cells during development. They are known to elicit either positive or negative signals in response to various stimuli: positive when they interact with membrane-bound adhesion molecules such as the N-CAMs or the extracellular matrix protein tenascin, leading to axonal growth and differentiation ; negative when they interact with Janusin (Restrictin) that induce neuronal repulsion (16). Interaction with RPTP-β requires the CAH domain. However, one knows little about the manner or direction in which the signal is transduced, to such an extent that the classical notion of ligands and receptors no longer applies. Both $R_\beta$ and contactin can exist in membrane-bound and soluble forms so that the soluble external domain of $R_\beta$ can act as a ligand for the contactin receptor just like the soluble form of contactin can serve as a ligand for the phosphatase receptor. And then, of course, the two receptors can come into contact with one another in the course of cell-cell interaction. As a result, the type and direction of the response could be conceivably switched from one to the other at various stages of neuronal development (15).

Likewise, intracellular protein tyrosine phosphatases (PTPs) display a great diversity of structures, either preceding or following a highly conserved catalytic core (17). These are undoubtedly involved in their regulation and localization. Some PTPs have segments homologous to cytoskeletal proteins such as band 4.1, ezrin or talin or SH2 domains that would allow them to interact with growth factor receptors. Tyrosine phosphatases were also found as the gene product of virulence plasmids from bacteria of the genus Yersinia (such as *Y. pestis* responsible for the bubonic plague that wiped out in times past a good segment of the human population). The enzyme is involved in the pathogenicity of the organism because when a dead enzyme is introduced in the bacterium, it is no longer pathogenic. It is also found in vaccinia virus and participates in its pathogenicity.

Described below are a couple of examples as to how intracellular PTPs may use their non-catalytic domains to regulate their activity, thereby eliciting a positive or negative response. The first concerns the family of SH2-containing phosphatases (18). Deficiency of one of these (PTP 1C or SHP-1) results in a severe if not lethal autoimmunity and immunodeficiency disease: the motheaten syndrome with focal abscesses of the skin giving a "motheaten" appearance (19,20). It is accompanied by an impaired lymphocytic development, migration of prothymocytes to the thymus and T-cell response to mitogens. Also, there is an increase in macrophage and macrophage-like cells in lymphoid tissue and bone marrow and erythropoiesis. Which suggests that under normal conditions, when the enzyme is present, it serves as a negative determinant in repressing lymphocytic development.

When expressed in E. coli or mammalian cells, these enzymes are almost inactive toward protein substrates. Activity can be increased ca. 20-fold if one chops off the c-terminus and 50-fold if one removes the SH2 domains, indicating some kind of interaction between the two even though the native enzyme contains no bound phosphate. On the other hand, they are activated more than a 1,000-fold by acidic phospholipids including phosphatidic acid (18). They contain several tyrosyl residues having the concensus sequence recognized by the adapter proteins Grb 2 (YXNX) and, indeed, 3 of these can be phosphorylated by activated growth factor receptors (18). The view is, then, that these enzymes (such as SHP-2) which exist in an inactive conformation in the cytoplasm can be recruited to the plasma membrane following cell stimulation by interaction of their SH2 domains with phosphotyrosyl residues having the proper recognition sequence (such as Tyr 1009 in the PDGF-R). Interaction between the two would unfold the phosphatases, allowing them to become phosphorylated by the receptor kinase, then to bind Grb 2 and initiate one of the signaling cascade such as the MAPK pathway. This would result in a positive response. Alternatively, these enzymes could act as negative determinants by causing the dephosphorylation and inactivation of other receptors or surface molecules that would not catalyze their phosphorylation. Acidic phospholipids would contribute to the latter function.

Finally, to examine the role the localization/regulatory domain might have in the function of TC-20, a human T-cell phosphatase (21), the properties of the wild-type enzyme were compared to those of a molecule in which the entire regulatory segment was deleted. What remains is fully active but now is soluble within the cytosol whereas the full-length enzyme localyzes on the endoplasmic reticulum. But deletion of the regulatory/localization domain induced large differences in behavior as seen, for instance, when the two forms were overexpressed in a highly tumorigenic BHK cell line and the transfected cells injected into nude mice. Within a month, tumors were produced in animals injected with the control cells containing the vector alone but, contrary to what one could have expected, the tumors were more fully developed (and highly vascularized due to angiogenesis) with cells overexpressing the wild-type phosphatase. By contrast, tumor formation was greatly reduced, if not abolished, with BHK cells transfected with the truncated form lacking the localization domain; in fact, several animals had no detectable tumor.

Similar data were obtained with embryonic Rat-2 cells transformed with the viral oncogene v-fms (22). Rat-2 cells are non-transformed and display a cobblestone morphology but acquire a spindle-shape, stringy appearance when transformed with v-fms. Cells overexpressing the full-length enzyme maintained the same stringy, transformed, phenotype, in contrast to cells containing the truncated enzyme which had reverted to the original, cobblestone morphology. To achieve a high level of expression of the tyrosine phosphatases, the enzymes were packaged in retroviruses with which the cells were infected. All spindle-shape cells grew readily on soft agar and produced large tumors when injected into nude mice. By contrast, tumor formation was abolished in cells transfected with the truncated enzyme, indicating that they had reverted to the non-transformed state (22).

The above data, and others not described herein indicate the following. First, phosphatases are not scavenger enzymes, only there to remove the phosphate groups introduced by the kinases. They cannot be viewed simply as providing an "off" switch in an "on/off", kinase/phosphatase system. These enzymes do not carry out one way - and opposing - reactions. The same phosphatases, depending on where they localize within the cell - or the molecule with which they interact, can serve either as positive or negative determinants in controlling cell behavior. In many instances, they can act synergistically with the kinases to enhance the phosphorylation reaction. Second, the factors that determine whether phosphatases would enhance or oppose a kinase reaction would seem to depend less on their state of activity than on their subcellular localization. This would suggest that if one wanted to call upon these to control transformation, one should try to tamper with their localization segments - or whatever binding proteins they might be attached to - rather than with their catalytic domains. Displacement of these enzymes from where they are meant to bind would seem a more promising approach than trying to modulate their catalytic activity. Finally, their architectural characteristics are so basically different from that of the kinases - with receptor tyrosine phosphatases displaying the structural motifs of cell adhesion molecules - that they must have a mission of their own in cell development, survival and death, quite apart from that of the kinases.

**References**

1. Tonks, N. K., Diltz, C. D., and Fischer, E. H. (1988) *J. Biol. Chem.* **263**, 6731-6737

2. Tonks, N. K., Diltz, C. D., and Fischer, E. H. (1988) *J. Biol. Chem.* **263**, 6722-6730

3. Charbonneau, H., Tonks, N. K., Walsh. K. A., and Fischer. E. H. (1988) *Proc. Natl. Acad. Sci. USA* **85**, 7182-7186

4. Thomas, M. L. (1989) *Annu. Rev. Immunol.* **7**, 339-369

5. Pingel, J. T. and Thomas, M. L. (1989) *Cell* **58**, 1055-1065

6. Koretzky, G., Picus, J., Schultz, T., and Weiss, A. (1991) *Proc. Natl. Acad. Sci. USA* **88**, 2037-2041

7. Trowbridge, I. S. and Thomas, M. L. (1994) *Annu. Rev. Immunol.* **12**, 85-116

8. Edelman, G. M. (1985) *Annu. Rev. Biochem.* **54**, 135-169

9. Edelman, G.M. (1993) *Cell Adhesion and Communication* 1, 1-7.

10. Brady-Kalnay, S. M., Rimm, D. L., and Tonks. N. K. (1994) *Adv.Prot.Phosphatases* **8**, 227-257

11. Zontag, G. C. M., Koningstein, G. M., Jiang, Y.-P., Sap, J., Moolenaar, W. H., and Gebbink, M. F. B. (1995) *J. Biol. Chem.* **270**, 14247-14250

12. Krueger, N. X. and Saito, H. (1992) *Proc. Natl. Acad. Sci. U. S. A.* **89**, 7417-7421

13. Levy, J. B., Canoll, P. D., Silvennoinen, O., Barnea, G., Morse, B., Honegger. A. M., Haung, J.-T., Cannizzaro, L. A., Park, S.-H., Druck, T., and et al (1993) *J. Biol. Chem.* **268**, 10573-10581

14. Milev, P., Friedlander. D., Sakural. T., Karthikeyan, L., Flad, M., Margolis. R.K. Grumet, M. and Margolis, R.U. (1994) *J. Cell Biol.* **127**, 1703-1715.

15. Peles, E., Nativ, M., Campbell, P. L., Sakurai, T., Martinez. R., Lev, S., Clary, D. O., Schilling, J., Barnea, G., Plowman, G. D., and et al (1995) *Cell* **82**, 251-260

16. Vaughan, L., Weber, P., D'Alessandri. L., Zish, A. H., and Winterhalter. K. H. (1994) *Persp. on Developm. Neurobiol.* **1**, 43-52

17. Hunter, T. (1995) *Cell* **80**, 225-236

18. Zhao, Z., Shen, S.-H., and Fischer, E. H. (1995) *Adv.Prot.Phos.* **9**, 297-317

19. Ahn, N. G., Seger, R., and Krebs, E. G. (1992) *Curr. Opin. Cell. Biol.* **4**, 992-999

20. White, M. F., Livingston, J. N., Baker, J. M., Lauris, V., Dull, T. J., Ullrich, A., and Kahn, C. R. (1988) *Cell* **54**, 641-649

21. Cool, D. E., Tonks, N. K., Charbonneau, H., Walsh, K. A., Fischer, E. H., and Krebs, E. G. (1989) *Proc. Natl. Acad. Sci. USA* **86**, 5257-5261

22. Zander, N. F., Cool, D. E., Diltz, C. D., Rohrschneider, L. R., Krebs, E. G., and Fischer, E. H. (1993) *Oncogene* **8**, 1175-1182

# Binding studies on the neuronal isoform of the non-receptor protein tyrosine kinase pp60$^{c\text{-}src}$, pp60$^{c\text{-}srcN}$ and potential target proteins

Christian Schoeberl[1], Edith Ossendorf[1], Joachim Kremerskothen[1], Charlotte Brewster[2], Steve Dilworth[2] and Angelika Barnekow[1]
[1]Dept. of Exp. Tumorbiology, University of Muenster, Badestr. 9, D-48149 Muenster, Germany
[2]Dept. of Chem. Pathology, Royal Postgrad., Med. School, Du Cane Road, London W12 ONN, U.K.

## 1. Introduction

Tyrosine kinases of the non-receptor type are enzymes that lack transmembrane and extracellular domains. As known so far, their target proteins are membrane receptor proteins, membrane-associated proteins and cytoplasmic proteins (Pawson and Hunter, 1994).

The non-receptor protein tyrosine kinases of the *src*-family are involved in important cellular processes like proliferation, differentiation and neoplastic growth (Barnekow et al., 1992, Rodrigues and Park, 1994). The *src*-kinases contain SH2- and SH3-domains that recognize special motifs within their target proteins. SH2-domains specifically bind to phosphotyrosine residues and SH3-domains recognize poly-proline sequences within their target proteins (Pawson, 1995).

Compared to pp60$^{c\text{-}src}$, the neuronal isoform pp60$^{c\text{-}srcN}$ contains six additional amino acids within the SH3-domain (Levy et al., 1987, Martinez et al., 1987). Therefore we were interested to investigate possible alterations in target protein binding specificities.

We constructed fusion proteins containing the SH3- and SH3N- domains of pp60$^{c\text{-}src}$ and pp60$^{c\text{-}srcN}$ (mouse) and the Glutathione-S-Transferase protein (GST). Binding studies with these fusion proteins and protein extracts of SH-SY5Y (neuroblastoma) cells revealed a 26 kDa protein and a 30 kDa protein that specifically bind to the SH3N- but not to the SH3-domain. Further characterization of these proteins are in progress.

## 2. Methods and Results

After induction of expression of mouse cDNA, coding for the SH3-/SH3N domains (Ala90 to Tyr157, subcloned in pGEX-2T), with 0.5 mM IPTG, the fusion proteins containing the GST protein (glutathione-S-transferase) linked to the SH3- or SH3N domains were purified on glutathione-Sepharose 4B columns. Subsequently the proteins were studied by Western Blot analyses using the monoclonal antibodies Mab 327 (Lipsich et al., 1983) and Mab V7B11 (Fig. 1).

NATO ASI Series, Vol. H 102
Interacting Protein Domains
Their Role in Signal and Energy Transduction
Edited by Ludwig Heilmeyer

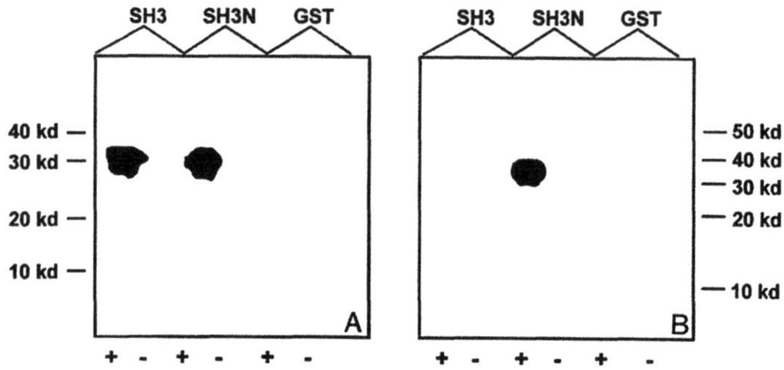


**Fig. 1: Western blot analysis of purified fusion proteins**
Purified GST fusion proteins (GST-SH3 and GST-SH3N, 2.5 µg) were loaded onto a 12 % polyacrylamide gel. After electrophoresis and transfer onto nitrocellulose, fusion proteins were detected by the monoclonal antibodies Mab 327 reacting with both isoforms **(A)** and Mab V7B11 recognizing pp60$^{c\text{-}srcN}$ **(B)**. The secondary antibody was an anti mouse conjugated to horseradish peroxidase. Finally the filters were washed and developed using the Amersham ECL system.
[SH3 stands for GST-SH3 fusion protein; SH3N stands for GST-SH3N fusion protein; +/- : with or without primary antibody].

As expected the Mab 327 (recognizing viral and cellular pp60$^{src}$) reacts with both, the SH3 and SH3N domain, whereas the V7B11, prepared against the six amino acid neuron-specific insertion, specifically binds to the SH3N domain.

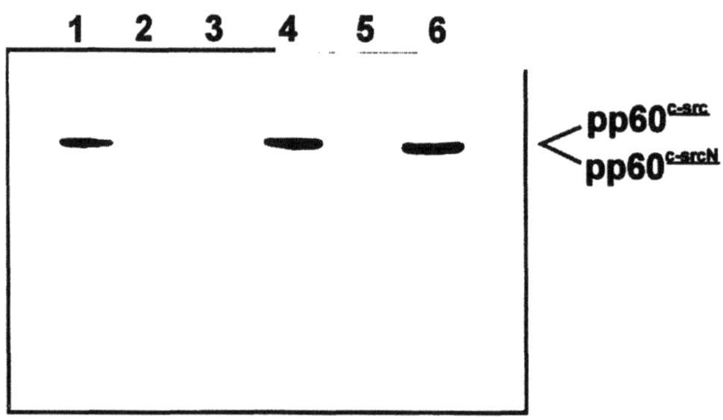


**Fig. 2: Competition experiments with GST-SH3 (lanes 3, 6) and GST-SH3N (lanes 2, 5)**
20 µg purified synaptic vesicle extract was separated on a 11 % PAGE and the proteins blotted onto PVDF-membrane. Mab327 (lanes 1-3) and MabV7B11 (lanes 4-6) were preincubated with 20 µg of GST-SH3 or GST-SH3N for 15 min. at 30 °C and the blotting membrane incubated with the Mab-GST-SH3/ SH3N mixture. pp60$^{c\text{-}src}$ and pp60$^{c\text{-}srcN}$ were detected using an enhanced chemiluminescent technique (Pierce, Rockford, USA). [lanes 1 and 4: Mab 327 and Mab V7B11 without preincubation with the GST-SH3/ SH3N].

The results shown in figure 2 confirmed the specificity of the heterologously expressed SH3- and SH3N-domains. Preincubation of pp60$^{\underline{c\text{-}src}}$ or pp60$^{\underline{c\text{-}srcN}}$ specific monoclonal antibodies with SH3- and SH3N-domains prevents the binding of the antibodies to pp60$^{\underline{c\text{-}src}}$ or pp60$^{\underline{c\text{-}srcN}}$ of synaptic vesicles, respectively.

To detect potential target proteins, the SH3-/SH3N domains were incubated with $^{35}$S-methionine labelled cell extracts of human neuroblastoma cells (SH-SY5Y). As shown in figure 3, SH3 and SH3N binding to cellular SH-SY5Y proteins resulted in different binding patterns. Mainly two proteins of molecular weight sizes 26 kD and 30 kD were detected using SH3N as binding partner. These proteins did not react with the SH3 domain. The nature of these proteins is still unknown.

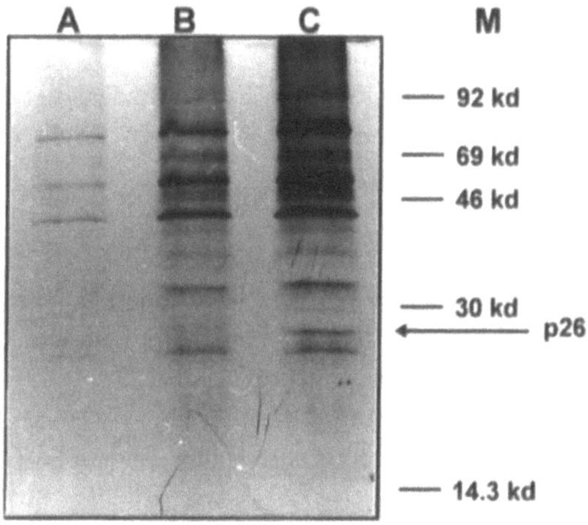


**Fig. 3: Binding of GST (A), GST-SH3 (B) and GST-SH3N (C) to cellular proteins from SH-SY5Y neuroblastoma cells**

1.2 mg of $^{35}$S-Met-labelled cellular protein extract from SH-SY5Y neuroblastoma cells were preincubated with 200 µg GST in 200 µl RIPA buffer for 2.5 h at 4°C. Then the protein solution was clarified by affinity chromatography filtration on glutathione-sepharose-4B columns to remove unspecific GST-bound cellular proteins. Subsequently the clarified protein solution was incubated with 120 µg of the fusion proteins (GST-SH3/GST-SH3N) for 2.5 h at 4°C in the presence of 100 µl glutathione-sepharose beads (1:1 slurry). After centrifugation at 10.000 x g the pellets were washed four times in RIPA buffer. Finally bound cellular proteins were eluted by soluble glutathione and separated on a 12 % SDS-PAA gel. Specific signals of eluted proteins were detected by fluorography.

**3. Summary and Conclusion**

While it was reported by Weng et al. (1993) that SH3/SH3N have similar binding patterns in fibroblasts, an additional set of proteins was detected in SH-SY5Y cells which specifically binds to SH3N.
We demonstrated that GST-SH3N binds to a 26 kD and to a 30 kD protein in extracts of the human neuroblastoma cells (SH-SY5Y) and that these proteins are not bound by GST-SH3.
From our data we conclude that the SH3 and SH3N domains of $pp60^{\underline{c\text{-}src}}$ and $pp60^{\underline{c\text{-}srcN}}$ display different binding behaviour in SH-SY5Y neuroblastoma cells, possibly due to the six amino acid insert within the SH3N-domain.

**Acknowledgements**
We thank S. Pahlman, Uppsala, for the SH-SY5Y cells, J.S. Brugge, Cambridge MA, for Mab 327. This work was supported by grants from the FONDS Chem. Industrie and Deutsche Forschungsgemeinschaft (Ba876/1-3) to A.B.
Chr. Schoeberl is a recipient of a grant from Evang. Studienwerk e.V. This article contains part of the Ph.D. thesis of Chr. Schoeberl.

**References**

**Barnekow, A., Ossendorf, E., Rosing, M., Titzenthaler, G., Wend, U. and Noeldeke, S.** (1992) Studies on the expression of tyrosine kinase-encoding protooncogenes in differentiating neuronal cells. *Neuroprotocols* **1**, 215-223

**Levy, J.B., Dorai, T., Wang, L.-H. and Brugge, J.S.** (1987) The structurally distinct form of $pp60^{\underline{c\text{-}src}}$ detected in neuronal cells is encoded by a unique *c-src* mRNA. *Mol. Cell. Biol.* 7, 4142-4145

**Lipsich, L.A., Lewis, A.J. and Brugge, J.S.** (1983) Isolation of monoclonal antibodies that recognize the transforming proteins of avian sarcoma viruses. *J. Virol.* **48**, 352-360

**Martinez, R., Mathey-Prevot, B., Bernards, A. and Baltimore, D.** (1987) Neuronal $pp60^{\underline{c\text{-}src}}$ contains a six-amino acid insertion relative to its non-neuronal counterpart. *Science* **237**, 411-415

**Pawson, T.** (1995) Protein-tyrosine kinases. Getting down to specifics. *Nature* **373**, 477-478

**Pawson, T. and Hunter, T.** (1994) Oncogenes and cell proliferation. *Curr. Opin. Gen. Dev.* **4**, 1-4

**Rodrigues, G.A. and Park, M.** (1994) Oncogenic activation of tyrosine kinases. *Curr. Opin. Gen. Dev.* **4**, 15-23

**Weng, Z., Taylor, J.A., Turner, C.E., Brugge, J.S. and Seidel-Dugan, C.** (1993) Detection of *src* homology 3-binding proteins, including paxillin, in normal and *v-src*-transformed balb/c 3T3 cells. *J. Biol. Chem.* **268**, 14956-14963

# The switch cycle of the Ras protein and its role in signal transduction

Alfred Wittinghofer

Max-Planck-Institut für molekulare Physiologie, Postfach 10 26 64,

44026 Dortmund, Germany

The Ras gene was discovered as an oncogene in mammalian tumors 15-17 years ago and its central importance in many signalling pathways has since been established. Point mutations in the three different Ras genes (H-, K- and N-Ras) had been found in approximately 25-30 % of human tumors and it had been established that Ras oncogenes are apparently involved in the early steps of cancerogenesis. Microinjection experiments with neutralizing antibodies, which specifically block Ras activity, had shown the requirement of functional Ras for several different growth factors. Oncogenic Ras genes induce growth factor independent proliferation in NIH3T3 fibroblasts and nerve growth factor independent differentiation in neuronal PC12 cells.

A mammalian cell contains an estimated 50-100 different GTP-binding proteins which function as molecular switches involved in regulating at least as many different biological processes ranging from protein biosynthesis, membrane trafficking to communicating signals from the outside of the cell to its interior. Ras is the founding father for the Ras-related superfamily of GTP-binding proteins and its properties can be considered a paradigm for the others. It is a 21 kDa protein which binds guanine nucleotides very tightly and cycles between an inactive GDP-bound state and an active GTP-bound state. (Fig. 1)

It becomes activated when protein-bound GDP is released and GTP, which is more prevalent in the cell than GDP, rebinds. In the GTP-bound state it reacts with downstream targets or effectors, which communicate the signal generated by growth factor receptors on the plasma membrane to other partners located further downstream in the reaction cascade. Effectors are defined as proteins which interact much tighter with the GTP-bound form than the GDP-bound form of the GTP-binding protein. This interaction is terminated by the GTPase reaction, which restores the GDP-bound form and terminates effector interaction. The intrinsic GTPase reaction is usually very slow (half life of the order of hours to minutes) and can be stimulated by several orders of magnitude (half-life seconds to milliseconds) by GTPase-

NATO ASI Series, Vol. H 102
Interacting Protein Domains
Their Role in Signal and Energy Transduction
Edited by Ludwig Heilmeyer

Activating proteins (GAP). In the resting state these proteins stay in the GDP-bound state and are activated by the action of guanine-nucleotide exchange factors (GEF) which increase the rate of nucleotide release by several orders of magnitude. In principle, however, the GTP-binding proteins can also be activated by the inhibition of GAP activity, as has been observed for Ras in T cells or adipocytes. A number of reviews dealing with various aspects of the biochemistry, biophysics and biology of Ras have appeared and are listed in the Reference section.

The Ras signalling is conserved in all higher organisms, and is also found in unicellular organism, although the function there is different from that in higher organisms. Mostly from genetic data the following sequence of events could be outlined for *Saccharomyces cerevisiae*: In a still unknown fashion nutritional signals cause the product of CDC25 to exchange GDP for GTP on the proteins RAS1 and RAS2, the two homologs of Ras, thereby converting them into the active form. The activated forms of RAS stimulate adenylate cyclase encoded by CYR1 through the bridging molecule encoded by SRV/CAP. The GAP homologous proteins encoded by IRA1 and IRA2 negatively regulate yeast RAS activity. cAMP causes the dissociation of the regulatory (BCY1) and catalytic subunit (TPK1,2,3) of the cAMP-dependent Kinase, which therefore can promote altered transcription and cell division through phosphorylation of multiple proteins.

It was originally hoped that the description of the signalling pathway in yeast *Saccharomyces cerevisiae* would help to define the signalling pathway of mammalian Ras. This was not the case since Ras does not stimulate adenylate cyclase of higher organisms. The situation appeared even worse when it was found that the Ras gene product of the fission yeast *Schizosaccharomyces pombe* is again involved in a different pathway.

However, in the last few years tremendous progress has been made towards our understanding of the structure, biochemistry and molecular biology of mammalian Ras in higher organisms including mammals, and the genetic analysis of the Ras signal transduction pathway in *C.elegans* and *D.melanogaster* has been one of the major breakthroughs. The sevenless pathway in the fruitfly *Drosophila melanogaster* determines the cell fate of a photoreceptor cell and leads to the development of a compound eye. In the worm *C. elegans* the Ras homologue let-60 is involved in the differentiation of gonadel cells and leads to the development of a vulva. Both signalling pathways transmit a signal from a receptor tyrosine kinase (RTK), the sevenless gene in drosophila and the let-23 gene in *C. elegans*. Both pathways work via activation of Ras and Raf kinase to initiate transcription of specific genes in the nucleus. It turns out that the main players in this game are conserved between worm, fly and man and are involved not only in the control of differentiation but also of cell growth control.

## 1. STRUCTURAL ASPECTS OF RAS FUNCTION

Another advance in understanding the molecular mechanism of Ras came by solving the X-ray structure of the protein in complex with either GDP or GTP-analogue. Gly-12 is the most frequently mutated residue in tumors and it has been shown that its mutation to any other amino acid except proline renders the protein oncogenic. These and other oncogenic mutations at position 13 and 61 cause a 10 fold decrease in the intrinsic hydrolysis rate of GTP. Double mutants involving Gly-12 and Ala-59 (A59T) have been found in the retroviral p21 proteins of the Kirsten and Harvey strain of MuSV. Since the protein GAP is no longer able to activate the GTPase activity of the oncogenic mutants, the difference in GTPase activity between normal and oncogenic p21 is actually very large (greater than $10^5$ fold) and is responsible for keeping the latter in the active, GTP-bound conformation. On the other hand, oncogenic p21 is still able to bind to GAP, and, as with wildtype p21, strong binding is dependent on the presence of GTP.

According to the three-dimensional structure Gly-12 is situated in a tight loop (loop L1) which wraps around the ß-phosphate group of GDP and the GTP-analogue GppNHp. This loop is found in many other mononucleotide (adenine and guanine nucleotide) binding proteins, such as adenylate kinase, was originally described by Walker et al. and has been called the P (phosphate binding)-loop. The three-dimensional structure analysis of Gly-12 mutants has shown that the mutations do not change the structure of the P-loop itself, or any other part of the 3D-structure, but rather perturb its spatial interaction with the phosphates, the effector loop L2 and the loop L4 containing glutamine 61. However, the overall structural changes between cellular and mutant proteins remain very small, as supported by data obtained by NMR.

Gln-61 is the second most frequently mutated residue of Ras in mammalian tumors. This residue is found in close proximity to glycine 12 and the g-phosphate on loop L4. It has been postulated that the role of this highly conserved glutamine is to activate the nucleophilic water molecule for attack on the γ-phosphate. Leucine at position 61 would least be able to perform the same function and in fact has the highest transformation efficiency of all Ha-Ras mutants in this position. The 3D-structure of the Q61L mutation is also the least perturbed as compared to wildtype supporting the idea of a catalytic role or Gln-61. Similar activating mutations in heterotrimeric G proteins have been found in certain pituitary tumors. The residues mutated were Gln-227 of $G_{sa}$ or Gln-205 of $G_{ia}$, which functionally correspond to glutamine 61 of Ras. The structure of $G_a$ subunits in the presence of GDP and $AlF_4^-$ has shown that Gln, and most likely Gln61 of Ras also, is involved in the stabilization of the transition state of the GTPase reaction.

Ser-17 is the most critical residue involved in contacting $Mg^{2+}$. The Ras mutants S17A and S17N have very fast nucleotide dissociation rates and are believed, although it has not been proven biochemically, to bind tightly to the exchange factor of Ras. Consequently Ras(S17N) has found wide application as a dominant negative inhibitor of normal Ras in vivo. It is not totally clear, whether this biological effect can be explained simply by the high nucleotide dissociation rates, the preferential affinity for GDP, the improper $Mg^{2+}$ coordination or by a blockade in the activation through the nucleotide exchange factor.

## 2. LIPID MODIFICATION

Ras proteins are only active in cellular signal transduction when they are localised at the inner surface of the plasma membrane after attachment of fatty acids. The posttranslational modification involves four consecutive enzymatic steps. First, the fourth cysteine from the C-terminus is farnesylated via a thioether linkage by a farnesyl-protein transferase, which recognises the so-called CaaX-box (C = cysteine, A = Aliphatic, X = any amino acid). Second, the three C-terminal residues are proteolytically removed, third, the C-terminus is carboxymethylated and fourth, $p21^{Hras}$, $p21^{Nras}$ and $p21^{Kras(A)}$ are palmitoylated on one or more cysteine residues within the C-terminal hypervariable region. In the case of $p21^{Kras(B)}$, a polybasic domain close to the C-terminus seems to cooperate with the farnesyl moiety in membrane attachment. The X position in the CaaX motif is occupied in Ras by either serine or methionine, which specifies it to be farnesylated. In Rap1A proteins the C-terminal leucine specifies geranylgeranylation of the cysteine. Cytosolic enzymes responsible for polyisoprenylation and therefore critical for the membrane association of Ras proteins have been identified from mammalian sources and from *S. cerevisiae*. Farnesyl-protein transferase consists of two distinct subunits, each of them approximately 50 kDa in molecular mass. In rat brain the ß-subunit confers specificity for the Ras C-terminal sequence, the a-subunit appears to bind the farnesyl donor farnesyl pyrophosphate and is shared between farnesyl and geranylgeranyl transferase. The homologous proteins of yeast, products of the RAM1/DPR1 and RAM2 genes, were identified independently as suppressors of the heat shock sensitivity of activated RAS2 and of a mating factor dysfunction, since the mating factor a also contains a CaaX-box.

The C-terminal modification of Ras is absolutely required for the biological function. Since it involves an enzymatic reaction that has been thoroughly tested in vitro it was also chosen as a possible target reaction to inhibit Ras action in human tumors. Several compounds have indeed been identified that are very potent in inhibiting the C-terminal modification and can therefore potentially inhibit the biological function of Ras. How and whether this compounds work in vivo and whether they are cytotoxic to the cell remains to be established since other proteins of the cell contain CaaS or CaaM motif and are also farnesylated . However, results with mice carrying a v-Ras transgene has established the high potency of such compounds to reduce tumors.

## 3. GUANINE NUCLEOTIDE RELEASE FACTORS AND ACTIVATION OF RAS

The genetic analysis of the CDC25 gene from *S. cerevisiae* initiated a concept, according to which CDC25 is an upstream element which promotes RAS activation by accelerating the exchange of GDP by GTP. CDC25 mutant cells are deficient in adenylate cyclase activation. Reduced cAMP production arrests growth and is accompanied by high heat shock resistance, hyperaccumulation of glycogen and the inability to grow on nonfermentable carbon sources. The mutation RAS2(G19V) (analogous to G12V in mammalian Ras) could compensate this defect, presumably because this protein, since it cannot hydrolyse GTP, is permanently activated and thus independent of GDP/GTP exchange. In normally growing yeast cells GDP/GTP exchange is the limiting step in the activation cycle, because RAS proteins are predominantly found in the GDP-bound state. The model was confirmed by results from the mutation RAS2(T152I), which has a normal GTPase but shows a 50 fold higher dissociation rate for GDP. This mutant restored normal cAMP production in CDC25 mutant strains as well, but for another reason than RAS2(G19V). The CDC25 gene was cloned and its disruption was shown to be is lethal, thus the conclusion was drawn that its gene product must act on both the RAS1 and the RAS2 protein. Nevertheless, it is still unclear how the nutritional signals are sensed by the CDC25 gene product before further signal transduction via RAS proteins.

Cytoplasmic as well as membrane bound guanine nucleotide exchange factors (GEF) have been described for mammalian cells. In the study of the signal transduction cascade of the sevenless gene which results in the development of the R7 photoreceptor cell of the compound eye of drosophila the son of sevenless gene (SOS) was identified to work downstream of the sevenless RTK and upstream of the drosophila Ras gene. The SOS gene contains a stretch of about 50 amino acids which are highly homologous to the catalytic domain of the CDC25 gene. The rest of the sequences of SOS on one hand and CDC25/SDC25 appear to be unrelated and thus represent different type of GEF proteins for Ras.

Molecular cloning of the first mammalian GEF, named Ras-GRF, was achieved from a rat brain cDNA library by the use of degenerate oligonucleotides designed to code for peptides present in the most conserved region. In a different approach a mouse cdc25 = $CDC25^{Mm}$ homolog was isolated by complementation of CDC25 deficiency in *S. cerevisiae*. Altogether the predicted protein sequences of seven mouse CDC25 homologs share the conserved catalytic domain (35 % amino acid identities over 196 residues) and encode proteins with 666 to 1260 amino acids, which show distinct tissue-specific expression patterns. Recombinant $CDC25^{Mm}$ protein has been shown to accelerate nucleotide exchange on Ras specifically by a factor of more than 1000 and to have a slight preference for the GDP complex. Additionally the longer versions of $CDC25^{Mm}$ contain the so-called dibble homology (DH) domain of approximately 250 amino acids. This region is found in Bcr, the breakpoint cluster protein, in the proto-oncogene products of vav and dbl (pronounced dibble), and the yeast gene CDC42. This domain is believed to be responsible for guanine nucleotide exchange on rho and rac and CDC24 proteins which are Ras-related GTP binding proteins engaged in the control of cytoskeletal organisation. That Ras-GRF contains also a SH3 domain (for src-homology region 3) which is also found in many cytoskeletal proteins (see below) suggests that the Ras pathway might be connected to a similar pathway involving the rho/rac proteins. This might explain the changes in cytoskeletal structure observed upon growth factor treatment.

Mammalian homologues of the Drosophila son of sevenless gene have also been cloned. They contain a stretch of about 250 amino acids homologous to CDC25 and GRF shown to be sufficient for the catalytic activity. The mammalian and drosophila SOS protein sequences contain a proline-rich region located at the C-terminal end of the polypeptide chain, behind the catalytic domain. A proline-rich domain protein, 3BP-1, has been identified as a ligand of the SH3 domain (src-homolgoy region 3) of several proteins such as the abl and src kinase. SH3 and SH2 regions are found in many signal transduction molecules and are elements responsible for protein-protein interactions. The first member of this family of proteins, which consists only of SH2 and SH3 domains, has first been identified as the sem-5 gene which is involved in a signal transduction cascade via the Ras gene let-60 in *C. elegans*. This process results in the formation of a vulva by differentiation from precursor cells. A mammalian homologue of the sem-5 gene product has identified and have been shown earlier to bind, via their SH-2 domains, to activated receptor tyrosine kinases. SH-2 domains are structural modules composed of approximately 100 amino acids which bind very tightly to phosphorylated tyrosines in a sequence-specific manner. The 3D-structure of a number of these domains have been determined and their binding to phospho-tyrosine residues is now well understood. The SH3 domain is another structural module for protein-protein interaction. It consists of about 50 amino acids and the tertiary structure of a number of different SH3 domains has been solved by X-ray and NMR, in the presence and absence of polyproline ligands. The presence of a proline-rich region in Sos immediately suggested a mechanism of activation for Ras involving the SH-2 and SH-3 domains of these adapter proteins that has subsequently been experimentally verified and is described below.

It has been shown that the mechanism of action of the GEF on Ras or Ras-related GTP binding proteins involves the formation of a binary complex between GEF and GNB without guanine nucleotide, which appears to be very tight in the absence of nucleotide and can be rapidly dissociated by guanine nucleotides. This is reminiscent of the interaction between elongation factors EF-Tu and EF-Ts or between Gα proteins and hormone receptors which also form a tight "empty" complex. The dominant inhibitory effect of the S17N mutation on the function of normal Ras has been assumed to be due to the tight binding to the Ras-GEF. Tight binding between CDC25 and Ras(S17N) has been indirectly demonstrated by a two-hybrid system in yeast. Similarly it could be shown that the effect of Ras(S17N) could be overcome by overexpression of the SCDC25 catalytic domain. It has also indirectly been shown by the inhibition of the exchange activity of SOS on normal Ras. However, it has sofar not been shown in the test tube whether the inhibitory mutant really binds much tighter to any Ras-GEF, the CDC25 or the SOS type.

The genes for the human and rat homologues of the drosophila Ras exchange factors SOS were cloned and shown to be expressed in all tissues, in contrary to the CDC-25 type exchange factors, also called Ras-GRF. Their proline-rich regions are probable motifs for protein-protein interactions involving SH3 domains. Therefore the search for proteins which contain SH3 domains and are involved in signal transduction concentrated on the so-called adapter proteins Grb-2, Ash, Nck, Shc, which are all homologues of the *C. elegans* sem-5 gene and contain SH-2 together with SH-3 domains. It turns out that Grb-2 and possibly the other proteins as well bind to SOS in unstimulated cells, as can be shown by co-immune precipitation and in vitro exchange factor analysis of cell extracts. The same interaction has been shown genetically and biochemically for the corresponding Drosophila proteins. The affinity between Grb-2 and SOS appears to be very high since it survives immune precipitation. Both SH-3 domains of Grb-2 are necessary for tight binding and the binding of both SH-3 domains seems to be cooperative. It is interesting that the inactivating mutations which have been found in the *C. elegans* sem-5 gene interrupt the Sos-Grb-2 interaction. Stimulation of the cells by the binding of growth factors to their respective receptors leads the trans-phosphorylation of the cytoplasmic domains of the receptors on tyrosine residues which can now interact with the SOS-Grb-2 complex via the SH2 domain of the adapter. This in turn leads to the activation of Ras, which becomes complexed to GTP and therefore activated. It is not directly obvious how this Ras-activation is accomplished since the exchange factor activity does not appear to be altered by the growth factor stimulation. It is rather likely that the relocation of Sos to the plasma membrane via the RTK-GRb-2 connection is responsible for Ras-activation.

While the activation of Ras via the SOS-type exchange factor can be modelled, the activation of Ras by Cdc25-type Ras exchange factors, which are much more highly expressed expressed in brain than in other tissues, is less clear. In one report it has been shown that $Ca^{2+}$ induced calmodulin binding is involved in Ras activation in cultured cortical neurons treated with ionomycin. However, neither ionomycin treatment nor calmodulin binding stimulate in vitro exchange activity of Ras-GRF. Recently it has been shown that activation of Ras by heptahelical muscarinic receptors occurs via $\beta,\gamma$ subunits of hetrotrimeric G proteins, and that an increase of GEF activity of the Cdc25 type protein can be demonstrated in vitro.

## 4. GAP PROTEINS IN SIGNALLING AND DOWNREGULATION OF THE RAS PATHWAY

During studies on the maturation promoting effect of Ras in *Xenopus oocytes* Trahey and McCormick discovered GAP as a protein which maintains the cellular H-Ras or N-Ras proteins in a biologically inactive state without affecting the oncoproteins. GAP seems to be necessary for the inactivation of Ras in order to avoid unregulated growth of the cell. It interacts preferentially with the active GTP-complex and accelerates the otherwise slow intrinsic GTP-hydrolysis rate of Ras (0.028 $min^{-1}$ at 37°C) by a factor of $10^5$. The gene corresponding to GAP was cloned from bovine brain and human placenta. It possesses two splice-forms, which code for proteins with calculated molecular masses of 100 kDa and 116 kDa (p120-GAP), which differ by a highly hydrophobic region at the N-terminus of p120-GAP. Sequence analysis revealed several signal transduction modules such as SH2, SH3 and PH domains. Additionally a region homologous to the calcium binding (CalB) region of phospholipase A2 has been identified. Deletion studies clearly defined the C-terminal amino acids from 702 to 1044 of p120-GAP as the catalytic domain, although a 20 fold decrease in GTPase stimulation accompanies truncation. However, the minimal GTPase activating domain consisting of 226 amino acids has be obtained by limited proteolysis from the another Ras-GAP neurofibromin, which is the gene product of a tumor suppressor gene which has been found to be frequently mutated in patients with the disease neurofibromatosis type I but also, albeit less frequently, in solid tumors. A mammalian homologue ($GAP1^{m}$) of the Drosophila GAP1 gene has been described as the third form of Ras-GAP. Recently, two other members of the $Ras^{GAP}$ family, GAPIII and $GAP1^{IP4BP}$ have been isolated, the latter of which has GTPase stimulating activity towards both Ras and Rap.

$p120^{GAP}$ and NF1 protein can be distinguished with respect to their catalytic properties. $p120^{GAP}$ increases the GTPase reaction of $p21^{H\text{-}ras}$ more than $10^5$-fold, $k_{cat}$ = 19 $sec^{-1}$, with a $K_M$ of 9.7 μM for Ras•GTP, whereas NF1 protein has been reported to have a lower $k_{cat}$ of 1.4 $s^{-1}$ but a higher affinity ($K_M$ of 0.3 μM). In studies with N-Ras the difference in affinity was similar, but much smaller in $k_{cat}$. Certain lipids were found to inhibit the GAP activity of NF1 protein at micromolar concentration without having an effect on $p120^{GAP}$ activity, although the latter binds membranes in a $Ca^{2+}$-dependent fashion in vitro.

Furthermore, it is likely that the biological role of these proteins are different: $p120^{GAP}$, in addition to its catalytic domain, contains several other independent functional domains such as SH2, SH3, PH, CaLB, all of which seem to be involved in interactions with other signalling molecules. However, no obvious sequence correlation with other signalling molecules outside the catalytic domain has yet been identified in the primary sequence of NF1. There is some evidence that $p120^{GAP}$ might be more than just an inactivator of Ras and may in fact be an effector. NF1 and $GAP1^{m}$, on the other hand, seem to be more negative regulators of Ras, since, outside the catalytic domain, they display a high homology to the *S. cerevisiae* IRA gene products and to the *Drosophila melanogaster* protein GAP1, which have genetically been found to be negative regulators of the Ras signal transduction pathway. In agreement with the concept of a negative Ras regulator embryonic neurons from NF1 knockout mice survive in the absence of neurotrophin, which signals via Ras. In addition, the NF1 gene has characteristics of a tumor suppressor gene, as evidenced from the abnormal regulation of Ras activity in neurofibromatosis type I patients.

Cell lines derived from Schwannoma from NF1 patients contain very little neurofibromin, but normal levels of functional Ras and p120-GAP. In these cells Ras displayed elevated levels of bound GTP, which could be reverted together with the tumorigenic phenotype by introduction of the catalytic domain of p120-GAP. The data were interpreted to support the idea of NF1 acting as a tumor-suppressor whose function is to reduce the level of p21-GTP. But in this context it is important to remember that in cells of neuronal origin, such as Schwann cells, Ras can transfer signals both for the proliferative response as well as for pathways leading to differentiation. Therefore it was argued that if NF1 was an effector molecule for the establishment of differentiation, its loss might promote the proliferative response through the accumulation of Ras·GTP. The role of neurofibromin as a negative regulator is also from the fact that it has been found mutated in solid tumors and that this mutation knocks out the catalytic activity. No such mutations have as yet been found for GAP itself, despite intensive search. Recently it has been shown by that tumors of neural crest derived cells other than Schwannomas, such as neurofibromas or melanoma, there is no correlation between loss of the NF1 gene and the amount of GTP bound to Ras in these cells suggesting that other regions outside the catalytic domain of neurofibromin are important for the tumorigenic phenotype.

A major clue to GAP function came from the observation that upon epidermal growth factor (EGF) or platelet derived growth factor (PDGF) stimulation, 10 % of p120-GAP becomes phosphorylated on tyrosine 460 by the corresponding growth factor receptor kinases and that the phosphorylation is responsible for the association with the kinase. This association is mediated by the interaction of the SH2 domains of p120-GAP with the phosphorylated tyrosine kinases. The same phenomenon was observed in cells transformed with oncogenic nonreceptor tyrosine kinases, such as v-src, v-abl and v-fps, for which Ras had already been suspected to be a downstream target. How this association with the kinases would lead to Ras activation could not be elucidated. Possible models are that GAP would either loose its GTPase stimulatory activity on phosphorylation or that GAP, being the Ras effector, would become competent for transferring the mitogenic signal. From the biochemical analysis of GAP activities, phosphorylation as a regulatory possibility is unlikely.

The first assay for any function of the Ras-GAP-complex was developed using a reconstituted membrane system. Currents through atrial potassium channels are initiated through coupling to the G protein GK after muscarinic receptor activation. The complex of Ras-GTP and GAP prevents current formation, probably by interfering with G protein and receptor coupling. The single components have no effect. In a different biological test system the effect of only the SH2-SH3 domains of GAP on a fos reporter plasmid was measured. Whereas full-length GAP was silent, the SH2-SH3 region induced gene expression in a cooperative fashion with Ras. This result apparently supports the notion that GAP consists of a down-regulating and an effector part. For the insulin stimulated germinal vesicle breakdown in *Xenopus oocytes* Duchesne et al. the SH3 domain of p120-GAP appears to be sufficient for signalling. Monoclonal antibodies directed against this region or peptides resembling this region interrupt this process associated with oocyte maturation.

The mechanism by which GAPs accelerate the GTPase reaction has been a matter of considerable debate and has centered on Ras, the impaired GTPase of which in oncogenic Ras mutants contributes to tumor formation in 25-30% of all cancer patients. It has been argued that Ras itself is an efficient GTPase machine but in order to achieve the fast GTPase reaction GAP favors the attainment of a GTPase competent conformation by catalyzing the rate-limiting isomerization reaction. Evidence for such an isomerization reaction in Ras and for

GAP catalyzing it has been presented but disputed by others. No similarly detailed kinetic analysis has been performed with GAPs regulating other subfamily GTP-binding proteins. In another model the actual chemical cleavage step is modified by GAP, most likely by GAP supplying residues for GTP hydrolysis on Ras and thereby stabilizing the transition state of the reaction. This model is sometimes called the arginine-finger hypothesis. It has recently been shown by the use of fluorescence spectroscopic measurements that Ras-GAPs stabilize $AlF_4^-$ binding to Ras, which based on evidence from $G_a$ proteins is believed to mimic the transition state of the GTPase reaction, and that stoichiometric amounts of GAP are needed for this effect. These experiments seemed to favor the second hypothesis. By using other GTP-binding proteins such as Cdc42, Ran and Rap and their respective GAPs it has become likely that the role of different GAPs to stabilize the transition state of the GTPase reaction is a general principle for many if not all GTP-binding proteins.

## 5. RAS INTERACTION WITH EFFECTORS

### 5.1 Identification of effector binding regions

10 years ago, extensive mutagenesis of the mammalian H-Ras oncogene identified a region of the Ras protein that is essential for biological activity: mutations within this region (amino acids 32-40) destroyed transforming activity without affecting overall protein integrity (estimated by guanine nucleotide binding and stable protein expression) and without affecting plasma membrane localization. This region was referred to as the 'effector binding region'. At this time, it seemed likely that Ras proteins would have a single effector function, just as the better characterized G-proteins $G_{s\alpha}$ and transducin-α appeared to regulate single effectors (adenylyl cyclase and phosphodiesterase). Now we know that Ras proteins interact with multiple effector proteins, each activating distinct signaling pathways and that the Ras mutants that were made to identify potential effector contacts actually perturbed several effector interactions simultaneously and to varying degrees, resulting in complex changes in Ras action. Current efforts focus on identification of relevant effector proteins that Ras proteins utilize to transform cells, the precise sites of binding of these effectors on Ras proteins, and the biological consequences of their activation.

The notion that Ras proteins may have multiple effector functions first arose from analysis of RAS action in *S. cerevisiae*: loss of functional RAS had a more severe phenotype than loss of its effector, adenylyl cyclase. This suggested additional effector functions necessary for survival. Clear evidence for multiple effectors in mammalian cells came on the heels of the discovery that the c-Raf-1 kinase is a major Ras effector: a mutant of H-Ras (E37G) was identified that failed to bind c-Raf-1 and was defective in transformation. A second mutant, T35S bound c-Raf-1 normally, but also failed to transform. However, the combination of the two mutants produced complementary activity, resulting in restoration of transformation. It was concluded that Ras uses at least two effectors (c-Raf-1 and unidentified effectors) together to transform cells. Similarly, it was discovered that Ras depends on synergistic activation of at least two pathways, the c-Raf-1/MEK/ERK pathway and the Rac and Rho pathway, for full transforming activity. The direct effector of the latter pathway is not yet clear, but PI(3)-kinase is an interesting candidate. This protein binds to Ras at the effector binding domain in a GTP-dependent manner, and is necessary for Rac activation. However, several other candidate Ras effectors have been identified, and efforts are currently focused on their biological relevance. The approach used by White and coworkers, also applied to Rac and Rho, in which Ras mutants are made with restricted effector interactions, should be useful in dissecting the contributions that these putative effectors make to Ras biological activity.

## 5.2 Ras sequences involved in Raf binding and activation

The best characterized Ras effector from mammalian cells, in terms of biological relevance as well as structural detail, is the c-Raf-1 serine/threonine kinase. An 81 amino acid domain from the Conserved Region 1 at the amino terminus of this protein is an independent folding domain (the RBD, Ras Binding Domain) and is sufficient to bind to Ras in a GTP-dependent manner with reasonable affinity. The complex of RafRBD with Rap1A, a close homologue of Ras which is likely to bind to RafRBD in a very similar manner than Ras, has been crystallized and the three-dimensional structure of the complex has been solved by X-ray crystallography (Fig. 2). It revealed the interactions of RafRBD with the residues D33, I36, E37, D38, S39 of Rap, which are conserved between Rap and Ras. These contacts are consistent with mutagenesis studies referred to above, and, equally satisfying, are part of a region of Ras proteins referred to as Switch I. This region changes significantly in conformation between the GDP- and the GTP-states. These two features of the interaction (binding at the 32-40 region, and binding to a GTP-sensitive region) were precisely anticipated for Ras-effector interactions.

It has not yet been possible to co-crystallize full length c-Raf-1 with Ras or Ras-related proteins, and there is clear evidence of contacts between Ras and c-Raf-1 outside the RBD, although the contact sites are controversial. Mutations at residues 31, 57, 59 of H-Ras inhibit binding to full length c-Raf-1 and mutations at positions 26, 29, 39, 40, 41, 44, 45, have been implicated as necessary for Ras-dependent Raf activation. It therefore appears that binding of the 32-40 region of Ras to the RBD is the major site of GTP-dependent binding, but other contacts contribute to binding and, significantly, to kinase activation.

## 5.3 Ras Sequences involved in other effector interactions

Mammalian Ras proteins bind to several classes of proteins in a GTP-dependent manner, and at least some of these are likely to have effector functions. These include the Raf proteins discussed above (c-Raf-1, A-Raf and B-Raf), regulatory subunits of PI(3)-kinases, the Ral GDS proteins, the GAPs as well as others with unknown biochemical activities. Mutagenesis studies reveal that each of these binds t the 32-40 Switch I region (as above), and each may have contacts outside this region. Within Switch I, significant differences exist between Raf interactions, and these other effector interactions. For example Akasaka et al. have shown by the yeast two-hybrid assay that the mutant Y32F is unable to bind to Raf, but still binds Byr2, whereas D38N and T58A outside the effector region show the opposite behaviour. Also, the G60A mutation has a weak effect on the binding to Raf, but the RalGDS interaction is more severely affected. There is also a hundredfold higher affinity of Ras for Raf-RBD than for Ral-GDS, and this difference is exactly the opposite for the affinities of these effectors for Rap. The Switch 2 region, the second region of Ras that differs between GDP-and GTP-states, but is not involved in Raf-RBD interaction, but maybe involved in the interaction with the cysteine-rich domain of Raf. However, GAP interacts with Switch II: perhaps this difference is related to the unique role of GAP as a regulator of Ras GTPase activity, although GAP and effector binding are mutually exclusive.

## 5.4 Ras-Raf interaction: functional aspects

Ras binding to c-Raf-1 has two clear consequences: c-Raf-1 translocates to the plasma membrane, and its kinase activity is turned on. These facts lead to the hypothesis that the sole function of Ras in the process of c-Raf-1 activation is to recruit c-Raf-1 to the membrane. According to this hypothesis, kinase activation occurs when c-Raf-1 is presented to activating components in the membrane, but the activation event per se does not involve Ras directly. To test this, c-Raf-1 mutants were made that would be targeted directly to the plasma membrane through C-terminal sequences derived from Ras itself (Raf-CAAX) or amino terminal myrsistoylation sequences from v-src. These mutant proteins localize to the plasma membrane constitutively and become activated in a Ras-independent manner. These results support the recruitment hypothesis, but do not prove it, since alternative interpretations cannot be excluded, as discussed below. However, these data prompted a search for membrane activators of c-Raf-1, and led to the discovery that tyrosine phosphorylation of c-Raf-1 results in kinase activation. An attractive model can then be formulated in which Ras binding to c-Raf-1 brings the kinase to the plasma membrane, at which site tyrosine phosphorylation results in kinase activation. This model does not assume that tyrosine phosphorylation is the only event involved in c-Raf-1 activation, but suggests it is a significant part.

A number of observations suggest that this model is not an accurate reflection of events leading to c-Raf-1 activation during normal signaling. Most problematic, Raf does not become phosphorylated on tyrosine residues at detectable levels in normal cells. Also, Ras activates mammalian c-Raf-1 in *S. cerevisiae*, which is unlikely to express activating tyrosine kinases.

This leaves open two possible models for authentic c-Raf-1 activation by Ras: in one model, the Ras-recruitment model is correct, but the relevant activating events remain to be identified. In the other, the recruitment model is fundamentally incorrect. Support for the latter interpretation comes from the discovery that one member of the Raf family, B-Raf, can be activated by H-Ras in its GTP state in a membrane-free system, through direct binding of the two proteins, in the presence of 14-3-3 protein. Post-translational modification of H-Ras was essential to this process, suggesting that H-Ras farnesylation may be involved in effector activation rather than membrane recruitment. This is reminiscent of earlier data showing that Ras proteins need to be post-translationally modified to activate adenylyl cyclase in cell-free extracts of *S. cerevisiae*, and suggests that B-Raf-1 activation by H-Ras is through an allosteric activation process involving a conformational change in B-Raf induced by H-Ras binding. This is analogous to the activation of the PAK65 kinase by direct binding of the Ras-related proteins Rac or CDC42 except that the post-translational processing is not required for the latter activating events.

In contrast to B-Raf, direct binding of purified c-Raf-1 to H-Ras does not result in kinase activation. While it is possible that c-Raf-1 activation by H-Ras is a fundamentally different process, it seems more likely that failure to activate c-Raf-1 in vitro is due to trivial technical difficulties, or missing co-factors. Also consistent with the possibility that the recruitment model is incorrect, Ras mutants have been identified that retain normal c-Raf-1 binding, but cannot activate c-Raf-1 effectively. Such properties would not be anticipated of the mutants if binding played the sole role in the activation process. Furthermore, Ras mutants that are farnesylated but remain cytoplasmic retain significant biological activity: this is not consistent with an essential role for recruitment in biological action.

The simplest model for Ras-dependent Raf activation is that Ras binding causes a conformational change in Raf that results in elevated kinase activity. Ras binding would involve the GTP-dependent Switch I region of Ras binding to the c-Raf RBD, as well as the other interactions described above, including direct interaction of the farnesyl group with c-Raf-1, possibly at the cysteine-rich Zinc finger region.

## 6. OTHER RAS-RELATED PROTEINS AND THEIR EFFECTORS

Individual Ras proteins can interact with several effectors at overlapping sites, as we have discussed. The 32-40 region of Switch I is identical in the mammalian H-Ras, K-Ras, and N-Ras proteins, and in *S. cerevisiae* RAS1 and RAS2, *S. pombe* Ras1, and Drosophila Ras1. Consistent with this region being a major effector binding site, all of these proteins are biologically interchangeable, to a first approximation. For example, truncated versions of *S. cerevisiae* proteins transform mammalian cells, and mammalian H-Ras can replace RAS proteins in *S. cerevisiae*. Cousins of this closely related family also share identical or virtually identical, 32-40 regions: however, these proteins have distinct biological activities. For example, Rap1A is 50% identical to H-Ras overall and has an identical 32-40 effector binding region and binds to the same set of effector proteins as the H-Ras family listed above. However, Rap1a does not transform mammalian cells, or substitute for RAS proteins in *S. cerevisiae*. Rap1a actually interferes with oncogenic transformation and Map kinase activation by Ras oncogenes. In *S. cerevisiae*, the Rap-related protein Rsr1 (Bud1) plays a role in bide site selection and is not involved in the same effector pathways as RAS.

Although qualitative methods such as the two-hybrid assay show that members of the Ras subfamily of proteins bind to the same set of effectors, quantitative measurements show that there are large differences in the respective affinities. In the case of Ras/Rap interacting with

Raf-RBD, the difference in affinity is due to a charge difference in residue 31, which is Lys in Rap and Glu in Ras, and the three-dimensional structure of a Rap(E30DK31E) double mutant shows that the interface is now electrostatically much more favorable for an interaction. This means that sequence differences outside the actual core effector domain (residue 32-40) are responsible for modulating the interactions between GTP-binding protein and effector.

# 7. SUMMARY

Thus, in summary, it appears on first glance, that the proteins involved in the transmitting a signal from the cell surface to the nucleus have been identified. However this may be only one of many signals, since it is becoming increasingly evident that signal transduction via Ras, Raf and Map kinase is not a linear process. Just to show a few of many observations it has been mentioned that full activation of Raf kinase in insect cells requires v-src and oncogenic Ras co-expression. On the other hand the activation of Map kinase can be achieved by coexpression with v-Ras and Raf-1 but not by coexpression with v-src. It has also been shown that activation of map kinase ($p44^{erk1}$), which in the linear pathway shown acts downstream of Raf or Ras, can be achieved either in a Raf-dependent or in a Raf-1 independent pathway. In PC12 cells v-Ras activates MAPK but v-Raf does not, whereas in other cell lines v-Ras fails to activate MAPK. Raf-1 independent pathways of MAP kinase activation must exist in certain cell lines, because transformation of PC12 cells by v-Ras, but not by v-Ras, triggers MAP kinase activity. Finally there is the old observation that Ras seems to be involved in regulating protein kinase C, which is not in the linear pathway shown. Also there is increasing evidence that at least one other activator of the MEK kinase, formerly called MAP kinase kinase kinase-MAPKKK which appears not to be activated by Ras but rather by heterotrimeric G proteins.

Taken these and many more recent observations together we will surely witness, in the near future, the detection of more signal transduction pathways that run parallel to the one described here and are likely interconnected with these. It is also likely that different cells have different signalling networks. Such a system might be expected to be much more suitable for the many different growth and differentiation programs that occur in an multicellular organism.

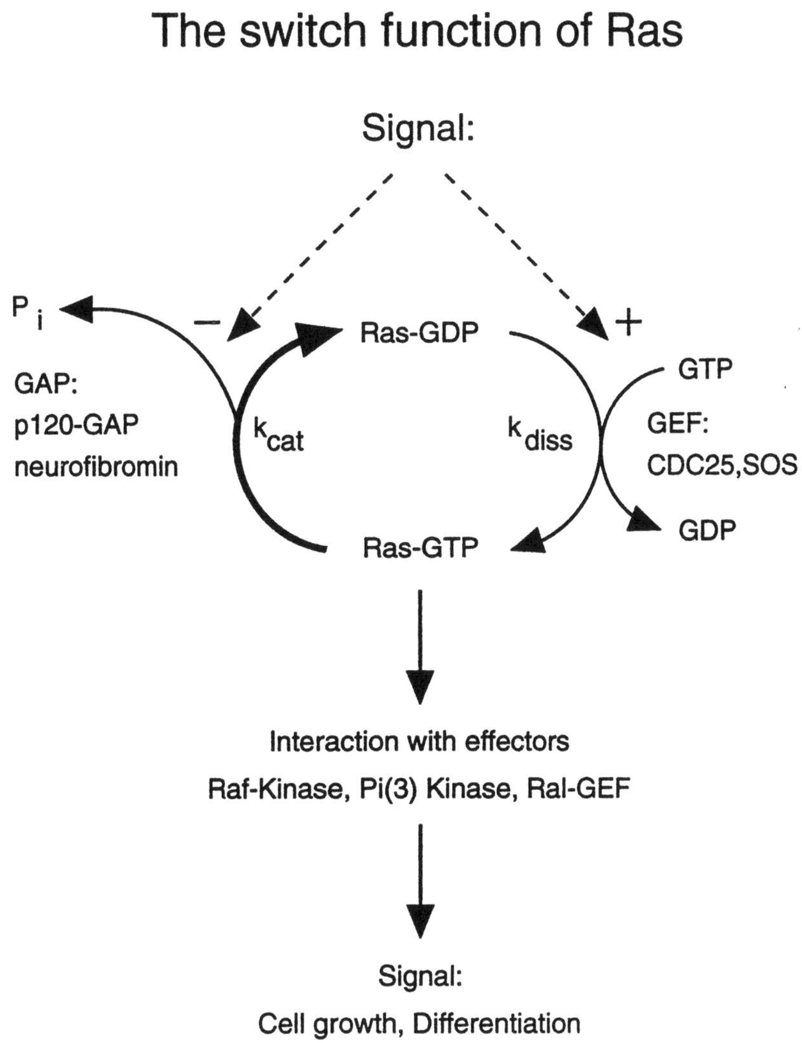


**Fig. 1:** Scheme of the switch function of Ras, which cycles between the GTP-bound ON-state and the GDP-bound OFF-state. It becomes loaded by either the activation of GEF, the specific guanine nucleotide exchange factor which increases the rate of dissociation of GDP, or by the inhibition of the GTPase-Activating protein GAP. Only in the GTP-bound form can it react efficiently with the effector protein and transmit a signal. Several effectors of Ras have been identified, the most important of which are the protein kinase Raf, phosphatidylinositol-3 kinase (PI(3)kinase) and the GEF for the Ras-like protein Ral, RalGEF.

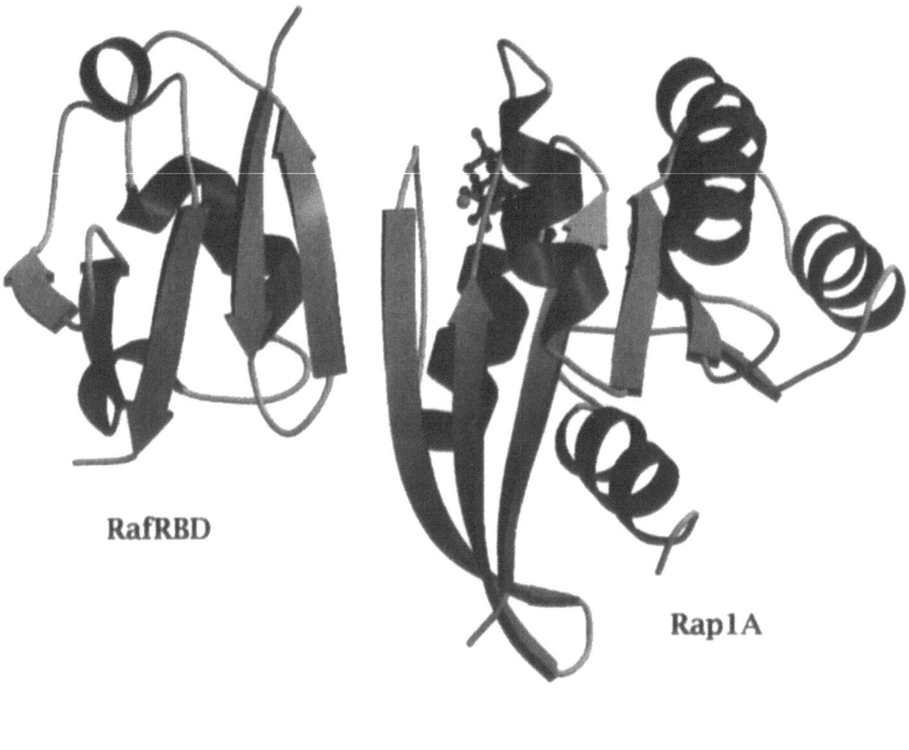


**Fig. 2:** Ribbon plot of the three-dimensional structure of the complex of Raps (a mutant of Rap that looks to the effector like Ras) and the Ras-Binding Domain (RBD) of C-Raf-1. The two proteins make an apparent continuous inter-protein ß-sheet. GTP (or rather the analogue GppNHp) is shown as a ball-and-stick model and is not located in the interface. Since the residues of Raps that make contact with RafRBD are conserved between Ras and Raps, the complex with Ras is considered to be identical.

RELEVANT REVIEWS:

Gibbs, J.B. and Marshall, M.S.
The ras oncogene - and important regulatory element in lower eucaryotic organisms.
Microbiological Reviews 53, 171-185 (1989)

Bourne, H.R., Sanders, D.A. and McCormick, F.
The GTPase superfamily: conserved structure and molecular mechanism.
Nature 349, 117-127 (1991)

Wittinghofer, A. and Pai, E.
The structure of ras protein: a model for a universal molecular switch.
TIBS 16, 383-387 (1991)

Boguski, M.S. and McCormick, F.
Proteins regulating Ras and its relatives.
Nature 366, 643-654 (1993)

Lowy, D.R. and Willumsen, B.M.
Function and regulation of Ras.
Annu. Rev. Biochem. 62, 851-891 (1993)

Feig, L.A.
Guanine-nucleotide exchange factors: a family of positive regulators of Ra and related GTPases.
Current Opinion in Cell Biology 6, 204-211 (1994)

Pronk, G.J. and Bos, J.L.
The role pf p21ras in receptor tyrosine kinase signalling.
Biochimica et Biophysica Acta 1198, 131-147 (1994)

Burgering, B.M.T. and Bos, J.L.
Regulation of Ras-mediated signalling: more than one way to skin a cat.
TIBS 20, 18-22 (1995)

McCormick, F.
Ras-related proteins in signal transduction and growth control.
Molecular Reproduction and Development 42, 500-506 (1995)

Wittinghofer, A. and Herrmann, C.
Ras-effector interactions, the problem of specificity.
FEBS Lett. 369, 52-56 (1995)

Feig, L.A., Urano, T. and Cantor, S.
Evidence for a Ras/Ral signaling cascade.
TIBS 21, 438-441 (1996)

Marshall, C.J.
Ras effectors.
Current Opinion in Cell Biology 8, 197-204 (1996)

Wittinghofer, A. and Nassar, N.
How Ras-related proteins talk to their effectors.
Trends in Biochemical Sciences 21, 488-491 (1996)

# *In vivo* quantitative assessment of Ras/Raf interaction using the two-hybrid system

Birgit Jaitner, Jörg Becker, Alfred Wittinghofer and Christoph Block
Max-Planck-Institut für Molekulare Physiologie
Abteilung Strukturelle Biologie
Rheinlanddamm 201, D-44139 Dortmund, FRG

GTP-binding proteins of the Ras subfamily act as molecular switches which regulate cellular proliferation and differentiation. They are switched „ON" by binding of GTP and „OFF" by the hydrolysis of GTP to GDP. Ras acts as an oncogene when mutated to a constitutively activated form which reflects its crucial role in the control of cell proliferation and differentiation. The interaction between the GTPase switch Ras and the downstream effector protein kinase c-Raf-1 represents a paradigm for the control of mitogenic signalling. Ras, when activated, binds to and activates its downstream effector, Raf, which in turn translates the GTP-dependent „ON" signal into protein phosphorylation. This signal is propagated within a setup of kinases (MEK, ERK) which is commonly referred to as the „MAP-kinase module", finally leading to the induction of gene expression (reviewed in Block & Wittinghofer, 1995).

Recently the structure of the Ras homologue Rap1A bound to the Ras-binding domain (RBD) of Raf has been solved (Nassar et al., 1995) and the contribution of individual amino acids within the contact surface to this protein-protein interaction has been elucidated (Block et al., 1996; Nassar et al., 1996). Although the interaction between Ras and the RafRBD is now well understood, the contribution of other regions of the Raf regulatory domain to Ras binding and Raf activation is highly debated. It has been claimed that residues within the Ser/Thr-rich CR2 domain of Raf are involved in Ras binding since mutations in Ras which affect Ras/Raf interaction in the two-hybrid system could be complemented by mutations within the RafCR2 domain (White et al., 1995).

Biochemical and structural characterization of the interaction of recombinant Raf proteins which exceed the RafRBD with Ras is severely hindered by the fact that these proteins have a very strong tendency to aggregate when expressed in *E. coli.* Measurement of binding constants without determination of active protein concentrations may therefore lead to erroneous results (Gorman et al., 1996; Block & Herrmann, unpublished observations). To cirumvent these limitations associated with bacterial expression of recombinant proteins we have established the use of the two-hybrid system as a semiquantitative method to assess the effect of point mutations on protein-protein interactions *in vivo*. Here we demonstrate that mutations within the RafCR2 domain do not affect Ras/Raf interaction. Using a Raf activation dependent reporter gene assay we show that mutations within the RafCR2 domain can activate Raf directly.

NATO ASI Series, Vol. H 102
Interacting Protein Domains
Their Role in Signal and Energy Transduction
Edited by Ludwig Heilmeyer

# 1 Materials and Methods

## 1.1 Yeast two-hybrid analysis

The yeast two-hybrid system we used is based on the system developed by Chevray and Nathans (1992) (kindly provided by R. Wolthuis). Wild type or mutant Raf-1-fragments (coding aa 1-275) were cloned into Gal4-transactivation domain fusion vector PC86. H-ras G12V (aa 1-166) was cloned into Gal4-DNA-binding domain fusion vector PC97.

For binding studies with Raf-1 and Ras-G12V, competent Y190 yeast cells prepared according to the Klebe protocol (1983), were cotransformed with 1 µg of each vector and grown on synthetic complete medium (SC medium) lacking leucine, tryptophane and histidine, containing 25 mM 3-amino-1,2,3-triazole (3-AT;Sigma), and glucose as carbon source. β-galactosidase assay was performed according to Breeden and Nasmyth (1985) as modified by Chevray and Nathans (1992). Blue staining occured within 3 h upon incubation with 1 mg/ml X-gal (5-bromo,4-chloro,3-indolyl-b-galactopyranoside) as substrate at 37°C (Miller 1972).

## 1.2 Liquid assay for β-galactosidase activity

Original transformants were restreaked on selective medium. From these plates single colonies were picked and inoculated in 15 ml SC medium lacking leucine and tryptophane. The liquid cultures were incubated at 30°C for 17 h. The following procedure was developed based on a method described by Miller (1972) with the following modifications: Sedimented cells were resuspended in 800 µl Z-buffer, permeabilized by addition of 30 µl chloroform and 20 µl 0.1 % SDS, followed by vortexing the cell suspension for 10 sec and incubating for 10 min at 30°C. Afterwards 160 µl ONPG (4 mg/ml in 0.1 M phosphate buffer, pH 7) were added and cells were incubated at 30°C until a yellow color developed. Reaction was stopped with 400 µl of 1 M $Na_2CO_3$. The suspension was centrifuged for 4 min in a microfuge to pellet the cell debris. The β-galactosidase activity was determined according to Miller (1972).

## 1.3 Reporter gene assay

The reporter gene assay demonstrating Raf kinase activity dependent gene expression was performed according to Block et al. (1996).

# 2 Results and Discussion

## 2.1 Correlation between *in vitro* and *in vivo* binding affinities

We used RafRBD interface point mutants, which we have characterized in our previous work (Block et al., 1996), to set up a semiquantitative assay system for protein-protein interaction using the yeast two-hybrid system. RafRBD mutants which display reduced binding affinity to Ras also show decreased binding within the two-hybrid system (Table 1). In contrast to the biochemical measurements using isolated proteins, these mutants could be tested using the two-hybrid system within the context of the complete regulatory domain of Raf. The correlation of affinities of isolated proteins *in vitro* and the two-hybrid reporter gene activity *in vivo* could not only be demonstrated using the qualitative X-gal staining but also by using a quantitative liquid assay for determination of β-galactosidase activity (Figure 1). This result clearly

**Table 1** Comparison of dissociation constants for the complexes between mant-Gpp(NH)p-Ras and wild-type or mutant Raf-RBD (Block et al., 1996) with the binding affinities measured *in vivo* for $Raf_{aa1\text{-}275}$ and Ras-G12V[a].

| Constructs | $K_D$ (μM) physiological ionic strength | estimates from the qualitative β-galactosidase-assay |
|---|---|---|
| wild-type | 0.13 | +++ |
| V69A | 0.95 | ++ |
| R59A | 3.8 | + |
| K84A | 14.0 | (+) |
| K84E | 65.0 | - |
| R89L | >100 | - |
| S257A | n.d. | +++ |
| S257L | n.d. | +++ |

[a] Ras-Raf interaction was qualitatively estimated using the filter assay for β-galactosidase.

Fig.1

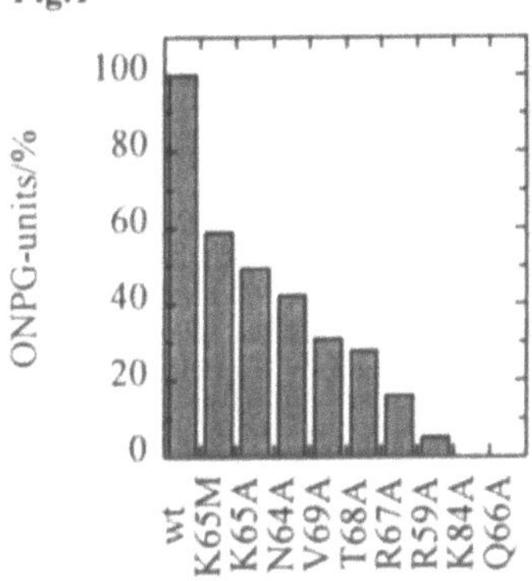


Fig.2

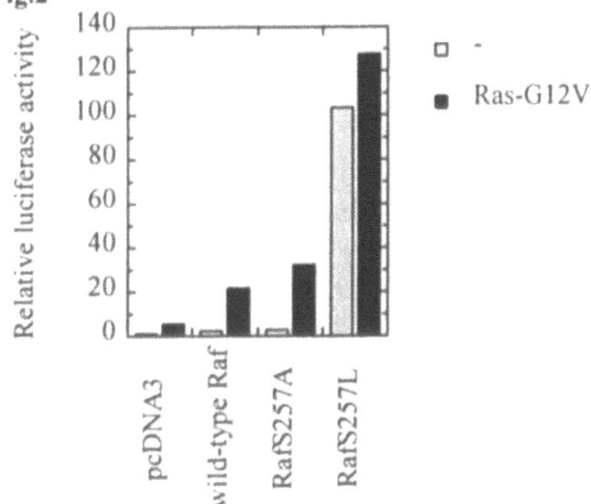


**Fig. 1** Strength of interaction between Ras-G12V and Raf mutants in the two-hybrid system. Data represent a typical β–galacatosidase experiment using liquid yeast cultures. β–galacatosidase activity was measured for $Raf_{aa1\text{-}275}$ and Ras-G12 V interaction with ONPG as substrate and determined as described in Material and Methods.

**Fig. 2** Effect of mutations within Raf-CR2 on Ras/Raf/MEK/ERK mediated transactivation. Transactivation induced by Raf or Raf-mutants was measured with a luciferase reporter construct driven by three E74 binding sites in transient transfection assays (described in Block et al., 1996).

demonstrates that the two-hybrid system is highly suitable for quantitative assessment of the effect of point mutations on protein-protein interactions especially when these proteins cannot be purified for biochemical measurements of binding affinity.

Since it has been claimed that point mutations within the RafCR2 domain (e.g. of Ser257) affect the interaction of Raf with Ras (White et al., 1995) we have tested the mutation of this residue in our assay system. Our two-hybrid analysis shows that RafCR2 mutants do not differ from the wild-type protein with respect to binding

(Table 1). This strongly suggests that this residue within the CR2 domain is not involved in the interaction of Raf with Ras.

**2.2 Raf activation by a point mutation within the RafCR2 domain**

Since an effect of the S257L mutation on Ras/Raf dependent cellular transformation was reported (White et al., 1995) we tested the effect of the S257A and S257L mutations on Raf activation dependent reporter gene expression. In the Raf activation assay the S257A mutation displays activation characteristics identical to wild-type Raf (Figure 2). In contrast to this, strong constitutive Raf activation was achieved by mutating Ser257 to leucine. These results, in combination with the results obtained using the two-hybrid interaction analysis, strongly suggest that the RafCR2 domain does not contribute to Ras/Raf interaction but plays an important role in the control of Raf activity.

Acknowledgement: We thank the DFG for financial support (Grant Nr. Bl 411/1-1).

# 3 References

Block, C. & Wittinghofer, A. (1995) Switching to Rac and Rho. Structure **3**, 1281-1284

Block, C., Janknecht, R., Herrmann, C., Nassar, N. & Wittinghofer, A. (1996) Quantitative structure-activity analysis correlating Ras/Raf interaction *in vitro* to Raf activation *in vivo*. Nature Struct. Biol. **3**, 244-251

Breeden L. & Nasmyth K. (1985) Regulation of the yeast HO gene. Cold Spring Harbor Symposia on Quantitative Biology. **50**,643-650

Chevray, P.M. & Nathans, D. (1992) Proteins interaction cloning in yeast: Identification of mammalian proteins that react with the leucine zipper of Jun. Proc. Natl. Acad. Sci. USA **89**, 5789-5793

Gorman, C., Skinner, R.H., Skelly, J.V., Neidle, S. & Lowe, P.N. (1996) Equilibrium and kinetic measurements reveal rapidly reversible binding of Ras to Raf. J. Biol. Chem. **271**, 6713-6719

Klebe R.J., Harriss, J.V., Sharp, Z.D. & Douglas, M.D. (1983). A general method for polyethylene-glycol-induced genetic transformation of bacteria and yeast. Gene. **25**, 333-341

Miller J.H. 1972. Experiments in Molecular Genetics. Cold Spring Harbor Lab. Cold Spring Harbor, New York. p351-355

Nassar, N., Horn, G., Herrmann, C., Scherer, A., McCormick, F. & Wittinghofer, A. (1995) The 2.2Å crystal structure of the Ras-binding domain of the serine/threonine kinase c-Raf1 in complex with Rap1A and a GTP analogue. Nature **375**, 554-560

Nassar, N., Horn, G., Herrmann, C., Block, C., Janknecht, R. & Wittinghofer, A. (1996) Ras/Rap effector specificity determined by charge reversal. Nature Struct. Biol. **3**, 723-729

White, M.A., Nicolette, C., Minden, A., Polverino, A., Van Aelst, L. & Wigler, M. H. (1995) Multiple Ras functions can contribute to mammalian cell transformation. Cell **80**, 533-541

# *in vivo* Analysis of C-Raf1 – 14-3-3 Interaction

**Javor P. Stolarov and Michael H. Wigler**

**Cold Spring Harbor Laboratory, Cold Spring Harbor, NY11724, and**
**Department of Genetics and Development, Columbia University**
**College of Physicians and Surgeons, N.Y.C., NY 10032**

## *Introduction*

Epistasis studies in model organisms and genetic and biochemical data in mammalian cell culture have unambiguously identified the c-RAF1 kinase as one of the immediate downstream effectors of RAS [10-16]. Thus, raf plays a critical role in cellular signal transduction [2,3,4,5,6,7,8,17,18,19]. It is also able to malignantly transform mammalian cells when activated constitutively [20]. This makes the study of its regulation significant.

Little is known about the mechanism of RAF regulation. Targeting full length RAF to the cell membrane gives it oncogenic properties [26,27]. Deleting the N-terminus of RAF also results in its activation [20]. Indeed, the v-RAF oncogene is a GAG-RAF fusion which is both targeted to the membrane and has the N-terminus of RAF deleted [1]. This and other observations have pointed to the autoinhibitory role of RAF's N-terminus.

Activation of RAF in vivo correlates with its phosphorylation. Phosphopeptide mapping has demonstrated that there are at least 6 sites on RAF which are phosphorylated either cotranslationally, or simultaneously with RAF activation [25]. RAF is deactivated in vitro by serine-threonine and tyrosine phosphatases, singly or in combination. Deactivation can be inhibited by prior addition of Hsp70 or 14-3-3z [44,45].

The recent interest in the 14-3-3 proteins is due to accumulating evidence that suggest a role for 14-3-3 in the regulation of the RAF kinase. The genetic data consistent with this hypothesis are:

i. A *ste11D* phenotype in *S. cerevisiae* (pheromone unresponsiveness and uninducibility of a FUS1 reporter) can be suppressed by either a dominantly activated RAF (N-terminal truncation) or overexpression of RAF and Ha-RAS, but not by RAF alone. Screening of a mammalian or yeast cDNA library for multicopy suppressors of *ste11D* in the presence of RAF yielded Ha-RAS and yeast (BMH1) 14-3-3 homologues [37].

ii. In a *ste11D, bmh1D* background, RAF is unresponsive to Ha-RAS activation while a dominantly activated RAF acts as a suppressor in a RAS-independent fashion [37].

iii. 14-3-3 and RAF interact in the two hybrid system [36-39].

The goal of this study is to determine whether the capacity of RAF proteins to interact with 14-3-3 is essential for RAF's biological activity.

We wish to use a genetic approach to address this question. We cannot make a 14-3-3 free cell, as there are at least nine abundantly expressed 14-3-3 isoforms in mammals [29], and the *S. cerevisiae* double knockout of the two yeast homologs BMH1&2 is dead. Therefore, we decided to use interaction suppression as an alternative. The idea is: RAF mutants incapable of interacting with 14-3-3, but wild type for all other known interactions ( RAS, MEK) are expected to have a loss of function phenotype. If only their inability to bind to 14-3-3 is responsible for this phenotype, then an interaction suppressor of 14-3-3 should restore wild type function when supplied in trans.

## *Results*

*Screen for 14-3-3 non interacting raf mutants.*

We have shown previously that in RAF there are three domains which can bind 14-3-3 by themselves: A (aa1-149); B (aa185-282); C (aa569-648). A and B are in the regulatory N-terminus of RAF (aa1-229), C is in the kinase domain (aa330-648). The isolated kinase domain containing only domain C can be activated by 14-3-3 in triton-disrupted mammalian cell extracts in a dose-dependent manner [38]. Others have shown that point mutations in domain B, which render RAF incapable of binding 14-3-3 *in vitro* and in baculovirus cells do not have an effect on RAF function [46]. We have shown that point mutations in domain C which abolish RAF interaction with 14-3-3 in the two hybrid, but do not affect it's interaction with RAS and MEK render it inactive in NIH3T3 focus formation assay.

The C domain of RAF (aa569-648) was mutagenized using PCR, and the resulting mutant fragment library was subcloned in the wild type kinase domain (DNRAF, aa330-648). The library was constructed in the LexAStop two-hybrid vector, creating a fusion protein with Lex A DNA binding domain N-terminally and RAF kinase domain C-terminally. Two-hybrid strain L40 was used (*MATa trp1 leu2 his3 LYS::(LexAop)4-HIS3 URA3::(LexAop)8-LacZ).*

*The 14-3-3 non binding raf mutants fall in two classes, depending on their ability to complement a ste 11 strain.*

NATO ASI Series, Vol. H 102
Interacting Protein Domains
Their Role in Signal and Energy Transduction
Edited by Ludwig Heilmeyer

As the capacity for interaction with 14-3-3 might be required for RAF's biological activity in mammalian cells, one needs to distinguish the above class mutants among the many 14-3-3 nonbinders. To address this problem, we decided to take advantage of the fact that wt RAF kinase domain (DNRAF) can complement a *ste11D* in the sex pathway of *S. cerevisiae*, provided that either MEK or STE7 is coexpressed. In yeast STE11, STE7 and FUS3/KSS1 form a conserved MAPK module. RAF, MEK, and ERK2 form an analogous module in mammalian cells. MEK and ERK2 are homologous to STE7 and FUS3/KSS1 respectively, STE11 and RAF are not homologous to one another. We hoped, that yeast, being different from mammalian cells will allow differences in biological activities among RAF mutants to be detected.

The strain SY1984 (*MATa ste11D his3D leu2 ura3 trp1 can1 FUS1-HIS3)* is prototrophic for histidine only when the sex kinase cascade is activated. Wild type DNRAF can activate the cascade, provided either a STE11gf allele or MEK is coexpressed.10 of the 32 mutants showed histidine prototrophy when MEK was coexpressed. 4 of them gave signal indistinguishable from wild type. The rest gave a weaker signal with varying intensity (Table I). None of the 32 mutants showed histidine prototrophy when STE7gf allele was coexpressed, even after > 10 days incubation. (TableI)

*ste11 suppressors and non-suppressors are positioned in distinct domains which may reflect structural features of raf's kinase domain.*

The catalytic domains of serine-threonine kinases consist of twelve subdomains showing high conservation. Regions between these domains vary to a much higher extent. Each subdomain contains several highly or absolutely conserved residues, which are believed to be essential for catalysis [21 and ref. therein]. The C 14-3-3 binding domain of RAF encompases subdomain XI. In this subdomain, there is one highly conserved V269 (V585 in RAF) - numbering is according to Hanks et al. for cAPK - and one absolutely conserved R280 (R596 in RAF). In the solved crystal structures of cAPKa, ERK2 and CDK2 this R280 forms a salt bridge with the absolutely conserved E208 in the APE motif in subdomain VIII. Alanine scan of TPK1 (yeast cAPK) shows that the R280A substitution results in a 45-fold (4500%) decrease in $k_{cat}$, while increasing Km by only 27%. It is likely that the RAF mutants incapable for suppressing *ste11* but still binding MEK as strongly as wild type fall in the same class. All solved crystal structures reveal a similar 3D structure of subdomain XI - two a helices ( H and I, see fig.1) connected by a loop containing the invariable R280. The borders of helix I and H are the same for all three kinases. Mutations in raf affecting interaction with 14-3-3 encompass a 42 aa fragment beginning 6 aa N-terminally of the C-terminus of helix H, spanning the entire intrahelcal loop and ending 15aa C terminally of helix I. 8 of the 10 ste11 suppressing mutants are in the loop, the remaining 2 are at the very borders of helices H and I flanking the loop.

*The 7 tested ste11 complementing mutants are biologically inactive in mammalian cells.*

One genetic assay for raf biological activity is the focus formation assay in NIH 3T3 cells. Wild type RAF kinase domain alone causes focus formation when overexpressed. The DNRAF mutants capable of suppressing *ste11D* were subcloned in a mammalian expression vector under the CMV promoter and transfected in NIH3T3 cells using the calcium phosphate protocol. Wild type RAF kinase domain gave an average of 107 foci per plate for five plates, SE=10.65. No statistically significant focus formation activity was observed for any of the mutants tested (six plates each). (Table II) This gives a difference in activity at least 600 fold.

*Lack of evidence for 14-3-3 stimulation of RAF in yeast*

While testing the SY1984 system for use with the RAF mutants, we attempted to repeat the original experiments showing stimulation of RAF by 14-3-3 in yeast [36,37]. We found that 14-3-3 was able to cause activation of the FUS1-HIS3 reporter in the absence of RAF (this control was omitted in [37]). We were not able to reproduce the results of 14-3-3 stimulation of RAF published in [36] (Table II and data not shown. For these experiments we used constructs sent to us by the authors as well as our own set.)

## *Discussion*

We are unable to make firm conclusions as to the role of 14-3-3 in RAF activation. In order to interpret our mutant study data we need a 14-3-3 interaction suppressor. Until such mutant is found, we cannot distinguish between two possible explanations of our results: i.) the mutations in RAF which abolish 14-3-3 binding also mechanistically impair the kinase function; ii.) 14-3-3 binding is required for RAF function and the mutations are specific in that they only affect 14-3-3 binding, but leave the kinase function intact.The next task is to find a 14-3-3 interaction suppressor and attempt rescue of the loss of function phenotype. Only full rescue will unambiguously demonstrate that loss of 14-3-3 interaction is the sole cause for loss of biological activity. Lack of rescue or partial rescue will leave ambiguity as to whether other functions (binding to other unknown proteins or kinase activity), besides the inability of RAF to interact with 14-3-3 are responsible for the loss of function phenotype.

The lack of stimulation of RAF by 14-3-3 in yeast, argues for the possibility that 14-3-3 is not a RAF activator. 14-3-3 may have an auxiliary role in protecting RAF from deactivation by phosphatases [44,45].

## TABLE I

**Two hybrid interaction profiles and biological activity of 14-3-3 non interacting cRaf1 mutants**

| # | mutation | times recovered | L40 two-hybrid interaction with 14-3-3: LacZ reporter | HIS3 reporter | SY1984 ste11 suppression: STE7gf coexpressed | MEK1 coexpressed |
|---|---|---|---|---|---|---|
| | wt | | +++ | +++ | +++ | +++ |
| 71 | R583W | 1 | - | - | - | - |
| 121 | C588S | 2 | - | - | - | + |
| 6 | V599M | 1 | - | + | - | +++ |
| 107 | V599A | 2 | - | +/- | - | ++ |
| 918 | K593E | 1 | - | N.D. | - | +++ |
| 916 | E595G | 1 | - | - | - | + |
| 904 | L598P | 1 | - | + | - | ++ |
| 45 | F599S | 2 | - | - | - | +++ |
| 123 | F599L | 1 | - | +/- | - | + |
| 70 | P600L | 1 | - | - | - | - |
| 87 | L603Q | 1 | - | - | - | - |
| 47 | S605P | 4 | - | - | - | - |
| 924 | I606T | 1 | - | - | - | +++ |
| 903 | L609H | 1 | - | + | - | - |
| 38 | L609R | 1 | - | - | - | - |
| 52 | I616T | 1 | - | + | - | - |
| 97 | R618W | 1 | - | - | - | - |
| 81 | S621P | 4 | - | - | - | - |
| 906 | E622G | 1 | - | + | - | - |
| 29 | L625S | 1 | - | - | - | - |

## TABLE II

**Biological activity of 14-3-3 non interacting cRaf1 mutants in NIH3T3 cells: focus formation frequency**

| RAF | FOCI/PLATE | N | S.E. |
|---|---|---|---|
| WT | 107 | 5 | 10.6 |
| C588S | 0.0 | 5 | 0.0 |
| V589M | 0.1 | 6 | 0.4 |
| V589A | 0.0 | 3 | 0.0 |
| K593E | 0.0 | 5 | 0.0 |
| E595G | 0.0 | 5 | 0.0 |
| F599S | 0.3 | 8 | 0.6 |
| I606T | 0.3 | 6 | 0.8 |

TABLE III

**Effects of 14-3-3 overexpression on c-Raf1 biological activity in yeast**

| OVEREXPRESSED | GENES: | | SUPPRESSION OF ste11Δ: |
|---|---|---|---|
| 14-3-3Hβ | cRaf1 | STE7gf | + |
| 14-3-3Hζ | cRaf1 | STE7gf | + |
| HaRas | cRaf1 | STE7gf | + |
| 14-3-3H | - | STE7gf | + |
| 14-3-3H | - | STE7gf | + |
| HaRas | - | STE7gf | - |
| 14-3-3H | cRaf1 | - | - |
| 14-3-3H | cRaf1 | - | - |
| HaRas | cRaf1 | - | - |
| | cRaf1 | STE7gf | - |

# References:


1. Rapp, U.R. et al.,1983, PNAS **80**: 4218-4222
2. Simon, M.A. et al.1991, Cell**67**:701-716
3. Perrimon,N. 1993, Cell **74**:219-222
4. Brand, A. H. et al. 1993, Genes Dev.**8**:629-639
5. MacNicol, A.M. et al. Cell 1993 **73** : 571-583
6. Kizaka-Kondoh, S. et al. , 1992 MCB **12/11**:5078-5086
7. Kyriakis, J.M. et al. 1992 Nature **358**:417-421
8. Howe, L.R. et al. 1992 Cell **71**: 335-342
9. Chardin et al. 1993, Science **260**:1338-1343
10. Voitek A.et al. 1993 Cell,**74**:205-214
11. Van Aelst L. et al. 1993 PNAS **90** :6213-6217
12 Zhang, Z-f. et al. 1993 Nature **364**: 308-313
13. Warne, P.H. et al. 1993 Nature **364**: 352-355
14. Moodie, S.A. et al. 1993 Science **260**: 1658-1661
16. White, M.A. et al. 1995 Cell **80**: 533-541
17. Dent, P. 1992, Science **257**:1404-1407
18. Macdonald, S.G. et al. 1993, MCB **13/11**:6615-6620
19. Daum, G. et al. 1994, TIBS **19**:474-480
20. Heidecker, G. et al.1990, MCB **10/6**:2503-2512
21. Hardie, G., Hanks, S. The Protein Kinase Factsbook, AP 1995
22. Fabian, J.R. 1994, PNAS **91**:5982-5986
23. Kolch et al. 1993, Nature **364**: 249-252
24. Marx, J. 1993 Science **262** : 989-990
25. Morrison, D.K. et al. 1993 JBC **268** :17309-17316
26. Leevers, S.J. 1994 Nature **360**:411-414
27. Stokoe, D. 1994 Science **264** : 1463-1466
28 Aitken A., 1995, TIBS 95-97
29. Martin, H. 1993, FEBS Lett. **3** :296-303
30. Izobe T. et al 1992, FEBS Lett. **2** : 121-124
31. Reuter , G.W. et al. 1994 Science **266** : 129-133
32. Robinson, K. et al. 1994 Biochem J. **299**: 853-861
33. Zupan, L.A. et al. 1992, JBC **267**:8707-8710
34. Herskowitz, I. 1995, Cell **80** :187-197
35. Marshall, C. 1995, Cell **80** : 179-185
36. Freed, E. et al.1994, Science **265**: 1713-1716
37. Irie, K. et al. Science **265**: 1716-1719
38. Li, S. et al. 1995 EMBO J. **14** :685-696
39. Fu, H. et al. 1994, Science **266**: 126-129
40. Fields, S. and Song, O. 1989 Nature **340** :245-246
41. Cadwallader, K.A. et al. 1994, MCB **14/7**: 4722-4730
42. Yao. B. et al. 1995 Nature **378** 307-310
43. Rugin, G., pers. comm.
44. Dent, P. et al. 1995 MCB **15/8**: 4125-4135
45. Dent, P. et al. 1995 Science **268**: 1902-1906
46. Michaud, N. et al. 1995 MCB **15/6**: 3390-3397
47. Elion, E. 1995 Trends in Cell Biol. **5**: 322-327

# Signal Transduction through the MAP kinase Pathway

Natalie G. Ahn[1,2], Scott C. Galasinski[3], Kristen K. Gloor[1], Timothy S. Lewis[1], Donna F. Louie[1,2], Sam J. Mansour[3], Theresa Stines Nahreini[1,2], Katheryn A. Resing[1], Paul S. Shapiro[1], and Anne M. Whalen[1]

[1] Department of Chemistry and Biochemistry,
[2] Howard Hughes Medical Institute,
[3] Department of Molecular, Cellular and Developmental Biology,
University of Colorado, Boulder, CO 80309

## 1. Introduction

An intracellular signal transduction pathway, called the MAP kinase cascade is rapidly stimulated in response to growth factors (1). Enzymes in this pathway include pp90 ribosomal S6 kinase, mitogen-activated protein kinase (ERK), and MAP kinase kinase (MKK), which form three tiers of a protein kinase cascade in which pp90rsk is phosphorylated and activated by ERK, and ERK is phosphorylated and activated by MKK. MKK is phosphorylated and activated by at least three protein kinases, Raf-1, MEK kinase, and Mos, and thus is a convergence point for diverse signalling pathways triggered upon cell surface receptor activation. The work of many laboratories has led to the definition of key connections between growth factor receptors and Raf-1. Some of these involve activation of Ras by association of the Ras-guanine nucleotide exchange factor with cell surface receptors via linker proteins and subsequent interaction of Ras with Raf-1 at the plasma membrane. Several cellular protooncogenes are components of this pathway. Furthermore, many transcription factors are downstream targets for ERKand pp90 ribosomal S6 kinase, both of which translocate to nuclei following cell stimulation. Thus, the MAP kinase cascade is a key pathway by which external mitogenic signals control cell growth at the level of transcription and plays an important role in mediating oncogenic cell transformation.

## 2. Phosphorylation sites on MAP kinase kinase

Electrospray ionization mass spectrometry was used to identify sites on MKK1 phosphorylated by immunoprecipitated v-Mos (2). Two phosphorylation sites were identified as $Ser_{218}$ and $Ser_{222}$ in the activation loop between subdomains 7 and 8 of the consensus kinase sequence for MKK; these were identical to the regulatory sites targeted by Raf-1 and MEK kinase, as determined by other groups (3,4). In addition, three sites of MKK autophosphorylation were found which were enhanced following activation by v-Mos. Two of these sites occurred within the sequence $Ser_{298}$(P)-Ser-$Tyr_{300}$(P) in a unique 40 amino acid insert between subdomains 9 and 10 of MKK. This sequence is similar to that found in ERK to be

NATO ASI Series, Vol. H 102
Interacting Protein Domains
Their Role in Signal and Energy Transduction
Edited by Ludwig Heilmeyer

phosphorylated by MKK [$Thr_{183}$(P)-Glu-$Tyr_{185}$(P)]. Residues surrounding this sequence are proline rich, suggesting a possible function of this domain in protein-protein interactions. A minor autophosphorylation site is located at $Thr_{23}$. Because active MKK is unable to phosphorylate a catalytically inactive MKK mutant, autophosphorylation occurs by an intramolecular mechanism.

The finding that the regulatory sites on MKK phosphorylated by v-Mos are the same as those phosphorylated by Raf-1 and MEKK confirms a common mechanism of MKK activation. The identification of several intramolecular autophosphorylation sites elsewhere in the molecule indicates that various domains of the protein outside the catalytic loop are accessible to the active site. No differences were observed between wild type MKK and mutants containing autophosphorylation site substitutions with respect to activation by v-Mos in vitro. However, mass spectrometric analyses suggest that MKK is autophosphorylated to a high stoichiometry under physiological conditions.

## 3. Mutagenesis of MKK

We investigated the regulation of MKK1 and 2 by using site directed mutagenesis to generate constitutively active mutants (5-7). Substitution of $Ser_{218}$ and $Ser_{222}$ with aspartic or glutamic acid resulted in up to 80-fold enhancement of basal activity. Increasing the number of acidic residues in the activation loop from two to four increased the basal specific activity of MKK by up to 5-fold, suggesting an important electrostatic contribution by phosphorylation to the activation.

Interestingly, deletion of N-terminal residues 32-51 enhanced the basal activity of MKK by 40-fold. These residues show a weak identity with the A-helix secondary structure motif of cAMP dependent kinase suggesting that a similar helical structure might be found in MKK, although an X-ray differaction structure of this enzyme has not yet been solved. Truncation of residues 1-51 did not alter the properties of MKK, and synthetic peptides based on the deleted sequences did not inhibit MKK in trans, indicating that the deletion does not activate the enzyme by removal of an autoinhibitory domain. However, replacing residues 32-51 with alanine resulted in mutants with wild type basal activity that were completely activatable by Mos, suggesting that an $\alpha$-helical structural motif is important in retaining wild type MKK in its inactive conformation. In support of this hypothesis, replacing three consecutive amino acids within this domain with proline residues mimicked the constitutive activation seen upon deletion, again suggesting that perturbation of secondary structure is involved in the mechanism of activation by the N-terminal mutations.

Both the acidic substitutions and the N-terminal deletion increase $V_{max}$, $V/K_{m,ERK2}$ and $V/K_{m,ATP}$, as is also observed following phosphorylation of wild type MKK1 by Mos. A synergistic enhancement of these steady state rate parameters occurs upon combining the mutations. It appears that structural rearrangements within different regions of the molecule cooperate to switch the enzyme to an active state. We hypothesize that conformational changes induced by mutagenesis together mimic those seen upon phosphorylation.

## 4. Regulation of cell differentiation by MKK

By combining these mutations in various ways, constitutively activated MKK mutants were constructed with activities ranging from 10- to 2100-fold over basal wild type MKK. These constructs provide useful tools for examining signalling downstream of MKK and ERK. In collaboration with Dr. George Vande Woude's lab we found that these mutants transform NIH3T3 cells, as also observed by others (6-9). These findings underscore the positive regulation of cell growth by the MAP kinase cascade.

In contrast to its function in regulating cell growth, the MAP kinase pathway has been shown to play an important role in regulating embryonic development and cell differentiation. Activation of MKK or ERK is essential for vulval development in C. elegans (10), photoreceptor cell specification and anterior-posterior body patterning in Drosophila (11), mesoderm induction during Xenopus laevis embryo development (12), and positive T cell selection (13). In addition, one example of a differentiation program induced in cultured cells by the MAP kinase cascade occurs in the rat pheochromocytoma (PC12) line, where activation of ERK by constitutively active MKK mimics features of neuronal differentiation observed in response to nerve growth factor (8).

The K562 erythroleukemia cell line is a model system for studying mechanisms regulating lineage commitment of hematopoietic stem cells. Hemin, aphidicolin, or 1-β-D-arabinofuranosylcytosine commit K562 cells towards an erythroid lineage (14). In contrast, phorbol 12-myristate 13-acetate (PMA) induces differentiation into megakaryocytes, with concomitant loss of monocyte- and erythroid-specific markers (15). Phorbol esters stimulate the MAP kinase pathway in many cells, through PKC-dependent as well as PKC-independent mechanisms, raising the possibility that ERK activation in K562 cells may be involved in specifying the megakaryocyte lineage.

Expression of constitutively active MKK in these cells resulted in cell adhesion and spreading, increased cell size, inhibition of cell growth, and induction of the platelet-specific integrin $\alpha_{IIb}\beta_3$, all hallmarks of megakaryocytic differentiation (16). The MKK inhibitor, PD98059, blocked MKK/ERK activation and cellular responses to phorbol ester, indicating that activation of MKK is necessary and sufficient to induce a differentiation program along the megakaryocyte lineage. Interestingly, constitutively active MKK also suppressed the hemin-induced expression of an erythroid marker, α–globin. Thus, the MAP kinase cascade, which promotes cell growth and proliferation in many cell types, is able to inhibit cell proliferation and initiate differentiation in K562 cells, suppressing the erythroid lineage at the same time that it promotes the megakaryocyte lineage. These results establish a now model system to investigate the mechanisms by which this signal transduction pathway specifies cell fate and developmental processes.

## 5. Mass spectrometric approaches to signal transduction

Electrospray ionization mass spectrometry (ESI-MS) enables determination of protein or nucleic acid masses to accuracies of 1 part in 10,000 Da (17). Experimental approaches used in our laboratory include (i) liquid chromatography/mass spectrometry (LC/MS), which separates molecules by high performance liquid chromatography followed by mass determination, (ii) tandem mass spectrometry (MS/MS, LC/MS/MS), which fragments ions by collision-induced dissociation, allowing peptide and nucleic acid sequences and specific sites of covalent modification to be determined, and (iii) deuterium exchange coupled to mass spectrometry, which measures solvent accessibility of protein backbone amides.

LC/MS and LC/MS/MS are being used to characterize phosphorylation sites on MKK and other proteins. In addition, deuterium exchange measurements are being carried to explore conformational differences between inactive and active forms of MKK. This technique involves analyzing the rates of exchange between backbone amide hydrogens with deuterium from solvent by incubation in $D_2O$, rapid proteolysis, and measurement of mass increases in the resulting proteolytic fragments. By comparing wild type and mutant forms of MKK, specific regions on the molecule were identified that underwent substantial increases in exchange rates upon enzyme activation (18). Enzyme activation led to conformational changes resulting in decreased exchange rates within the glycine rich ATP binding loop in β2 and in regions including the catalytic cleft and activation loop of the predicted kinase structure. These were accounted for solely by phosphorylation site mutations, and could be attributed to motions involving domain closure or electrostriction. Enzyme activation also led to increased exchange rates in the rest of the N-terminal ATP binding domain, including the conserved 5 stranded beta sheet (β1-β5) and the intervening Cα-helix. Increased exchange rates in this region were observed with both activation loop and N-terminal deletion mutants and were enhanced upon combining the mutations, accounting for synergistic effects of these mutations on activity. Evidently, activation involves enhancement of flexibility in the ATP binding domain, in order to accommodate changes in conformation during ATP binding, phosphoryl transfer, and/or ADP release.

Mass spectrometry is also being used to examine masses of components within large protein/nucleic acid complexes. Recent studies have established methods to analyze masses of protein components within mammalian 40S ribosomal subunits (19). The results indicate ribosomal proteins that are post-translationally modified, and in many cases the chemistry of modification can be deduced from differences in mass. Several proteins, however, have masses that cannot be accounted for by known modifications, suggesting the possibility of novel chemistries. In addition, proteins not accounted for by known ribosomal protein sequences are apparent, suggesting new protein constituents of the 40S subunit. Studies are underway comparing these masses with those of ribosomal proteins from transformed or growth factor treated cells or in cells responding to stress, with the aim of identifying novel protein post-translational modifications under signal transduction control.

## Citations:

1. Seger, R. and E.G. Krebs. 1995. The MAPK signaling cascade. *FASEB J.* 9:726-735.
2. Resing, K.A., S.J. Mansour, A.S. Hermann, R.S. Johnson, J.M. Candia, K Fukasawa, G.F. Vande Woude, and N.G. Ahn 1995. Determination of v-Mos catalyzed phosphorylation sites and autophosphorylation sites on MAP kinase kinase by ESI/MS. *Biochemistry* 34:2610-2620.

3. Alessi D.R., Y. Saito, D.G. Campbell, P. Cohen, G. Sithanandam, U. Rapp, A. Ashworth, C.J. Marshall, and S. Cowley 1994. Identification of the sites in MAP kinase kinase-1 phosphorylated by $p74^{raf-1}$ *EMBO J.* 13:1610-1619.
4. Gardner, A.M., R.R. Vaillancourt, C.A. Lange-Carter, and G.L Johnson 1994. MEK-1 phosphorylation by MEK kinase, Raf, and mitogen-activated protein kinase: Analysis of phosphopeptides and regulation of activity. *Mol. Biol. Cell* 5:193-201.
5. Mansour, S.J., J.M. Candia, J. Matsuuda, M. Manning, and N.G. Ahn (1996) Dual mechanisms of activation of mitogen-activated protein kinase kinase 1 as revealed by mutagenesis and rate analysis. (1996) *Biochemistry* in press.
6. Mansour, S.J., Matten, W.T., Hermann, A.S., Candia, J.M., Rong, S., Fukasawa, K.. Vande Woude, G.F., and Ahn, N.G. (1994) Gain-of-function mutations in MAP kinase kinase promote mammalian cell transformation. *Science* 265:966-970.
7. Mansour, S.J., J.M. Candia, K.K. Gloor and N.G. Ahn. 1996. Constitutively active mitogen-activated protein kinase kinase 1 (MAPKK) and MAPKK2 mediate similar transcriptional and morphological responses. *Cell Growth Diff.* 7:243-250.
8. Cowley, S., H. Paterson, P. Kemp and C.J. Marshall. 1994. Activation of MAP kinase kinase is necessary and sufficient for PC12 differentiation and for transformation of NIH 3T3 cells. *Cell* 77:841-852.
9. Brunet, A., G. Pages and J. Pouyssegur. 1994. Constitutively active mutants of MAP kinase kinase (MEK1) induce growth factor-relaxation and oncogenicity when expressed in fibroblasts. *Oncogene* 9:3379-87
10. Sundaram M. and M. Han. 1996. Control and integration of cell signaling pathways during C. elegans vulval development. *Bioessays* 18:473-480.
11. Perrimon, N. 1994. Signalling pathways initiated by receptor protein tyrosine kinases in Drosophila. *Curr. Op. Cell Biol.* 6:260-266.
12. Umbhauer, M., C.J. Marshall, C.S. Mason, R.W. Old, and J.C. Smith.1995. Mesoderm induction in Xenopus caused by activation of MAP kinase. *Nature* 376:58-62
13. Aberola-Ila, J., K.A. Forbush, R. Seger, E.G. Krebs, and R.M. Perlmutter. 1995. Selective requirement for MAP kinase activation in thymocyte differentiation. *Nature* 373:620-623.
14. Rutherford, T., J.B. Clegg, D.R. Higgs, R.W. Jones, J. Thompson and D.J. Weatherall. 1981. Embryonic erythroid differentiation in the human leukemia cell line K562. *Proc. Natl. Acad. Sci. USA.* 78:348-352.
15. Alitalo, R. 1990. Induced differentiation of K562 leukemia cells: a model for studies of gene expression in early megakaryoblasts. *Leuk. Res.* 14: 501-514.
16. Whalen, A.M., S.C. Galasinski, P.S. Shapiro, T.S. Nahreini, and N.G. Ahn 1996. Megakaryocytic differentiation induced by constitutive activation of the MAP kinase pathway. *Mol. Cell. Biol.* Submitted.
17. Mann, M. and M. Wilm 1995. Electrospray mass spectrometry for protein characterization. *Trends Biochem. Sci.* 20:219-224.
18. Resing, K.A. and N.G. Ahn 1996. Structural dynamics of MAP kinase kinase analyzed by deuterium exchange mass spectrometry. In preparation.
19. Louie, D.F., K.A. Resing, T.S. Lewis, T.S. and N.G. Ahn 1996. Mass spectrometric analysis of 40S ribosomal proteins from Rat-1 fibroblasts. *J. Biol. Chem.* 271:28189-28198.

# Identification and Characterization of MKKX, a novel mammalian MAP kinase kinase

Pamela M. Holland[1,2] and Jonathan A. Cooper[2]

[1] Department of Biochemistry, University of Washington, Seattle, WA 98109, USA
[2] Department of Basic Sciences, Fred Hutchinson Cancer Research Center, Seattle, WA 98109, USA

## 1. Introduction

The MAPK cascade has emerged as an evolutionarily conserved element in the transduction of a wide variety of extracellular signals. In mammals, at least three distinct pathways have been identified. These include p42 and p44 ERKs, activated by growth factors and many other mitogenic stimuli; p54 and p46 JNKs, strongly activated by stressful stimuli such as cytokines, UV radiation, and protein synthesis inhibitors; and p38 MAPKs, which are also responsive to stress-related stimuli (1).

Although the existence of distinct MAPK pathways has been suggested to provide a basis for signaling specificity, it remains unclear how this might be achieved at a molecular level. One shared characteristic of MAPKs is their requirement for phosphorylation on both threonine and tyrosine within a TXY motif for activation. These reactions are catalyzed by a family of dual specificity kinases termed MAP kinase kinases (MAPKKs). Multiple MAPKKs have been shown to share specificity for a particular MAPK and individual MAPKKs can activate multiple MAPKs.

In order to further understand the activation, localization, and substrate specificity of MAPKKs, we performed a yeast two-hybrid library screen using full-length human MKK1 as bait (1). Results from screening a day 9.5-10.5 mouse embryo library identified the C-terminal fragment of a putative novel MAPKK family member, referred to herein as MKKX.

## 2. Results

The cloning of MKKX demonstrates there are two isoforms, MKKXa and MKKXb, which differ in their N-terminal sequences. MKKXa encodes a 419 amino acid protein with a predicted molecular weight of 48 KDa. MKKXb is slightly shorter, encoded by 391 amino acids. MKKXa/b contains all the residues conserved among dual specificity kinases. In addition, both contain stretches of proline rich sequences in their shared N-terminal region. All data presented has been obtained with MKKXa, and will be referred to simply as MKKX.

A sequence comparison of known MAPKKs shows that MKKX has 56% identity to a recently identified *Drosophila* MAPKK, termed hemipterous (hep) (Fig. 1). Hemipterous is required for the establishment of the dorsal epidermis,and (hep) mutants fail to undergo dorsal closure (3). It is of interest to note that within the region of subdomains VII and VIII, only MKKX and Dhep contain a serine

NATO ASI Series, Vol. H 102
Interacting Protein Domains
Their Role in Signal and Energy Transduction
Edited by Ludwig Heilmeyer

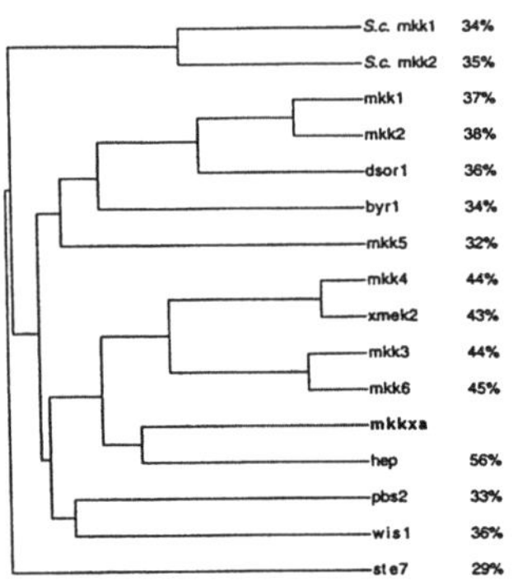


**Fig. 1.** The comparison was created by the PILEUP program (WGCG) using a pair wise-alignment. The indicated percent identity of the kinases with MKKX was calculated using the BESTFIT program.

residue at position 278 (Fig. 2). This may represent a unique regulatory phosphorylation site in MKKX and Dhep which is not present in other MAPKKs. In addition, preliminary complementation studies indicate that MKKX is capable of rescuing a *Drosophila* hep mutant (S. Noselli, pers. comm.). Among mammalian MAPKKs, MKKX is more closely related to MKK6, MKK3 and MKK4 than to MKK1 and MKK5 (4,5,6,7). A Northern blot analysis reveals that MKKX is expressed ubiquitously as a 4 kb transcript, with the greatest expression in muscle and brain (data not shown).

Previous studies in our lab have employed an in vitro association assay in order to establish the relationship between kinases and their substrates. Using this assay, we have determined that MKKX is capable of binding to JNK1, but not p38 or ERK2 *in vitro* (Fig. 3). Similar results are observed in a yeast two-hybrid assay (data not shown). No other candidate substrates have been identified to date.

Transient overexpression of MKKX in NIH3T3 cells has demonstrated that MKKX undergoes a mobility shift on SDS-PAGE in response to various stimuli, as detected by Western analysis (Fig. 4). UV irradiation and treatment of cells with anisomycin, but not growth factors such as PDGF, can induce a shift in MKKX. This mobility shift may be indicative of a change in the phosphorylation and activation state of MKKX.

Predicted phosphorylation site mutants of MKKX (8) have been generated (SA 272, TA 276 and SE 272, TE 276) and analyzed by transient overexpression in NIH3T3 cells (Fig. 5). Co-transfection of MKKX wild type with JNK1 potentiated the induction of JNK1 activity observed in response to anisomycin (5.8 fold). In addition, co-expression of MKKX wild type significantly increased JNK1 activity in either PDGF-stimulated or unstimulated cells. These results suggest that MKKX has a high level of basal activity when co-transfected with JNK1.
A similar pattern was also observed with the MKKX SA and MKKX SE mutant forms. The fact that MKKX SA and MKKX SE activities with respect to JNK1 were comparable indicates that neither mutant was fully inactive or active, respectively. It is particularly striking that co-expression of the MKKX SA mutant does not block induction of JNK1 activity. In contrast, a corresponding MKK4 SA mutant (SA 257, TA 261) is completely inactive and blocks

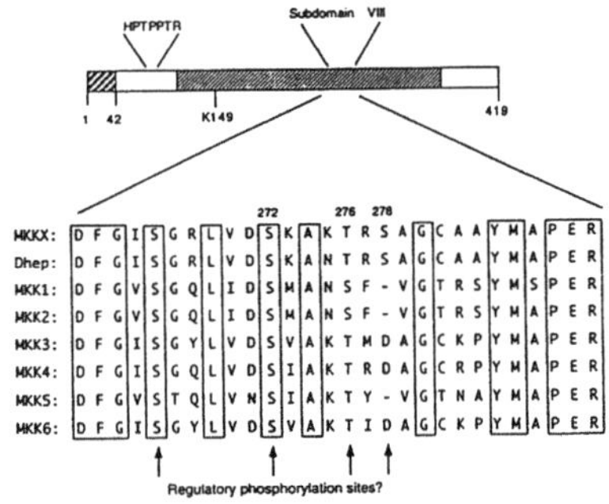


**Fig. 2.** Alignment of sequences within subdomains VII and VIII of mammalian MAPKKs identified to date. *Drosophila* hemipterous is also included. Proposed regulatory phosphorylation sites are indicated by arrows.

induction of JNK1 activity in this system(data not shown). It is possible that additional phosphorylation sites are involved in the regulation of MKKX. Further mutational analyses are required to clarify this issue.

In order to gain further insight into the biological role of MKKX, we have studied its subcellular localization in stably transfected Swiss 3T3 fibroblasts. Preliminary data indicates that MKKX is found diffusely throughout the cell, with areas of increased expression along the cell periphery (data not shown). It is possible that like Dhep, MKKX may play a role in regulating changes in cell morphology. Studies to determine whether various Rho family members also known to control cell morphology may regulate MKKX are ongoing (9).

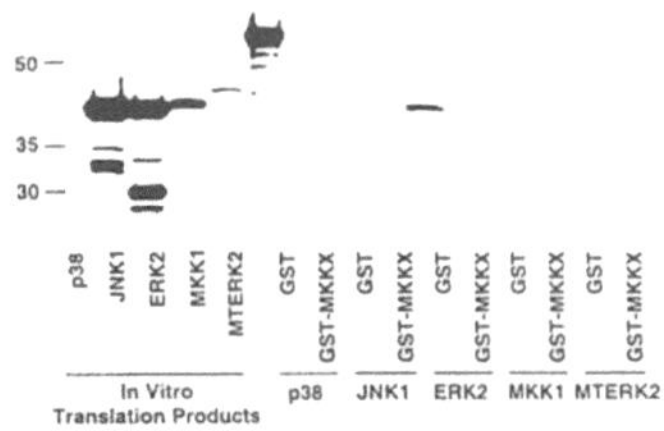


**Fig. 3.** **Recombinant GST-MKKX or GST alone was incubated at 4 °C for 1 hour with either $^{35}$S-Met labeled *in vitro* translated p38, JNK1 or ERK2 (Promega TNT Kit). Samples were washed and analyzed by SDS-PAGE and autoradiography.**

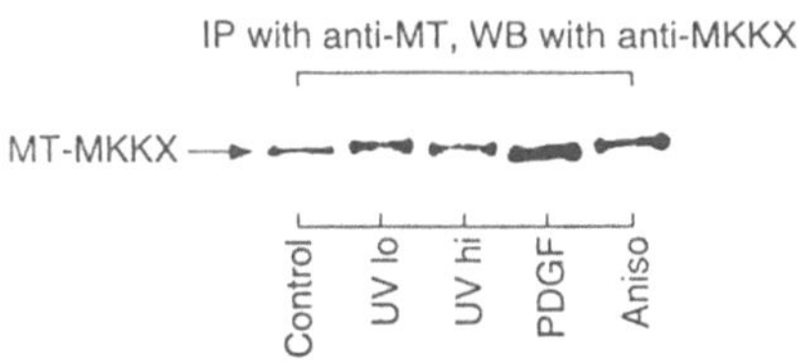


**Fig. 4.** NIH3T3 cells were transiently transfected with myc-tagged wild type MKKX. After 48 hours, cells were treated with either UV-C irradiation (50 J/m$^2$, 100 J/m$^2$), PDGF (5μg/ml, 20 min), anisomycin (10μg/ml, 20 min), or left untreated. After lysis, myc-tagged MKKX was immunoprecipitated with an anti-myc antibody (9E10), and visualized by western blotting with anti-MKKX antibody.

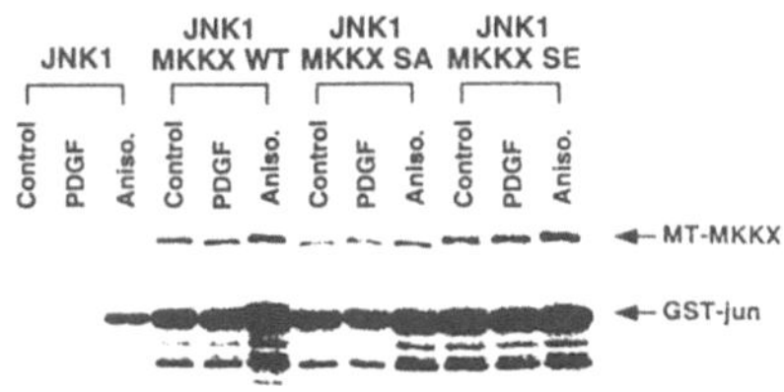


**Fig.5.** NIH3T3 cells were transiently co-transfected with myc-tagged JNK1 and either myc-tagged wild-type MKKX, MKKX SA (SA 272, TA 276), MKKX SE (SE 272, TE 276), or CS3MT vector alone. After 48 hours, cells were treated with either PDGF (5μg/ml, 20 min), anisomycin (10μg/ml,20 min) or left untreated. After lysis, myc-tagged JNK1 was immunoprecipitated with an anti-myc antibody (9E10), and immune complex activity was measured using GST-jun as substrate. Myc-tagged MKKX constructs were also present in the immunoprecipitate.

## 3. Summary

We have identified a novel MAPKK, termed MKKX, from a yeast two-hybrid screen using MKK1 as bait. MKKX is a novel MAPKK family member with homology to *Drosophila* hemipterous. MKKX is encoded by two isoforms, MKKXa and MKKXb, which differ in their N-terminal sequences. MKKX is expressed ubiquitously in adult mouse tissues, particularly in brain and muscle. MKKX binds to JNK1 *in vitro*, but not p38 or ERK2. Transiently overexpressed MKKX is capable of undergoing a mobility shift on SDS-PAGE in response to UV irradiation or anisomycin treatment, but not in response to PDGF. Co-expression of MKKX with JNK1 in NIH3T3 cells increases JNK1 activity. Immunostaining studies of

MKKX in Siwss 3T3 fibroblasts indicate that MKKX is distributed throughout the cell with areas of increased expression along the periphery.

References

1. J.M. Kyriakis and J. Avruch, *J. Biol. Chem.*, 271, 24313 (1996).
2. A.B. Vojtek and S. M. Hollenberg, *Meth. Enzymol.*, 255, 331 (1995).
3. B. Glise, H. Bourbon, S. Noselli, *Cell*, 83, 451 (1995).
4. J. Raingeaud et al., *Mol. Cell. Biol.*, 16, 1247 (1996).
5. A. Lin et al., *Science*, 268, 286 (1995).
6. G. Zhou, Z. Bao, J.E. Dixon, *J. Biol. Chem.*, 270, 12665 (1995).
7. J.M. English et al., *J. Biol. Chem.*, 270, 28897 (1995).
8. S.J. Mansour et al., *Science*, 265, 966 (1994).
9. A.B. Vojtek and J. A. Cooper, *Cell*, 82, 527 (1995).

# Interleukin-1 activates a novel p54 MAP Kinase Kinase in rabbit liver

[1]Michael Kracht, [2]Andrew Finch and [2]Jeremy Saklatvala

Cytokine Laboratory, Babraham Institute, Cambridge, CB2 4AT, UK
Institute for Molecular Pharmacology, Medical School Hannover, D-30623 Hannover, Germany

## 1.Introduction

The major inflammatory cytokine interleukin-1 (IL-1) exerts pleiotropic effects during acute or chronic inflammatory disease. On a cellular level it often increases expression of genes for many proteins involved in initiating or maintaining an inflammatory process, like cytokines (IL-2, IL-6, IL-8, G-CSF, GM-CSF), adhesion molecules (ICAM-1, ELAM-1), proteinases (collagenases, stromelysin, gelatinases), cyclooxygenases, acute phase proteins and others. However the molecular mechanisms of signal transduction induced by IL-1 are poorly understood. IL-1 enhances phosphorylation of intracellular proteins, including the epidermal growth factor receptor (EGF-R) and the small heat shock protein hsp27. Accordingly, IL-1-activated protein kinases can be detected in extracts of stimulated cells. Some of these have been identified as members of the mitogen activated protein kinase (MAPK) family (1,2). As yet no consistent pattern of activation of these kinases by IL-1 has been found amongst different cell lines (1,2,3). In order to identify protein kinases which are physiologically relevant to IL-1 signalling, we injected rabbits intravenously with IL-1 and prepared extracts from their livers. Using the EGF receptor peptide T669 as a substrate, we purified two IL-1-induced protein kinases by sequential chromatography to homogeneity and identified them as two isoforms of stress activated protein kinase (SAPK) $\alpha$ also called p54 MAP kinase $\alpha$ (4). Here we present evidence using recombinant protein kinases as substrates that IL-1 also activates a novel SAPK activator (a p54 MAP Kinase Kinase) in rabbit liver.

## 2.Results

Having identified IL-1 induced stress activated protein kinases *in vivo*, we tried to establish assays for their activators and to purify them. As shown in (Fig.1) IL-1 activates a p54 MAP kinase kinase about sixfold in liver which can be detected in cation exchange chromatography fractions. We also detected a kinase that phosphorylates (Fig.2.A) and activates (Fig.2.B) recombinant p38 MAP kinase in those fractions. The stimulated p54 MAP kinase kinase was partially purified as summarised in (Table I). Although the p54 MAPKK copurified with the p38 MAPKK, it did not coelute on MonoS, suggesting that the activators are different enzymes (Fig.3). Using antibodies against MKK3 and SEK1/MKK4 we were not able to identify proteins comigrating with p54 or p38 MAPKK in our purest material by western blotting or immunoprecipitation (data not shown).

NATO ASI Series, Vol. H 102
Interacting Protein Domains
Their Role in Signal and Energy Transduction
Edited by Ludwig Heilmeyer

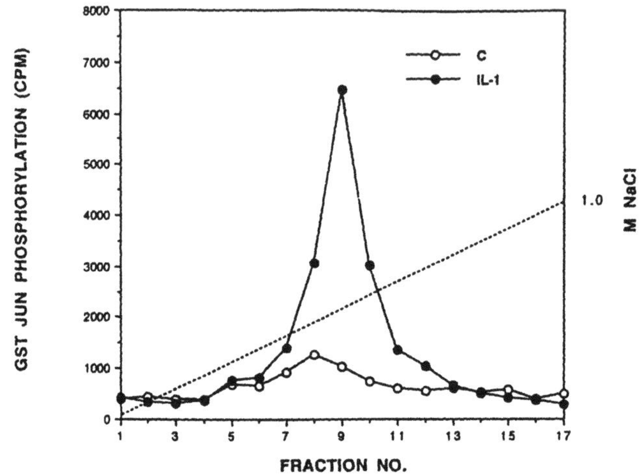


**Fig.1. IL-1 activates a p54 MAP Kinase Kinase activity in rabbit liver.** Rabbits were injected with recombinant human IL-1 α ( ) or vehicle ( ). After 8 min rabbits were killed with pentabarbitone and livers removed as described in detail in (4). An extract of soluble protein was prepared and loaded on to a 2ml FFS cation exchange chromatography column . Proteins were eluted with a linear salt gradient fom 0-1,0 M NaCl in 17 fractions and 1ml fractions collected. Aliquots of fractions were assayed for the ability to phosphorylate and activate recombinant bacterially expressed glutathion-S-transferase p54 MAPK β (a kind gift of J.R. Woodgett, Ontario Cancer Institute, Toronto, Canada). Activated GST p54 MAPK was purified from the reaction mixture with glutathion Sepharose beads and assayed for its activity on recombinant GST c-jun substrate (aa 1-135) using $^{32}$P-ATP. Phosphorylated GST c-jun was separated on SDS PAGE, detected by autoradiography and quantitated by Cerenkov counting excised gel slices.

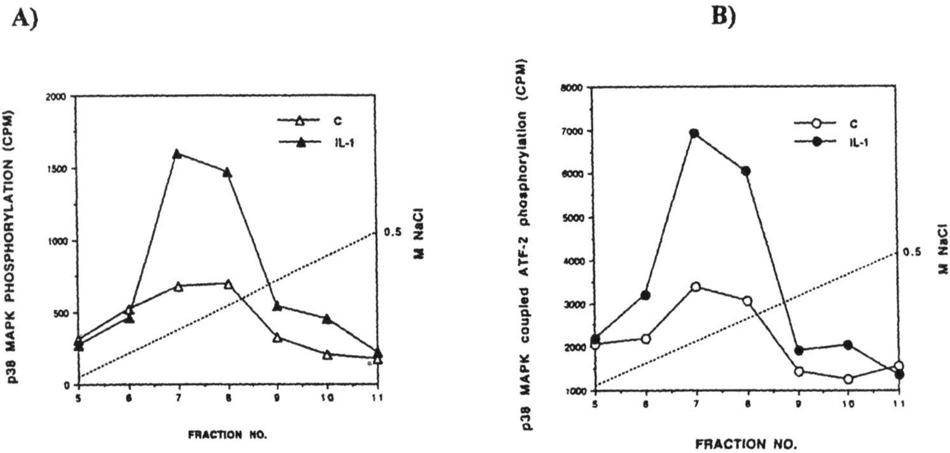


**Fig.2. Interleukin-1 activates a p38 MAP Kinase Kinase activity in rabbit liver.** Aliquots of fractions from a similiar experiment as in ( Fig.1.) were assayed for the ability to phosphorylate and activate recombinant human GST p38 MAP kinase in the presence of $^{32}$P-ATP and GST ATF-2 as substrates. Substrates were separated on SDS PAGE and detected by autoradiography.$^{32}$P incorporation into p38 MAP kinase (A) and coupled activity of the phosphorylated and activated enzyme on ATF-2 (B) is shown.

| | protein (mg) | volume (ml) | activity (CPMx10-6) | yield (%) |
|---|---|---|---|---|
| | | | **Control** | |
| S20 | 6460 | 380 | – | |
| FFS | 92 | 158 | 5 | |
| PS | 24 | 20 | 0.15 | |
| MQ | 7.4 | 20 | 0.13 | |
| | | | **IL-1** | |
| S20 | 6290 | 370 | 5.6 | – |
| FFS | 75 | 146 | 22 | 100 |
| PS | 22 | 20 | 4 | 18 |
| MQ | 10 | 20 | 2.7 | 12 |
| MS | 5.15 | 7 | 1.6 | 7 |
| Sup12 | 2.6 | 1.75 | 1.4 | 6 |

**Table I. Partial purification of a IL-1-stimulated p54 MAP Kinae Kinase from rabbit liver.** Total soluble protein from one liver of rabbits injected with IL-1 orvehicle was obtained by a 20,000xg centrifugation (S20). Proteins were then subjected to sequential chromatography on Fast Flow S-Sepharose (FFS), phenyl Sepharose (PS), MonoQ anion exchange chromatography (MQ), MonoS cation exchange chromatography, (MS) and superose 12 gel filtration chromatography (Sup12). Assays for p54 MAPKK were performed as in described for (Fig.1).

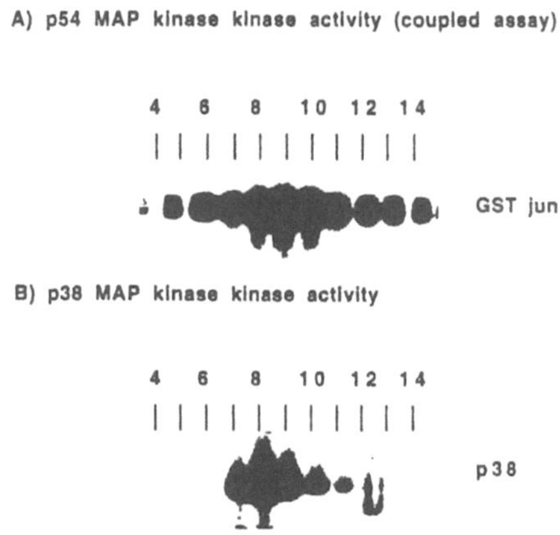


**Fig.3. MonoS purified rabbit liver p54 MAPKK does not coelute with p38 MAPKK.** Rabbit p54 MAPKK was purified on FFS, PS, MQ and MS (see Table I). Aliquots of the fractions were asssayed for p54 MAPKK activity (A) as described in (Fig.1.). The same fractions were also assayed for the ability to phosphorylate p38 MAPK (B)

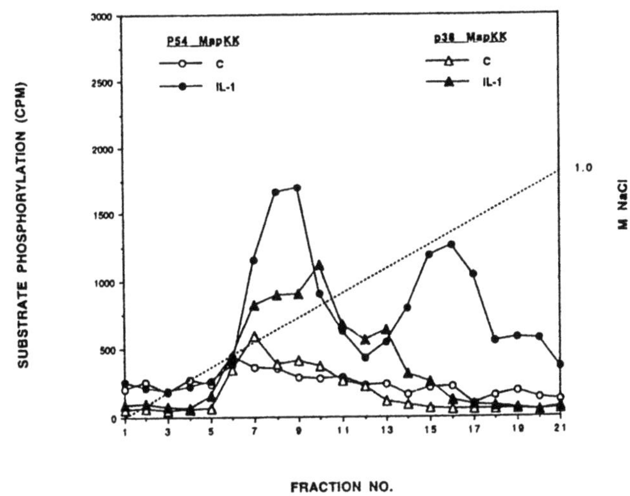


**Fig.4. IL-1 stimulates p54 MAP Kinases Kinases and p38 MAP Kinases Kinases in KB human epidermal carcinoma cells .** Extracts of KB cells stimulated with IL-1 for 15min or left untreated were chromatographed on a FFS chromatography column. and proteins eluted with a linear salt gradient as described in Fig. 1. Aliquots of fractions were assayed for the ability to activate recombinant p54 MAPK β or to phosphorylate p38 MAP Kinase.

**3.Discussion**

We have identified two forms of SAP kinase α as IL-1-stimulated protein kinases *in vivo*. Other studies have also shown that SAPK's or JNK's are activated by IL-1 in cell lines. The kinases phosphorylate transcription factors, but their physiological substrates after induction by IL-1 remain to be identified. The availability of recombinant SAP kinases enabled us to follow the pathway upstream. Although we can detect activators for p54 SAP kinase and for p38 MAPK we were not able to identify them yet. Three novel MAP kinase kinases have been cloned by homology cloning (MKK 3,4,6) and have been shown to activate SAPK's (MKK 4) or p38 MAPK (MKK 3,6) when overexpressed in cells. A specific p54 MAP kinase kinase has not been cloned so far. We were not able to detect activated MKK3 and MKK4 in rabbit livers. However, we are investigating the possibility that the p38 MAP kinase kinase is related to MKK6. SAPKK 3 and 6 have also been purifed from tissues and shown to be specific major activators of p38 MAPK *in vivo* (5). Several more activities have been characterised after MonoS fractionation of cellular extracts from stressed cells (6). We found two p54 MAP kinase kinases and one p38 MAP kinase kinase in KB cells (Fig.4.). Those data suggest, that on the level of the kinase kinases activation of the MAP Kinase pathways is complex depending on the stimulus and cell or tissue studied. Identifying the endogenous SAPKK's induced by IL-1 is a neccessary step to investigate how the activated IL-1 R complex is linked to the JNK/SAPK pathway.

**References**


**1.**Saklatvala, J., Rawlinson, L.M., Marshall, C.J., and Kracht, M. (1993) *FEBS Lett.* 334:189-192

**2.**Kracht, M., Shiroo, M., Marshall, C.J., Hsuan, J.J., and Saklatvala, J. (1994) *Biochem.J.* 302:897-905

**3.**Uciechowski, P., Saklatvala, J., von der Ohe, J., Resch, K., Szamel, M., and Kracht, M. (1996) *FEBS Lett.* 394: 273-278

**4.**Kracht, M., Truong, O., Totty, N.F., Shiroo, M., and Saklatvala, J. (1994) *J.Exp.Med.*. 180:2017-2025

**5.**Cuenda, A., Alonso, G., Morrice, N., Jones, M., Meier, R., Cohen, P., and Nebreda, A.R. (1996) *EMBO J.* 15: 4156-4164

**6.**Meier, R., Rouse, J., Cuenda, A., Nebreda, A.R., and Cohen, P. (1996) *Eur.J.Biochem.* 236:796-805

7.Raingeaud, J., Whitmarsh, A.J., Barrett, T., Derijard, B., and Davis, R.J. (1996) *Mol.Cell.Biol.* 16:1247-1255

# THE PROTEIN KINASE B FAMILY - STRUCTURE, REGULATION AND FUNCTION

*Roger Meier, Mirjana Andjelković, Matthias Frech and Brian A. Hemmings*
*Friedrich Miescher Institute*
*P.O. Box 2543*
*CH-4002 Basel*
*Switzerland*

## *Introduction*

The phosphorylation of signal transduction molecules is of central importance for growth factor-induced transduction of mitogenic signals, cell growth and differentiation. Stimulation of growth factor-receptor kinases by their ligands can activate several signalling modules depending on the cell type. Activation of receptor tyrosine kinase cascades can be either independent of a second messenger system or dependent on a second messenger.

The protein kinase B (PKB) family is apparently regulated by protein phosphorylation, which is in turn promoted by putative second messengers generated by growth factor-induced activation of phosphoinositide 3-kinase (PI 3-kinase). This review discusses recent advances in our understanding of the signalling modules involving growth factor-receptor tyrosine protein kinases, PI 3-kinase and PKB.

NATO ASI Series, Vol. H 102
Interacting Protein Domains
Their Role in Signal and Energy Transduction
Edited by Ludwig Heilmeyer

*The PKB protein kinase family*

The PKB family consists of three isoforms (α, β and γ) with an apparent molecular weight of about 60 kDa. They are very similar in structure (81% amino acid identity) except that PKBγ has a carboxyl-terminal deletion of 23 amino acids (which probably corresponds to a regulatory phosphorylation site - see below). The primary structure of the PKB family consists of an amino terminal pleckstrin homology (PH) domain attached to the kinase catalytic domain, and a carboxy terminal Ser/Thr-rich regulatory domain. The sequence of the catalytic domain is most closely related to the second messenger sub-family of protein kinases (Ingley et al., 1995).

PKB was initially identified by homology cloning using a cAMP-PK-specific probe (Jones et al., 1991a). PKBα (Coffer and Woodgett, 1991; Jones et al., 1991a), also known as c-Akt-1 (Bellacossa et al., 1991), represents the cellular homologue of the oncogenic v-Akt, the protein encoded by AKT-8, an acute transforming retrovirus which causes T-cell leukaemia's and lymphomas in mice (Staal et al., 1977). PKBβ (Jones et al., 1991b) is identical to Akt-2 (Cheng et al., 1992).

Interestingly, v-Akt is a fusion protein between the N-terminal region of the retroviral GAG capsid protein inserted 6 base pairs upstream of the PKBα (Akt) proto-oncogene full length sequence. Myristoylation of the retroviral GAG/PKBα chimera is thought to induce membrane localisation and subsequent activation of PKBα, which is presumably the reason for its oncogenic potential (Ahmed et al., 1993; Bellacosa et al., 1991).

Furthermore, PKBα and -β are amplified and overexpressed in ovarian (Cheng et al., 1992), gastric (Staal, 1987), pancreatic (Cheng et al., 1996) and breast carcinomas (Bellacosa et al., 1995). The third isoform, PKBγ (Konishi et al., 1995) has not yet been found to be implicated in any disease states.

### *The role of Ser/Thr phosphorylation in PKBα regulation*

PKBα is activated by many growth factors, such as PDGF, EGF, basic FGF, insulin, IGF-1 and phosphatase inhibitors (pervanadate, okadaic acid, calyculin A). Recent studies indicate that the activation of PKBα is triggered by receptor tyrosine kinase-mediated stimulation of PI 3-kinase. The activating signal is transmitted to the p110 catalytic subunit of PI 3-kinase by interaction of the inter-SH2 domain of the regulatory subunit of p85 with the appropriate SH2-binding site. Although it is not known how reception of the signal leads to activation of PKBα, several lines of evidence place PKBα downstream of PI 3-kinase: (1) activation of PKBα is prevented by the PI 3-kinase inhibitor wortmannin (Burgering and Coffer, 1995; Kohn et al., 1995; Andjelković et al., 1996), (2) dominant negative PI 3-kinase mutants abolish PKBα activation (Burgering and Coffer, 1995), (3) overexpression of constitutively active PI 3-kinase activates PKBα (Didichenko et al., 1996; Klippel et al., 1996), and (4) PDGF-receptor mutants that lack PI 3-kinase-binding sites fail to activate either PI 3-kinase or PKBα (Burgering and Coffer, 1995; Franke et al., 1995).

Recent studies on insulin and IGF-1-dependent PKBα activation (Alessi et al., 1996) revealed that the kinase is regulated by phosphorylation of two critical residues Thr-308 and Ser-473. Phosphorylation of both residues is necessary for full activation of PKBα, and although this can be inhibited by wortmannin (Alessi et al., 1996), the amino acid sequences of the Thr-308 (ATMK*T*FCGT) and Ser-473 (FPQF*S*YSAS) sites indicate their independent regulation. The Thr-308 sequence is conserved in all PKB isoforms whereas the Ser-473 site is only conserved in PKBα and -β. Interestingly, the Ser-473 site is apparently deleted from the γ isoform. This suggests a similar upstream

Thr-308 activator for all isoforms and a different regulation mechanism for the Ser-473 phosphorylation site.

### *The role of the PH domain in PKBα activation*

Although it is clear that activation of PKBα by phosphorylation is a major regulatory mechanism (Alessi et al., 1996), the key question now emerging is how the PH domain participates in the regulation of PKB. PKB was the first kinase to be recognised with a PH domain (Haslam et al., 1993). PH domains are often found in proteins involved in signal transduction and cytoskeletal organisation (Shaw, 1996). Their ability to bind inositol phosphates and phospholipids is of critical importance for the understanding of their regulation by PI 3-kinase. Since PKBα lies downstream of PI 3-kinase, recent studies have focused on the influence of phosphoinositides on the PH domain and kinase activity (James et al., 1996; Frech et al., 1997). The results indicate that the PH domain binds inositol phosphates and phospholipids with a high affinity ($K_d$ = 0.5-8 μM). The product of PI 3-kinase, phosphatidylinositol(3,4,5)-trisphosphate (PtdIns(3,4,5)P3), inhibits PKBα activity with half-maximal inhibition at 2.5 μM. In contrast, phosphatidylinositol(3,4)-bisphosphate (PtdIns(3,4)P2) stimulates kinase activity. The opposite effects of PtdIns(3,4,5)P3 and PtdIns(3,4)P2 may indicate that PKBα activity is regulated by changes in phosphoinositide synthesis. Effects of inositol phosphates and phospholipids on the activities on the β and γ isoforms have not yet been described.

Although it has been reported that PKBα does not require a PH domain for activation (Andjelković et al., 1996; Kohn et al., 1996), and that the PH domain is not a target for the activating phosphorylation (Alessi et al.,

1996), other reports indicate the importance of the PH domain in PKBα activation (Franke et al., 1995; Frech et al., 1997). For example, PDGF does not fully activate PKBα bearing a mutation in the PH domain (Franke et al., 1995; Andjelković et al., 1996). Furthermore, coexpression of ΔPH-PKBα with constitutively active p110 or membrane-associated p110 failed to activate ΔPH-PKBα (Klippel et al., 1996), suggesting that PI 3-kinase activates PKBα only if it contains a functional PH domain. The data so far obtained indicate that the PH domain localises PKBα at the cell membrane. This translocation step is, therefore, a prerequisite for the activation of the kinase by phosphorylation. An intriguing possibility in this model is that PtdIns(3,4)P2 induces a conformational change that facilitates phosphorylation by an upstream kinase.

***Membrane targeting and PKB activation***

The hypothesis that PKB needs to be recruited to the plasma membrane for subsequent activation by phosphorylation of Thr-308 and Ser-473 is supported by several lines of evidence. PKBα becomes constitutively activated upon attachment of an N-terminal Lck tyrosine kinase myristoylation/palmitoylation motif, which constitutively induces the translocation of PKBα to the plasma membrane (Andjelković et al., 1997). Furthermore, activation occurs in ΔPH-PKBα membrane-targeted mutants, indicating that the PH domain is not necessary under these conditions (Kohn et al., 1996, Andjelković et al., 1997). To elucidate a possible role for PI 3-kinase as direct activator of PKBα, Kohn et al. (1996) fused the inter-SH2 domain of the p85 regulatory subunit of PI 3-kinase to PKBα. The results indicate that the inter-SH2/PKBα chimera associates with p110, leading in turn to activation of the lipid kinase. Apparently, the activated p110 subunit subsequently promotes the activation of PKBα. However, the question still remains whether an upstream kinase is required for the activation of PKBα.

### *Possible functions of PKB kinases in cell proliferation*

A central question now emerging from the studies of PKB is the identification of the downstream signalling. The first physiological substrate identified for PKBα was glycogen synthase kinase-3 (GSK3) (Cross et al., 1995). In skeletal muscle, insulin stimulation results in wortmannin-sensitive, rapamycin-insensitive, phosphorylation and inactivation of GSK3. Since PKBα activates $p70^{s6k}$ in cotransfection assays, it has been proposed that PKBα lies upstream of this kinase (Burgering and Coffer, 1995; Franke et al., 1995). This idea is supported by the finding that pretreatment of NIH 3T3 cells with wortmannin abolishes PDGF-induced activation of both PKBα and $p70^{s6k}$ (Franke et al., 1995). In contrast, rapamycin inhibits $p70^{s6k}$ but not PKBα (Burgering and Coffer, 1995; Andjelković et al., 1996), indicating that PKBα activates $p70^{s6k}$ upstream of rapamycin inhibition.

### *Concluding remarks*

Over the past two years, we have seen rapid change in the perception of PKB. Recent data indicate it to be of central importance in receptor tyrosine-kinase activated PI 3-kinase signalling. Future studies will probably concentrate on the modulation of biological processes by PKB, with emphasis on the fact that this signal pathway is potentially oncogenic. Clearly, it will be of importance to identify downstream elements of PKB in order to understand this oncogenic potential.

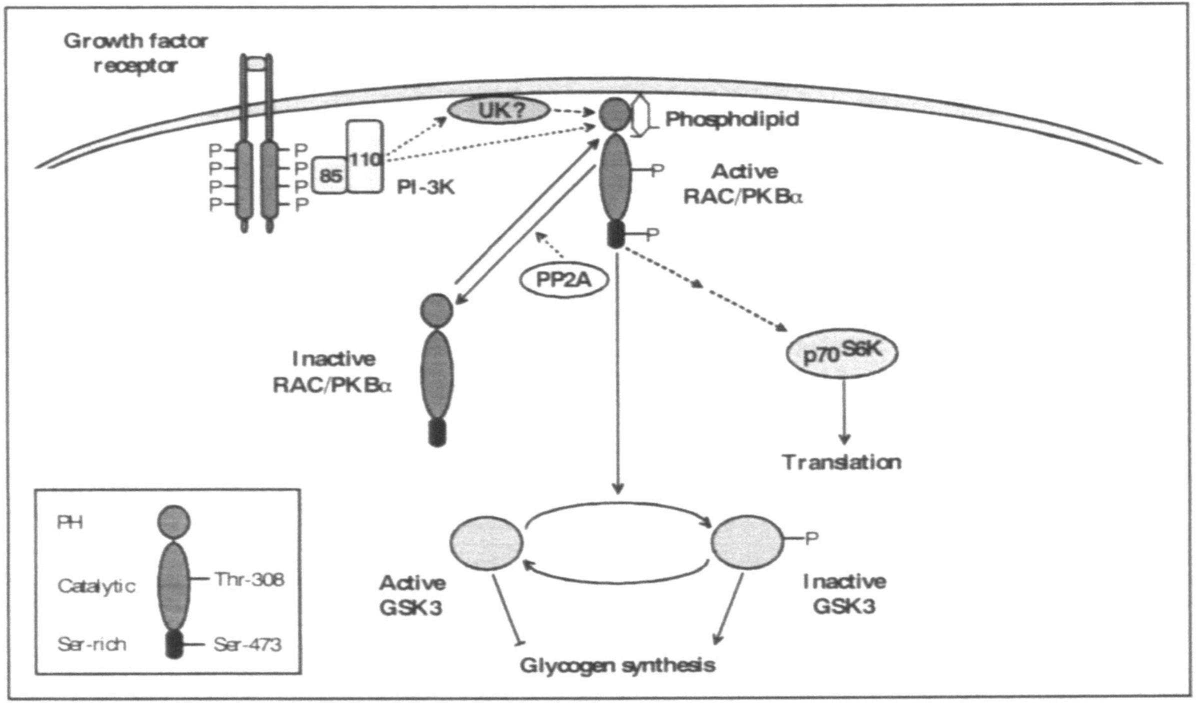


**Fig. 1: Model of PKBα regulation.**
Dashed arrows indicate possible steps. UK stands for Upstream Kinase and GSK3 for glycogen synthase kinase-3.

***References***


Alessi, D.R., Andjelković, M., Caudwell, B.F., Cron, P., Morrice, N., Cohen, P., and Hemmings, B.A. (1996). Mechanism of activation of protein kinase B by insulin and IGF-1. EMBO J., 15, 6541-6551.

Ahmed, N.N., Franke, T.F., Bellacosa, A., Datta, K., Gonzalez-Portal, M.E., Taguchi, T., Testa, J.R., and Tsichlis, P.N. (1993). The proteins encoded by c-akt and v-akt differ in post-translational modification, subcellular localization and oncogenic potential. Oncogene, 8, 1957-1963.

Andjelković, M., Jakubowicz, T., Cron, P., Ming, X.-F., Han, J.-W. and Hemmings, B.A. (1996). Activation and phosphorylation of a pleckstrin homology domain containing protein kinas (RAC-PK/PKB) promoted by serum and protein phosphatase inhibitors. Proc. Natl. Acad. Sci., 93, 5699-5704.

Andjelković, M., Alessi, D., Meier, R., Cron, P., Frech, M., Cohen, P. And Hemmings, B.A. (1997). Manuscript in praparation.

Bellacosa, A., Testa, J.R., Staal, S.P. and Tsichlis, P.N. (1991). A retroviral oncogene, akt, encoding a serine-threonine kinase containing an SH2-like region. Science, 254, 274-277.

Bellacosa, A., DeFeo, D., Godwin, A.K., Bell, D.W., Cheng, J.Q., Altomore, D.A., Wan, M., Dubeau, L., Scampia, G., Masciullo, V., Ferrandina, G., Benedetti, P., Mancuso, S., Neri, G. And Testa, J.R. (1995). Molecular alterations of the Akt2 oncogene in ovarian and breast carcinomas. Int. J. Cancer (Pred. Oncol.), 64, 280-285.

Burgering, B.M.Th. and Coffer, P.J (1995). Protein kinase B (c-Akt) in phosphatidyl-3-OH kinase signal transduction. Nature, 376, 599-602.

Cheng, J.C., Godwin, A.K., Bellacosa, A., Taguchi, T., Franke, T.F., Hamilton, T.C., Tsichlis, P.N., and Testa, J.R. (1992). Akt2, a putative oncogene encoding a member of a subfamily of protein-serine/threonine kinases, is amplified in human ovarian carcinomas. Proc. Natl. Acad. Sci., 89, 9267-9271.

Cheng, J.C., Ruggeri, B., Klein, W.M., Sonada, G., Altomare, D.A., Watson, D.K. and Testa, J.R. (1996). Amplification of Akt2 in human pancreatic cancer cells and inhibition of Akt2 expression and tumorigenicity by antisense RNA. Proc. Natl. Acad. Sci., 93, 3636-3641.

Coffer, P.J. and Woodgett, J.R. (1991). Molecular cloning and characterisation of a novel putative protein-serine kinase related to the cAMP-dependent and protein kinase C families. Eur. J. Biochem., 201, 475-481.

Cross, D.A.E., Alessi, D.R., Cohen, P., Andjelković, M. and Hemmings, B.A. (1995). Inhibition of glycogen synthase kinase-3 by insulin mediated by protein kinase B., Nature, 378, 785-789.

Didichenko, S.A., Tilton, B., Hemmings, B.A., Ballmer-Hofer, K. And Thelen, M. (1996). Constitutive activation of protein kinase B and phosphorylation of p47(Phox) by membrane-targeted phosphoinositide 3-kinase. Curr. Biol., 6, 1271-1278

Franke, T.F., Yang, S.-I., Chan, T.O., Datta, K., Kazlaukas, A., Morrison, D.K., Kaplan, D.R. and Tsichlis, P.N. (1995). The protein kinase encoded by the Akt proto-oncogene is a target of the PDGF-activated phosphatidylinositol 3-kinase. Cell, 81, 727-736.

Frech, M., Andjelković, M., Reddy, K., Falck, J.R. and Hemmings, B.A. (1997). High affinity binding of inositol phosphates and phosphatidylinositides to the pleckstrin homology domain of RAC/Protein Kinase B and their influence on the kinase activity. Submitted.

Haslam, R.J., Koide, H.B. and Hemmings, B.A. (1993). Pleckstrin domain homology., Nature, 363, 310-311.

Ingley, E. and Hemmings, B.A. (1995). In The Protein Kinase Facts Book, eds. Hardie, and Hanks, S. (Academic Press, London), pp.95-97.

James, S.R., Downes, C.P., Gigg, R., Grove, S.J.A., Holmes, A.B. and Alessi, D.R. (1996). Specific binding of the Akt-1 protein kinase to phosphatidylinositol 3,4,5-triphosphate without subsequent activation. Biochem. J., 315, 709-713.

Jones, P.F., Jakubowicz, T., Pitossi, F.J., Maurer, F. and Hemmings, B.A. (1991a). Molecular cloning and identification of a serine/threonine protein kinase of the second-messenger subfamily., Proc. Natl. Acad. Sci. USA, 88, 4171-4175.

Jones, P.F., Jakubowicz, T. and Hemmings, B.A. (1991b). Molecular cloning of a second form of rac protein kinase., Cell. Regulation, 2, 1001-1009.

Klippel, A., Reihard, C., Kavanaugh, M., Apell, G., Escobedo, M.-A. and Williams, L.T. (1996). Membrane localization of phosphatidylinositol 3-kinase is sufficient to activate signal-transducing kinase pathways. Mol. Cell. Biol., 16, 4117-4127.

Kohn, A.D., Kovacina, K.S. and Roth, R.A. (1995). Insulin stimulates the kinase activity of Rac-Pk, a pleckstrin homology domain containing ser/thr kinase. EMBO J., 14, 4288-4295.

Kohn, A.D., Takeuchi, F. and Roth, R.A. (1996). Akt, a pleckstrin homology domain containing kinase, is activated primarily by phosphorylation. J. Biol. Chem., 271, 21920-21926.

Konishi, H., Kuroda, S., Tanaka, M., Matsuzaki, H., Ono, Y., Kameyama, K., Haga, T. and Kikkawa, U. (1995). Molecular cloning and characterisation of a new member of the rac protein kinase family: Association of the pleckstrin homology domain of three types of rac protein kinase with protein kinase c subspecies and βγ subunits of G proteins., Biochem. Biophys. Res. Comm., 216, 526-534.

Shaw, G. (1996). The pleckstrin homology domain: an intriguing multifunctional protein module. BioEssays. 18, 35-46.

Staal, S.P., Hartley, J.W. and Rowe, W.P. (1977). Isolation of transforming murine leukemia viruses from mice with a high incidence of spontaneous lymphoma. Proc. Natl. Acad. Sci. USA, 74, 3065-3067.

Staal, S.P. (1987). Molecular cloning of the akt oncogene and its human homologues Akt1 and Akt2: Amplification of Akt1 in a primary human gastric adenocarcinoma. Proc. Natl. Acad. Sci. USA, 84, 5034-5037.

# Regulation of $p70^{s6k}$

Patrick B. Dennis and George Thomas

Friedrich Miescher-Institut, Maulbeerstrasse 66, 4058 Basel, Switzerland

This manuscript represents the first of two lectures concerning the $p70^{s6k}$ and translation. It is presented as closely as possible to the actual talk.

Our Group has had a long term interest in the mechanisms by which growth signals induce cells to proliferate, more specifically we have focused on the multiple phosphorylated substrate, ribosomal protein S6 and the kinase which modulates this response. The approach we have taken in asking how a growth factor induces cells to exit the $G_0$ state of the cell cycle, has been to begin with an obligatory step in this process, the activation of protein synthesis, and to work our way back along this pathway describing molecular components involved in this event. This approach very rapidly led us to S6, an integral 40S ribosomal protein which becomes multiply phosphorylated in response to oncogenic transformation and growth factors. More recently, we have also moved downstream to understand the role of this signalling pathway during mitogenesis in terms of the activation of protein synthesis. In this first lecture, we will focus on the mechanism by which kinase activation is brought about and the signalling pathway which controls this response. S6 resides in what is commonly referred to as a small head region of the 40S ribosome just opposed to the larger 60S subunit in a position involved in binding mRNA. We spent some effort in identifying the sites of phosphorylation and showed that these sites are

NATO ASI Series, Vol. H 102
Interacting Protein Domains
Their Role in Signal and Energy Transduction
Edited by Ludwig Heilmeyer

clustered at the carboxy end of the protein in serines 235, 236, 240, 244 and 247. Further studies, in collaboration with Richard Wettenhall (Melbourne, Australia), went on to show that there is a specific order in which these sites are modified, the first site of phosphorylation being 236 followed by 235, 240, 244 and then 247, and it appears to be these later sites of phosphorylation which are associated with the activation of protein synthesis. We took advantage of this sequence to generate an affinity column which eventually allowed us to purify the kinase to homogeneity. From this purified material, we obtained sequence data which we then used to clone the kinase, and from the cloning studies, we were able to generate specific antibodies. In brief, the cDNA clone that we obtained encoded a protein of 502 amino acids which had a very short amino terminus, that was quite acidic in nature, followed by a conserved catalytic domain, and finally a putative regulatory domain at the carboxy end of the molecule. During the cloning of this molecule, it became evident that there was a second isoform arising from the $p70^{s6k}$ which was identical in the common coding region but which contained a 23 amino acid extension at its amino terminus. We know from the work of Yi Chen and Sara C. Kozma (Basel, Switzerland), that both these transcripts are generated from a common gene and furthermore, that they are generated from the same mRNA transcript through the use of alternative initiation translational start sites. The first initiation start site, which encodes the larger $p85^{s6k}$ isoform has a poor Kozak consensus initiation start site. In most instances, there is read through and initiation at the second translational start site which displays a strong consensus initiation start site encoding the $p70^{s6k}$ isoform. Whether this process is regulated is not known at this point. If one focuses a little more closely on this 23 amino acid extension, one sees that it contains all the hallmarks of a nuclear targetting sequence. In fact, a few years ago, we showed that this form of the enzyme is constitutively targetted to the nucleus. In contrast, if we use an antibody which detects the $p70^{s6k}$ isoform, along with the nuclear form, one observes very strong cytoplasmic staining. By carrying out quantatative confocal mycroscopy, Ned Lamb and Ann Fernandez (Montpellier, France) were able to demonstrate that the $p70^{s6k}$ isoform of the protein appears to be exclusively cytoplasmic. Having these antibodies in hand also allowed us to examine the mechanism by which kinase activation is brought about through phosphorylation. Initially, we identified four sites of phosphorylation which lie within the putative regulatory end of the molecule. There are two striking features about this sequence. The first is that it has minimal but significant homology with S6 itself, suggesting that this sequence may serve as an autoinhibitory domain within the molecule. In fact, peptides covering this sequence inhibit the kinase in the low μM range. The second interesting feature is that all four sites of phosphorylation, S411, S418, T421 and S424, display S/TP motifs, a motif recognised by MAP kinase, leading to the initial speculation that activation of $p70^{s6k}$ is coupled to the activated receptor through activation of MAP kinase. However, we spent some effort in showing that the $p70^{s6k}$ was not an *in vivo* substrate for MAP kinase. In fact, we showed that dominant negative forms of either Ras or Raf, upstream signalling components on the MAP kinase pathway, block $p44^{mapk}$ activation in cotransfection studies but have no effect on $p70^{s6k}$ activation. These studies eventually led to the finding that these two signalling pathways bifurcate at the level of the receptor at the specific docking sites of the human PDGF receptor, 740 and 751 which are responsible for binding both PI3 kinase and NCK. In parallel, it was found that the fungal metabolite wortmannin, which is thought to be a specific inhibitor of PI3 kinase, also blocked the activation of the $p70^{s6k}$. This has led to a tentative model of the signalling pathway leading to the $p70^{s6k}$ which is initiated by recruitment of PI3 kinase to the activated receptor leading to the increased production of phosphoitylinositol 3, 4, 5 phosphate and in turn leading to the activation of the recently described kinase, PKB, which contains a phospholipid-binding pleckstrin homology domain. Further downstream, PKB would then lead to the activation of FRAP. FRAP is a large molecular weight protein of Mr 300,000 which has homology with lipid kinases and the property to self-phosphorylate, suggesting it could also play the role of a protein kinase. FRAP is also the target of the gain of function, inhibitory, rapamycin - FKBP12 immunophillin complex. As this inhibitory complex has no effect on either the activation of PI3 kinase or PKB, the ordering of the pathway in which PI3 kinase signals to PKB, which then signals to FRAP and then downstream to the $p70^{s6k}$ was proposed. However, when we analysed the individual point mutations of the PDGF receptor, we found when tyrosine 740, the strongest of the two docking sites, was changed to a phenylalanine, this altered form of the receptor was unimpaired in its ability to activate $p70^{s6k}$. Yet this mutation ablated PI3 kinase and PKB activation as well as failing to lead to an increase in the production of phosphotylinositol 3, 4, 5 phosphate in response to PDGF. Further studies, carried out by the group of Joseph Avruch (Boston, USA), showed that a double truncated form of the $p70^{s6k}$, which exhibits about 50% of the activity of the wild type enzyme, is totally rapamycin resistant, but still inactivated by wortmannin. These observations made it difficult to place the PI3 kinase block upstream of the rapamycin sensitive block.

At this point, we realised that to work out the signalling pathway upsteam, it was very important to establish the identity of the inhibitory target which was closest to the kinase. Therefore, we focused our efforts on the role of the rapamycin in regulating $p70^{s6k}$ activity. Given that we had just identified the phosphorylation sites associated with activation of the enzyme, the first question we wanted to ask was whether the inhibitory effects of rapamycin on cell growth were elicited through $p70^{s6k}$. To address this question we took advantage of the fact that others have shown, in many cases, that substitution of acidic residues for phosphorylation sites can mimmick the phosphorylation state and at the same time confer phosphatase resistance to the kinase. Therefore, we changed the four residues in the autoinhibitory domain to acidic residues ($D_3E$ $p70^{s6k}$) and then transiently expressed this construct in 293 cells in order to measure the activity of the enzyme. These experiments led to two unexpected results. First, the mutant was as active as the wild type enzyme in the presence of mitogens, and second, when we treated cells transiently expressing this construct with rapamycin this form of the kinase was inactivated to the same extent as a wild type enzyme. This latter finding suggested that rapamycin brought about $p70^{s6k}$ inactivation through a mechanism independent of phosphorylation, or conversely that there were other unidentified phosphorylation sites associated with kinase activation. When we concentrated on SDS-PAGE western blots of the $p70^{s6k}$ we observed band shifts with the mutant which were as extensive as those observed for wild type $p70^{s6k}$, and those band shifts collapsed into a single band when we pre-treated cells with rapamycin. These band shifts, or changes in electropheretic mobility of the protein are indicative of phosphorylation and dephosphorylation and suggested that there were additional sites of phosphorylation associated with kinase activation. In fact, a re-analysis of the phosphopeptide maps from quiescent and stimulated cells, in the absence and presence of rapamycin, led to the identity of three additional phosphopeptides whose phosphorylation levels were associated with kinase activation, denoted as b, c and d. We spent some effort in trying to identify these sites by taking advantage of the computer program which the group of Tony Hunter (San Diego, USA) had described for predicting the mobility of tryptic phosphopeptides in the two dimensional system. However, this approach failed, so at this point we His-tagged the kinase, transiently expressed it at high levels in human 293 cells, isolated this material, digested it with trypsin and then purified the phosphopeptides by reverse phase HPLC. Add back experiments, with endogenous $p70^{s6k}$, demonstrated that the purified peptides were equivalent to those which we had observed *in vivo* by carrying out phosphopeptide maps from material isolated from Swiss 3T3 cells. These sites were then identified by taking advantage of mass-spectrometry, phosphate release, N-terminal peptide sequencing and phospho amino acid analysis. All four procedures were required for unambiguous identification of the sites as T229, in the T loop and, T389 as well as S404, in the linker region coupling the catalytic with the putative autoinhibitory domain of the kinase. All three sites resided within atypical tryptic cleavage products which largely explained our difficulty in the initial identification. Because of the importance of T loop phosphorylation in kinase activation, we reasoned that T229 would be the principal target of rapamycin induced $p70^{s6k}$ inactivation. To test this possibility, we stimulated cells for a short time, 30 minutes, in the presence of $^{32}P$, then treated these cells with rapamycin and followed the time course of inactivation in parallel with the loss of phosphate from each one of the three residues through two-dimensional phosphopeptide analysis. The results showed that the kinase was inactivated by about 50% in 4 minutes and that by 15 minutes it was almost totally inactive, having largely returned to basal levels. If we look at the corresponding phosphopeptide maps, the surprise again was that the dephosphorylation of the site which followed inactivation of the kinase, was T389, followed by S404, whereas at 30 minutes we could still detect phosphate in T229, when kinase activity was totally abolished. To test the importance of each one of these sites in modulating kinase activity, we either changed these residues to an acidic amino acid or to an alanine. Conversion of S404 to a neutral or acidic residue, raised basal kinase activity but had no influence on the ability of mitogens to activate the enzyme, suggesting that this was not a principal target of acute kinase regulation, but instead was involved in modulating kinase activity. In contrast, if we applied a similar approach to T229, regardless of the residue which we substituted into this position, we ablated kinase activity making it difficult to use this approach to assess the importance of this phosphorylation site for kinase function. Similarly, if we substituted an alanine for T389, we ablated kinase activity, however, much to our pleasure, when we substituted an acidic residue at this site, the kinase displayed high basal activity, although stimulated activity could only be raised to 50% of the wild type level in the presence of the mitogen. We had previously found that when we substituted acidic residues into the autoinhibitory domain, basal kinase activity was raised, therefore we reasoned since T389 lies just upstream of the autoinhibitory domain, the 389 acidic mutation in the background of the acidic mutations in the autoinhibitory domain may be able to raise kinase activity equivalent to that of the wild type enzyme. Indeed, this is what we observed, raising the question that, if T389 is the principal target of rapamycin-induced $p70^{s6k}$ inactivation, we should be able to test this possibility by taking advantage of the acidic T389 mutation. Thus, the T389 acidic mutation has been placed either into the wild type background or into the $D_3E$ background. In both cases, parent constructs were inactivated by rapamycin treatment. However, in the two constructs harboring the acidic T389 mutation, the kinase activities were largely protected against rapamycin inactivation. These findings established T389 as an important site for kinase regulation and prompted us to look more closely at the motif surrounding this site. A novel feature of the phosphorylation site is that there are no charged residues upstream or downstream, suggesting that phosphorylation was carried out by a kinase of novel specificity. Furthermore, we found that this sequence, although not previously noted, was present in many members of the second messenger family of serine/threonine kinases. Most notably in all the PKC isoforms, $p90^{rsk}$ as well as RAC/PKB and in the putative homologues of $p70^{s6k}$ from arabadopsis and yeast, ATPK1 and YPK1, respectively. In all cases, there is an invariant phenylalanine in the -4 position, and the phosphorylated residue is surrounded by large, uncharged aromatic residues. There is also a propensity for glycine in the -2 position and valine in the +2 position. Most importantly, the distance that this motif in $p70^{s6k}$ resides from the site in the T-loop, T229, is conserved in all these kinases suggesting that this sequence is an extension of the catalytic domain.

In parallel, while these studies were being carried out, the group of Joseph Avruch (Boston, USA) took a second approach to try to determine the mechanism of kinase activation. In brief, they made a series of truncation mutants of the p70$^{s6k}$ in an attempt to identify important regulatory domains within the molecule. When they removed the N-terminus of the kinase, the ability of the kinase to be activated by mitogens was completely lost. In contrast, if they removed the carboxy tail, they reduced the ability to activate the kinase by about 25%, but conferred partial rapamycin resistance on the molecule though still retaining sensitivity to wortmannin. However, when they removed the carboxy tail in combination with the N terminal truncation, they could rescue the loss of activity caused by removal of the amino terminus alone. This form of the kinase was totally rapamycin resistant, but still sensitive to wortmannin. This finding suggested that wortmannin and rapamycin were bringing about p70$^{s6k}$ inactivation by independent mechanisms. However, when we examined the sites of phosphorylation which were sensitive to the two inhibitors of p70$^{s6k}$ activation, we found that the same sets of sites were ablated by both agents. This suggested that either we had missed critical sites of phosphorylation associated with inactivation by wortmannin, or instead, that these two inhibitors were bringing about kinase inactivation through two independent mechanisms. To test this possibility, we examined the resistance of the T389 acidic mutants to wortmannin treatment in relationship to rapamycin. The results showed that in either a post- or pre-treatment regime, the kinase is as resistant to wortmannin as it is to rapamycin, consistent with our previous conclusion derived from phosphopeptide map analysis. This observation suggested that difference observed in the truncation mutants might be inherent in those mutants rather than having anything to do with the regulation of kinase. To test this possibility, we generated similar mutants in the p70$^{s6k}$ as those described by Dennis Templeton

(Cleveland, USA) and collaboraters. The first 54 amino acids of the p70$^{s6k}$ were removed to generate an amino terminal truncation mutant, the last 104 residues were removed to generate a carboxy tail truncation mutant, and both truncations were combined to generate the double truncation mutant. In transient transfection studies, the activity of these mutants was very similar to those described by Joseph Avruch (Boston, USA) in the p85$^{s6k}$ isoform. The amino terminal truncation had low basal activity which could not be stimulated by mitogen treatment. The carboxy tail truncation also had low basal activity, but it could be activated to about 75% the level of the wild type enzyme. Following mitogen treatment, removal of the carboxy tail rescued kinase activity of the N terminal truncation, though the total activity rescued was only about 50% of the wild type enzyme. So removal of each domain led to a significant loss of kinase activity. We next analysed the sensitivity of each one of these constructs to either rapamycin or wortmannin, and again, very similar to the observations of Joseph Avruch (Boston, USA) and colleagues, we found that the carboxy tail truncation was totally sensitive to wortmannin as was the double truncation mutant, yet the carboxy tail mutant exhibited some rapamycin resistance, and the double truncation mutant was totally resistant to the rapamycin. Given the importance of phosphorylation in bringing about p70$^{s6k}$ activation, the question we first wanted to address was whether the inability to bring about kinase activation by mitogens in the amino terminal truncation was due to the inability of an upstream kinase to access specific phosphorylation sites. Therefore, when we compared the phosphopeptide maps of serum-stimulated amino terminal truncation mutant with the wild type, three rapamycin and wortmannin sensitive sites, T229, T389 and S404, were not phosphorylated in the truncation mutant. Given the importance of T389 and T229 phosphorylation in bringing about kinase activation, this finding is consistent with this form of the enzyme's failure to be activated by mitogens. If this model is correct, then one would reason that further removal of the carboxy tail, which leads to rescue of kinase activity, would also lead to the corresponding phosphorylation of T389 and T229. To test this possibility, we analysed phosphopeptide maps of the double truncation mutant in the absence and the presence of serum. Much to our surprise, in the basal state, along with phosphorylation of T367 and S371, we observed strong phosphorylation of T229, not seen in the wild type enzyme under these conditions. Addition of serum had very little effect on any one of these three sites, however, there was an acute increase in phosphorylation of T389. These findings suggest that removal of both the amino terminus and the carboxy end of the molecule allows T229 phosphorylation to take place in the absence of mitogens, suggesting that the kinase which regulates this site is constitutively activated in the cell, and its inability to phosphorylate the enzyme in the whole kinase is due to the fact that this site is inaccessible. Furthermore, because the double truncation mutant of $p70^{s6k}$ has low basal activity, despite the presence of T229 phosphorylation, phosphorylation of T229 is not sufficient to bring about kinase activation. Consistent with this conclusion, addition of serum brings about acute phosphorylation in T389 suggesting that this site is responsible for triggering kinase activation.

When we treat cells expressing the double truncation mutant with rapamycin and wortmannin we find that wortmannin blocks activation of the kinase, whereas it is activated in the presence of rapamycin. So we wanted to determine the state of phosphorylation in the double truncation mutant in the presence of the two inhibitory agents. The results show that in the presence of rapamycin, both T389 and T229 are phosphorylated, consistent with the inability of rapamycin to block kinase activation in this construct. More importantly, this result shows that the gain of function inhibitory complex, formed by FKBP12 and rapamycin, does not affect the kinases responsible for phosphorylating these two key residues, instead this inhibitory effect seems to be delivered to the amino terminus of the molecule through either stimulation of an inhibitor, or inhibition of a positive effector which interacts with this domain. Finally, if one analyses the effect of wortmannin on a double truncation mutant, again what one sees is loss of T389 phosphorylation with no effect on T229, despite the fact that T229 was identified as a potential target downstream of PI3 kinase activation. Taken together, a model begins to emerge for $p70^{s6k}$ activation (see Figure). In this model, one would predict that the amino terminus and the carboxy terminus of the molecule are interacting with one another, preventing kinases from phosphorylating specific residues involved in kinase activation. Mitogen stimulation would lead to the activation of some effector molecule which would operate on the amino termini of the molecule, disrupting the interaction of the carboxy and amino terminus and allowing kinases to phosphorylate specific residues. This would lead to the hyperphosphorylation of the S/TP sites within the autoinhibitory domain, as well as acute phosphorylation at T389. This would further drive the carboxy tail away from the amino terminus allowing access to T229 and, finally, generating a fully active form of the enzyme. However, what has to be noted is that T389, as well as the phosphorylation sites in the autoinhibitory domain, contribute to kinase activity and are not simply opening

# A Model for p70$^{s6k}$ Activation

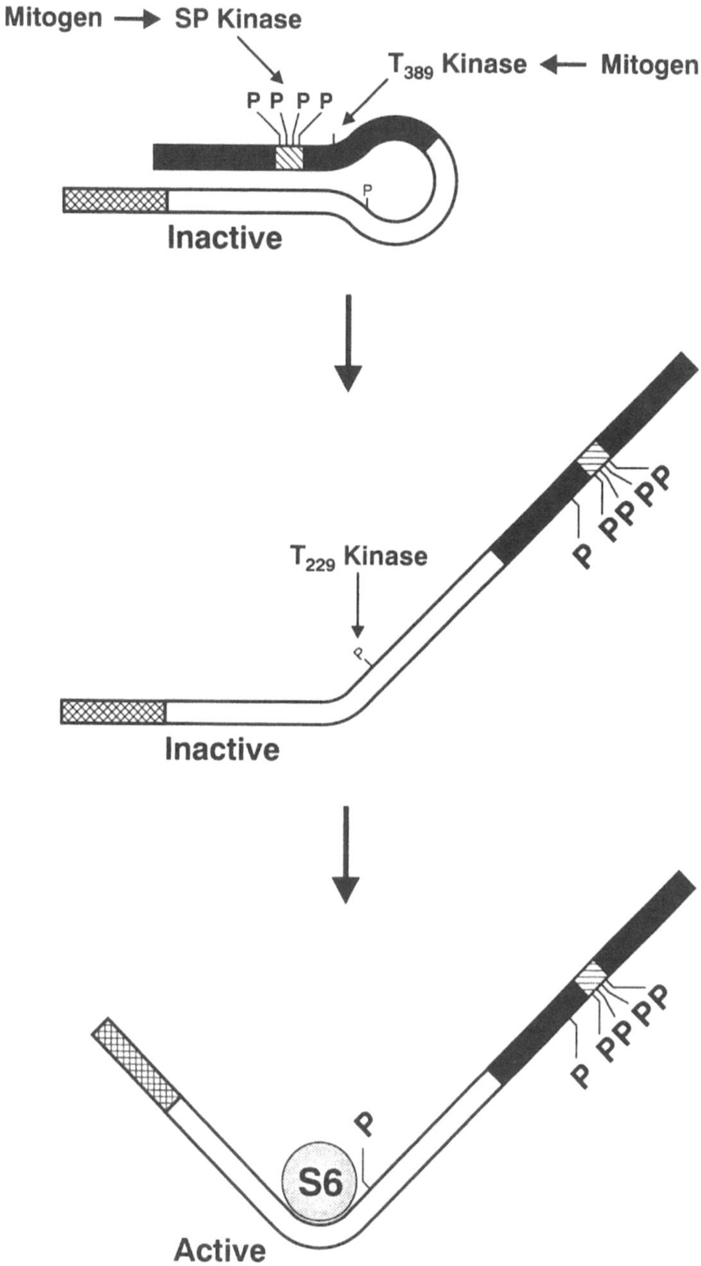

up the molecule such that T229 can become phosphorylated, as T229 phosphorylation is not sufficient to bring about kinase activity. Consistent with this finding, removal of the carboxy tail of the molecule does not raise kinase activity, however, if we substitute acidic residues in those positions, we do raise basal kinase activity and this carboxy terminal truncation has 75% of the activity of the wild type molecule. These results point to is the importance of the S/TP sites, as well as the T389 phosphorylation in bringing about kinase activation, and the importance of identifying the kinase which modulates these two sets of sites. Given the fact that the $p70^{s6k}$ does not lie downstream of the MAP kinase signalling pathway, it suggests that the $p70^{s6k}$ is regulated by a novel S/TP kinase as well as a kinase which phosphorylates residues of novel specificity which are flanked by large aromatic residues.

In the following lecture we will discuss how we have been able to take advantage of these mutations to elucidate about the role of the $p70^{s6k}$ in signalling downstream to translation.

Key References

1. Ferrari, S. and Thomas, G. 1994. S6 phosphorylation and the $p70^{s6k}/p85^{s6k.}$ Critical Reviews in Biochem. and Mol. Biol. **29**, 385-413.

2. Pearson, R. B., Dennis, P. B., Han, J-W., Williamson, N. A., Kozma, S. C., Wettenhall, R. E. H. and Thomas, G. 1995. The principal target of rapamycin-induced $p70^{s6k}$ inactivation is a novel phosphorylation site within a conserved hydrophobic domain. EMBO J. **21**, 5279-5287.

3. Weng, Q. P., Andrabi, K., Kozlowski, M. T., Grove, J. R. and Avruch, J. 1995. Multiple independent inputs are required for activation of the p70 S6 kinase. Mol. Cell. Biol. **15**, 2333-2340.

4. Chou, M. M. and Blenis, J. 1996. The 70 kDa S6 kinase complexes with and is activated by the Rho family G proteins cdc42 and rac1. Cell **85**, 573-583.

5. Dennis, P. B., Pullen, N., Kozma, S. C. and Thomas, G. 1996. The principal rapamycin-sensitive $p70^{s6k}$ phosphorylation sites, T-229 and T-389, are differentially regulated by rapamycin-insensitive kinase kinases. Mol. and Cell. Biol. **16**, 6242-6251.

# Rapamycin, FRAP and the Control of 5'TOP mRNA Translation

Stefano Fumagalli and George Thomas

Friedrich Miescher-Institut, Maulbeerstrasse 66, 4058 Basel

This manuscript represents the second of two lectures concerning the p70$^{s6k}$ and translation. It is presented as closely as possible to the actual talk.

In the previous lecture we explained that the goal of our laboratory has been to use the protein synthetic apparatus to understand the mechanism by which a growth factor acting on the outside of the cell is able to induce an essential metabolic response required for cell growth within the cell. The importance of protein synthesis in cell growth is made more evident by the observation that when a cell divides the two ensuing daughter cells receive the same amount of genetic information but they do not always receive the same amount of cytoplasm. As a consequence of this unequal cell division, when these cells are stimulated to proliferate, the larger cell progresses through the cell cycle much more rapidly than the smaller cell. If one analyses each stage of the cell cycle, the smaller cell has a prolonged $G_1$. Indeed, this cell must grow to a certain size before it can pass the G1/S boundary. In turn, to grow to this size the smaller cell has to activate its protein synthetic machinery. Interestingly, the major product that these cells are producing is more protein synthetic machinery which is required for subsequent cell divisions. Our focus on this model and the use of the protein synthetic machinery rapidly led us to the multiple phosphorylation of S6, which is an integral ribosomal protein of the small ribosomal 40S subunit, that in response to mitogenic signals and transformation becomes phosphorylated at 5 serines clustered at its carboxy terminus. The fact that S6 resides at the messenger RNA (mRNA)/transfer RNA (tRNA) binding site in the ribosome led to the speculation that its phosphorylation might somehow be involved in changing the affinity of the ribosome for mRNA. This prompted us to ask the question if, at the time when S6 is being phosphorylated we could observe changes in the pattern of protein synthesis

NATO ASI Series, Vol. H 102
Interacting Protein Domains
Their Role in Signal and Energy Transduction
Edited by Ludwig Heilmeyer

which were controlled at the translational level. To address this question, we pulse labelled either resting or serum stimulated cells with $^{35}$S-methionine and examined the pattern of protein synthesis by two-dimensional NEPHGE gel electrophoresis. The patterns were quite similar in both cases. However, there were some notable changes. We were able to identify in the order of 10 proteins where the amount of $^{35}$S methionine incorporated changed quite dramatically after stimulation. More importantly, we were able to show that increases in at least half of these proteins were controlled at the translational level. More recently we have been able to show that this selective regulation is conferred by an oligopyrimidine tract (5'TOP) at the transcriptional start site of the mRNAs encoding these proteins. The question we wanted to address at this point is whether S6 phosphorylation is involved in the upregulation of these transcripts and secondly what is the mechanism responsible for controlling this process.

Initially, we found that the synthesis of such proteins, for example Q49 which we later identifed as eukaryotic elongation factor eEF-1α (which is involved bringing the charged tRNA to the ribosome during the elongation phase of translation) was upregulated about 5-fold in terms of the amount of $^{35}$S-methionine incorporated into the protein following mitogenic-stimulation. If we corrected for the increase in protein synthesis, which increases in the order of 3- to 4-fold following mitogen stimulation, the increase in the amount of $^{35}$S- incorporated into the protein is 15- to 20-fold. When the experiment was performed in the presence of the transcription inhibitor actinomycin D we blocked increases in about half of the proteins, arguing that their increased levels were controlled at the transcription level. In contrast, such proteins as eEF-1α were still upregulated in the presence of actinomycin D arguing that their mRNAs were already present in resting cells but very inefficiently utilised by the translational apparatus until cells were stimulated to proliferate. To test this model we carried out northern blot analysis and found that the levels of eEF-1α remained constant for at least 3 hours after serum stimulation, whereas β-actin mRNA, whose transcription is known to be upregulated by mitogens, went up in the order of 5- to 10-fold during this same time period. There are two models used to explain the upregulation of such a transcript. The first model is referred to as Lodish, or competitive model of translation. According to this model, the translational apparatus would recognise the eEF-1α mRNA with very low efficiency in resting cells. Upon stimulation the increase of translational machinery available for protein synthesis would allow recruitment of eEF-1α mRNA on to polysomes. Messages which are efficiently recognised by the translational apparatus are already present on polysomes. Thus if one compares the amount of $^{35}$S-methionine incorporated into two such proteins one would see proportionally that the protein encoded for by the mRNA which was under-utilised in the quiescent cell would stand out much more strongly than the protein encoded for by the mRNA which was already being efficiently translated in the quiescent cell. The second model, which we refer to as a selective model of protein translation, suggests instead that alterations in the translational apparatus, like the phosphorylation of S6 or another translational factor would allow the selective upregulation of mRNAs such as eEF-1α. To distinguish between these two models, we analysed the distribution of eEF-1α, and as a control β-actin, in both resting and serum stimulated cells. The important thing to note, before we describe the results, is that in both quiescent or stimulated cells the mean polysome size remains constant. Thus, mitogen stimulation increases the amount of translational machinery but the average number of ribosomes bound to any single message remains constant during this time and most messages remain in the same place in the polysome profile. If the polysome distribution of actin is examined, for example, in quiescent Swiss 3T3 cells we found that actin mRNAs were present on polysomes of about 7 to 8 ribosomes per transcript. When cells were stimulated to proliferate, there was a large increase in the amount of actin message due to an increase in its transcription. However, the actin transcripts were always associated with polysomes of about 7 to 8 ribosomes per transcript, despite the fact that the number of ribosomes in polysomes had increased. In contrast to actin, eEF-1α mRNA in quiescent cells is associated with mRNP particles or monosomes/disomes being very inefficiently used by the translational apparatus. Upon mitogen stimulation, eEF-1α redistributes to polysomes of about 11 to 12 ribosomes per transcript, despite the fact that its size is very similar to that of actin mRNA. Indeed, we find that actin transcripts begin to accumulate in mRNPs at the same time that eEF-1α is being recruited into polysomes. In the competitive model of translation, the eEF-1α mRNA present in mRNP particles should have moved to monosomes/disomes, instead both populations move to very large polysomes, showing that this mRNA is selectively upregulated when we stimulate cells to proliferate. To ensure that eEF-1α transcripts on monosomes/disomes were actively being translated cells were treated with puromycin, and the eEF-1α mRNA transcripts were shown to run off of monosomes and disomes and accumulate in mRNP particles. eEF-1α falls into a family of mRNAs which have gained some notoriety in the last few years because they contain a very unusual structure at their 5' transcriptional start site. These mRNAs always begin with a cytosine whereas most mammalian mRNAs begin with an adenosine. The cytosine is followed by a stretch of polypyrimidines usually made up of uridines and can vary in length from 5 to 14 residues. This region of the transcript is known as the 5'TOP. They represent a very small family of transcripts, maybe only 100 to 200 different mRNAs, but they can make up to 20% of the total mRNA in the cell, with most of them encoding for components of the translational apparatus. Indeed, of the 27 mRNAs encoding mammalian ribosomal proteins in which primary extension has been carried out all begin with a cytosine and contain an oligopyrimidine tract. What we would like to know is whether S6 phosphorylation is involved in the selective translational upregulation of these mRNAs when cells are stimulated to proliferate. To address this question we have undertaken both a genetic and a biochemical approach. However, the immunosuppressant rapamycin allowed us to get some insight into this question. Rapamycin leads to the selective dephosphorylation and inactivation of the $p70^{s6k}$ when cells are stimulated to proliferate. In resting cells $p70^{s6k}$ activity is quite low. Following mitogen stimulation there is a large activation of the kinase that is totally blocked by pre-treatment of the cells with rapamycin. In quiescent cells, S6 is largely unphosphorylated with a small amount of phosphate in the first and second sites of phosphorylation. Upon serum stimulation S6 becomes highly phosphorylated. However, in the presence of rapamycin we totally ablate the increase in S6 phosphorylation. It is also known that post-treatment of cells with rapamycin leads to the rapid inactivation of the $p70^{s6k}$. If 3 hours after mitogen stimulation, cells are treated for 30 min, 60 min or 120 min with rapamycin, we observe a very rapid inactivation of $p70^{s6k}$. On the other hand S6 dephosphorylation proceeds much more slowly such that at 30 min the protein is still partially phosphorylated, and though it is largely dephosphorylated by 60 min, it only becomes totally dephosphorylated following 120 min treatment. The kinetics of this response are much slower than the kinase inactivation, strongly suggesting that S6 dephosphorylation is a consequence of the inactivation of the kinase. What then is the effect of rapamycin on global translation? When we examined the amount of $^{35}$S-methionine incorporated into protein we observed only slight inhibition in the presence of rapamycin, in the order of a 15% to 20%. Consistent with this finding, rapamycin had only a small effect on the recruitment of 80S ribosomes into polysomes. Next we examined the effect of rapamycin on 5'TOP mRNAs; to address this question we stimulated cells in the presence or absence of rapamycin or we added rapamycin after post-mitogen stimulation. Actin message was used as a control in these experiments and as described above, actin mRNA is upregulated transcriptionally but remains in the same place in the polysome profile following mitogen stimulation, whereas eEF-1$\alpha$ mRNA redistributed to very large polysomes. In the presence of rapamycin, the shift of eEF-1$\alpha$ mRNA to polysomes was largely blocked, while there was no effect on actin. Indeed, actin mRNA shifts to slightly larger polysomes. Approximately 30% of the eEF-1$\alpha$ mRNA was still upregulated in the presence of rapamycin, arguing that the signalling pathway leading to the translational upregulation of 5'TOP mRNAs is redundant and there is a second rapamycin insensitive mechanism which can lead to the upregulation of these mRNAs. If the experiment was performed such that rapamycin was added 3 hours post-mitogen stimulation, actin mRNA remained associated with polysomes, whereas eEF-1$\alpha$ redistributed from large polysomes back into mRNP particles and monosomes/disomes. Other mRNAs which contain a 5'TOP were qualitatively effected in a very similar way, whereas mRNAs which lacked the track were unaffected by rapamycin. This would suggest that the track is required for rapamycin to elicit this inhibitory effect. To address this question we took advantage of NIH3T3 cell lines stability expressing chimeric mRNAs where either the 5'UTR of a ribosomal protein in its wild type configuration or where the 5'TOP had been mutated to include five purines, were fused to the growth hormone messenger RNA. In quiescent cells the wild type chimeric mRNA was present in mRNPs and monosomes/disomes, very much as described for eEF-1$\alpha$. In contrast, the mutant chimeric transcript has largely present polysomes, arguing that the polymirimidine track acted as a suppressor in quiescent cells. Upon serum stimulation the wild type transcript, as well as the mutant transcript, moved to polysomes of about 7-8 ribosomes per transcript. In contrast, in the presence of rapamycin, only the wild type chimeric transcript moved back to mRNP particles and monosomes/ disomes, while there was no effect on the mutant transcript. This data demonstrates that rapamycin requires the polypyrimidine track to elicit its inhibitory effect. What is the role of $p70^{s6k}$ in the upregulation of these transcripts? To address this question we took advantage of a rapamycin resistant mutant $p70^{s6k}D_3E\text{-}E_{389}$, in which T389 and the 4 phosphorylation sites in the autoinhibitory domain had been changed to acidic residues. The mutant or the wild type $p70^{s6k}$ constructs were transiently transfected into 293 cells. After starvation, the cells were stimulated for 3 hours and then rapamycin was added for one hour. We then examined the distribution of the eEF-1$\alpha$

on polysomes and saw that in the case of the p70$^{s6k}$D$_3$E-E$_{389}$ mutant more eEF-1α mRNA was present on polysomes in comparison to cells overexpressing the wild type form of the kinase. This effect is probably due to the higher basal kinase activity of the p70$^{s6k}$D$_3$E-E$_{389}$ mutant. Rapamycin treatment caused eEF-1α mRNA transcripts to run off polysomes in cells expressing the wild type kinase, whereas in cells expressing the mutant construct the upregulation of the same transcript was completely protected. Thus, the inhibitory effects of the rapamycin on 5'TOP mRNAs are elicited through the p70$^{s6k}$. How does p70$^{s6k}$ exert this effect on 5'TOP mRNAs? We predict that it is through the phosphorylation of S6, but at this point we have no direct evidence to support this model (see Figure). We can speculate that S6 itself when phosphorylated could interact directly with the 5'TOP (i) or with proteins which bind to the tract (ii). Alternatively its phosphorylation might alter the mRNA binding site within the ribosome (iii).

While the studies above were being carried out, a second translational component was identified which becomes heavily phosphorylated when cells are stimulated to proliferate. The protein has been referred to as 4E-BP1 for the 4E binding protein 1. Its function is to act as a repressor of the initiation factor 4E which is responsible for recognising and binding the methyl-$^7$G of the mRNA cap structure and mediating the binding of 40S ribosomes to the initiation complex. Phosphorylation of 4E-BP1 leads to its release from 4E allowing 4E then to interact with the initiation complex. *In vitro* 4E-BP1 is a very good substrate for the MAP kinase, suggesting that the MAP kinase pathway was responsible for increased 4E-BP1 phosphorylation and the upregulation of translation. However, in all these initial studies, insulin was used as the mitogen. We know that insulin addition to mouse Swiss 3T3 cells and human 293 cells does not lead to MAP kinase activation. This suggests that 4E-BP1 does not have as much general importance as had been predicted, or that the MAP kinase is not the only kinase that leads to its phosphorylation. To distinguish between these possibilities we examined the level of p70$^{s6k}$ activation and 4E-BP1 phosphorylation in both human 293 and Swiss 3T3 cells. The results showed that upon addition of insulin, p70$^{s6k}$ was activated, as was previously shown, but also 4E-BP1 phosphorylation increased arguing that MAP kinase is not necessary for inducing the increase in 4E-BP1 phosphorylation. When we treated either cell type with rapamycin we blocked the increase of S6 phosphorylation and also the increase in 4E-BP1 phosphorylation. To address the question of whether

# S6 Phosphorylation and 5'TOP mRNA Translation

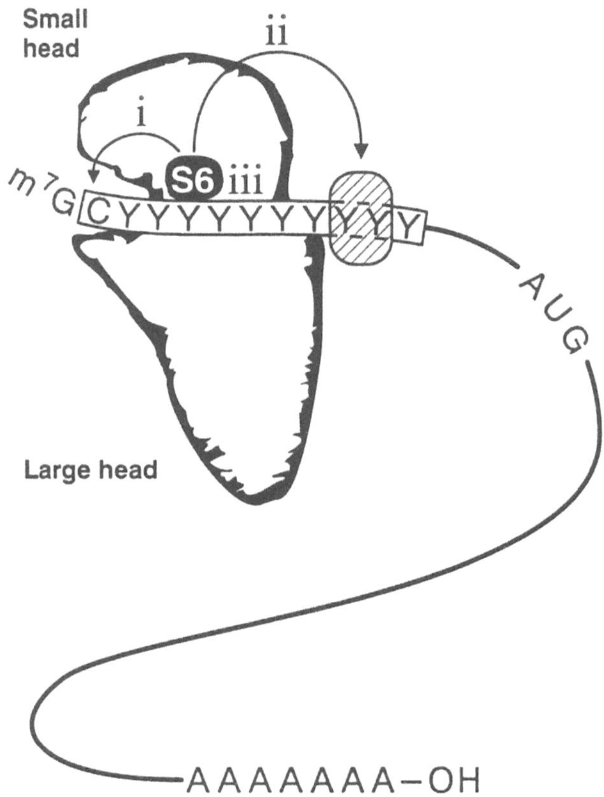

MAP kinase was sufficient to bring about 4E-BP1 phosphorylation we employed TPA, which through protein kinase C leads to a 20-fold activation of both MAP kinase and activates $p70^{s6k}$ to the same extent as insulin. Such treatment also led to an equivalent increase in 4E-BP1 phosphorylation. If cells were treated with TPA together with rapamycin there was no effect on MAP kinase activation but $p70^{s6k}$ activation and 4EBP1 phosphorylation were abolished. These experiments argue that MAP kinase is neither necessary nor sufficient for turning on this pathway and suggest instead that 4E-BP1 lies somewhere downstream of the FRAP/$p70^{s6k}$ signalling pathway. To test this possibility we took advantage of cells expressing the double point mutant of the human PDGF receptor, Y740F/Y751F that when stimulated to proliferate with PDGF fail to activate the $p70^{s6k}$, but not MAP kinase. In these cells 4E-BP1 failed to become phosphorylated arguing again, that $p70^{s6k}$ and 4E-BP1 lie on the same signalling pathway. The question then is, where does 4E-BP1 lie in respect of $p70^{s6k}$ in this signalling pathway? To address this question we have taken advantage of the rapamycin resistant form of the kinase and found that it does not protect 4E-BP1 from dephosphorylation induced by rapamycin, arguing that 4E-BP1 phosphorylation bifurcates upstream of $p70^{s6k}$.

Key References


1. Jefferies, H. B. J., Reinhard, C., Kozma, S. C. and Thomas, G. 1994. Rapamycin selectively represses translation of the "polypyrimidine tract" mRNA family. Proc. Natl. Acad. Sci. USA, **91**, 4441-4445.

2. Beretta, L., Gingras, A-C., Svitkin, Y. V., Hall, M. N. and Sonenberg, N. 1996. Rapamycin blocks the phosphorylation of 4E-BP1 and inhibits cap-dependent initiation of translation. EMBO J. **15**, 658-664.

3. von Manteuffel, S. R., Gingras, A-C., Ming, X-F., Sonenberg, N. and Thomas, G. 1996. 4E-BP1 phosphorylation is mediated by the FRAP-$p70^{s6k}$ pathway and is independent of mitogen-activated protein kinase. Proc. Natl. Acad. Sci. USA. **93**, 4076-4080.

4. Brown, E. J. and Schreiber, S. L. 1996. A Signalling pathway to translational control. Cell **86**, 517-520.

5. Jefferies, H. B. J. and Thomas, G. 1996. Ribosomal protein S6 phosphorylation and signal transduction. In: Hershey, J. W. B., Mathews, M. B., Sonenberg, N. (eds) Cold Spring Harbor, 389-409.

# Structure, Regulation and Targeting of Protein Phosphatase 2A

*Nataša Andjelković [1], Stanislaw Zolnierowicz [1,2], Regina E. Mayer-Jaekel [1,3] and Brian A. Hemmings [1]*

[1] Friedrich Miescher-Institut, P.O. Box 2543, CH-4002 Basel, Switzerland;
[2] Current Address: Department of Biochemistry, Faculty of Biotechnology, Medical University of Gdansk, Debinki 1, 80-211 Gdansk, Poland;
[3] Current Address: Arbeitsgruppe Zellbiologie-Tumorbiologie, Universität Konstanz, Postfach 5560, 78434 Konstanz, Germany

## *Introduction*

Reversible phosphorylation of proteins on serine, threonine and tyrosine residues is widely accepted as a principal mechanism by which eukaryotic cells respond to extracellular signals. This process is mediated by two classes of enzymes, protein kinases and protein phosphatases. The protein kinases have been extensively studied at both the biochemical and molecular level, which has led to the identification of almost 400 distinct protein kinases involved in complex signal transduction pathways (Hardie & Hanks, 1995).

Protein phosphatases (PPs) found in eukaryotic cells are structurally and functionally diverse enzymes that can be classified by their substrate specificities into phosphoserine/threonine (PSTPs) and phosphotyrosine phosphatases (PTPs). Serine/threonine specific phosphatases belong to the two distinct gene families, PPP and PPM (reviewed in Wera & Hemmings, 1995; Barford, 1996; Zolnierowicz & Hemmings, 1996). The PPP family encompasses the three major eukaryotic subfamilies of Ser/Thr phosphatases termed PP1, PP2A and PP2B, as well as some less abundant but structurally homologous enzymes, such as PPQ, PPV, PPX, PPY, PPZ, SIT4 and PP5.

NATO ASI Series, Vol. H 102
Interacting Protein Domains
Their Role in Signal and Energy Transduction
Edited by Ludwig Heilmeyer

The PPM family members include PP2C and a related mytochondrial pyruvate dehydrogenase phosphatase. Phosphotyrosine-specific and dual specific protein phosphatases represent another large group of enzymes with the catalytic and structural properties different to those of the phosphoserine/threonine specific phosphatases (reviewed in Denu et al, 1996; Tonks & Neel, 1996). The crystal structures of the catalytic subunits representing both serine/threonine and tyrosine/dual specific phosphatases have been solved, contributing largely to our knowledge of the catalytic mechanisms and substrate binding properties of these enzymes (reviewed by Barford, 1996; Fauman & Saper, 1996).

The major question that now emerges is: how is the function of protein phosphatases precisely regulated? Some answers to this question can be provided by the example of protein phosphatase 2A (PP2A), one of the most complex and best characterized multisubunit protein phosphatases found in eukaryotic cells. PP2A is a serine/threonine specific protein phosphatase that performs a multitude of functions in eukaryotic cells. It is essential for the control and modulation of many cellular processes, including metabolism, signal transduction, transcription, protein synthesis, cell cycle progression, cell transformation, growth and development (reviewed in Mumby & Walter, 1993; Mayer-Jaekel & Hemmings, 1994). PP2A can also act as a tyrosine phosphatase, especially after stimulation with phosphotyrosine phosphatase activator protein (PTPA), which suggests that PP2A could also be a dual specificity enzyme (reviewed in Van Hoof et al, 1994). While enzymatic properties of PP2A have been discussed elsewhere (reviewed in Wera & Hemmings, 1995), this review will focus on some of the recent findings that contribute to our understanding of the complex structure and regulation of this universally important enzyme.

***Structure of PP2A holoenzymes***

PP2A encompasses a family of trimeric holoenzymes which universally consist of a 'core dimer' composed of a 36-kDa catalytic subunit (PP2Ac/C) bound to the constant regulatory subunit of 65 kDa (PR65/A). The core dimer further complexes with a third or variable type of regulatory subunit to form the physiologically relevant PP2A heterotrimer. Several different types of PP2A variable subunits have been identified from purified heterotrimers where detailed analysis of individual subunits resulted in a comprehensive understanding of the complex molecular architecture of PP2A (reviewed in Kamibayashi & Mumby, 1995). The three gene families that encode the variable regulatory subunits of PP2A are termed PR55/B, PR61/B′ and PR72/B″ (Figure 1). The possible complexity of PP2A holoenzymes is further emphasized at the genetical level, where all mammalian PP2A subunits exist in multiple isoforms encoded by different genes (Figure 1).

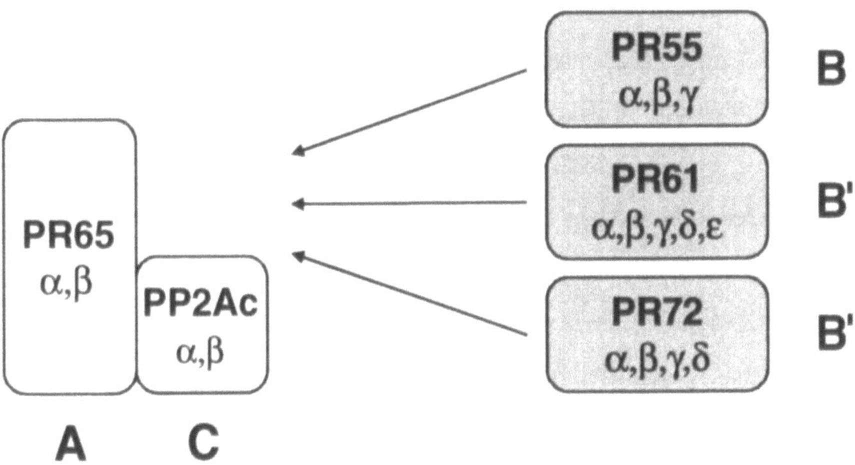


**Figure 1.** Three gene families encompassing more than 10 mammalian genes encode variable regulatory subunits of PP2A. Greek letters indicate the name of each gene for the respective subunit.

The current number of genes encoding the variable subunits in mammals is 12, some of which are alternatively spliced, making the number of PP2A holoenzymes that could theoretically be assembled beyond all previous expectations.

***Regulation of PP2A function by variable subunits***

Given the numerous cellular processes involving PP2A, we sought to identify novel PP2A holoenzymes by employing an affinity based purification strategy (Figure 2). The key feature of this approach is the characterization of PP2A-containing fractions with a large panel of subunit-specific antibodies (described in Zolnierowicz et al, 1995). This approach resulted in the isolation of two novel PP2A holoenzymes (Zolnierowicz et al, 1996) and additional PP2A-associated proteins (Andjelković et al, 1996).

The two purified PP2A heterotrimers contained variable regulatory subunits with apparent molecular masses of 61 kDa and 56 kDa, respectively (Zolnierowicz et al, 1996). Both holoenzymes displayed low basal phosphorylase phosphatase activity which could be further stimulated by protamine. The elution profile from DEAE-Sepharose corresponded to that of the previously described $PP2A_0$ holoenzyme (Tung et al, 1985). Molecular cloning revealed that both the 61-kDa and 56-kDa proteins belong to the same subunit family (PR61/B′) that comprises at least three genes, one of which gives rise to several splicing variants. Comparisons of these sequences with the available databases identified one more human gene and predicted another one based on a rabbit cDNA-derived sequence, thus bringing the number of genes encoding PR61/B′ family members to five (McCright et al, 1995; Csortos et al, 1995; Tanabe et al, 1996; Tehrani et al, 1996; Zolnierowicz et al, 1996).

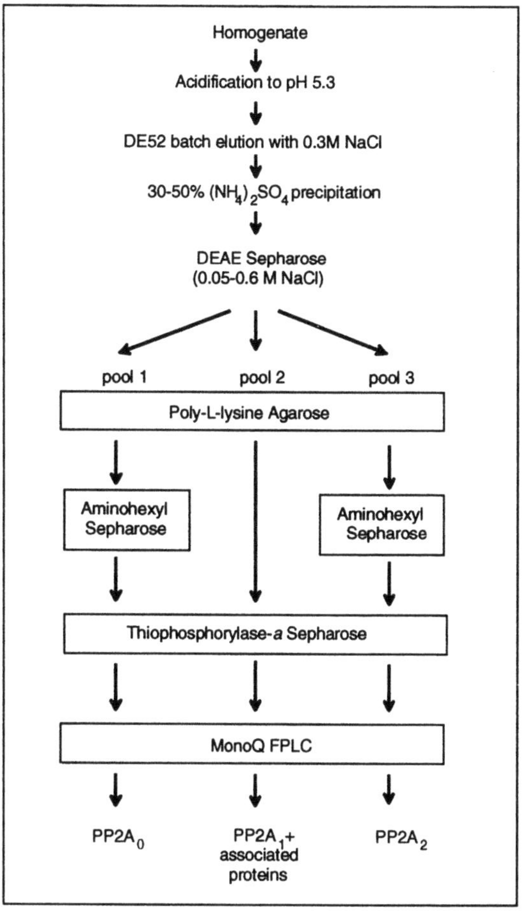


**Figure 2.** Schematic outline of the purification protocol used to isolate different PP2A holoenzymes from rabbit skeletal muscle.

Several PP2A regulatory subunits belonging to the PR61/B′ family contain consensus sequences for nuclear localization and may therefore target PP2A to nuclear substrates. Recent report by McCright et al. (1996) demonstrated that α, β and ε isoforms of PR61/B′ are localized to the cytoplasm, while γ1, γ3 and δ isoforms are concentrated in the nucleus. The same work further shows that almost all of the PR61/B′ family members are phosphoproteins. Significantly, the α and β isoforms of PR61/B′ associate with cyclin G in a p53-regulated manner (Okamoto et al, 1996).

The variable regulatory subunits appear to play the crucial role in determining the function of various PP2A holoenzymes *in vivo*, by directing subcellular localization, activity and substrate specificity. Our laboratory has shown that the disruption of the gene encoding the 55-kDa regulatory subunit of PP2A in *Drosophila* leads to a lethal mitotic phenotype described as $aar^1$ (*a*bnormal *a*naphase *r*esolution). This PP2A subunit is required for progression through mitosis and the most severe mitotic abnormalities are observed in proliferating cells of the larval brains (Mayer-Jaekel et al, 1993). Different alleles of the PR55 gene, termed *twins*, were identified due to cell fate abnormalities leading to duplications of imaginal disk pattern and sensory bristles (Uemura et al, 1993; Shiomi et al, 1994). We have subsequently analyzed the protein levels and localization of PP2A subunits, as well as the changes in PP2A activity towards various substrates in different mutant larval brains compared to the wild-type flies. The results obtained suggest that the mutants with decreased levels of PR55 protein poorly dephosphorylate $p34^{cdc2}$ phosphorylated substrates. Since the levels of the catalytic and PR65 subunits are normal, and no cell cycle specific changes in PR55 localization can be observed, this implies that that the defects in progression through anaphase are due to the reduced amount of PR55-containing holoenzyme and lack of dephosphorylation of certain specific substrates (Mayer-Jaekel et al, 1994). *Saccharomyces cerevisiae cdc55* mutants are also deficient in the 55-kDa regulatory subunit and, while still viable, display abnormal cytokinesis when grown at low temperatures (Healy et al, 1991). Overexpression of the PR65 regulatory subunit in mammalian fibroblasts also affects cytokinesis (Wera et al, 1995), which all points to an essential role that PP2A plays during mitosis.

***Regulation and targeting of PP2A by interacting proteins***

Recently, further analysis of purified preparations of PP2A holoenzymes has revealed additional proteins that interact with PP2A. These proteins appear to be much less abundant than the previously characterized variable subunits. By protein microsequencing we identified one of these proteins as eRF1 (*e*ukaryotic *r*elease *f*actor 1), a protein that functions in translational termination as a polypeptide chain release factor (Frolova et al, 1994; reviewed in Nakamura et al, 1996). By a number of criteria we have shown that eRF1 interacts with PP2Ac (Andjelković et al, 1996). In the yeast two-hybrid system human eRF1 interacts with PP2Ac, but not with the PR65 or PR55 subunits. By deletion analysis the binding domains are found to be located within the 50 N-terminal amino acids of PP2Ac, and between amino acid residues 338 and 381 in the C-terminal part of human eRF1. This association also occurs *in vivo*,

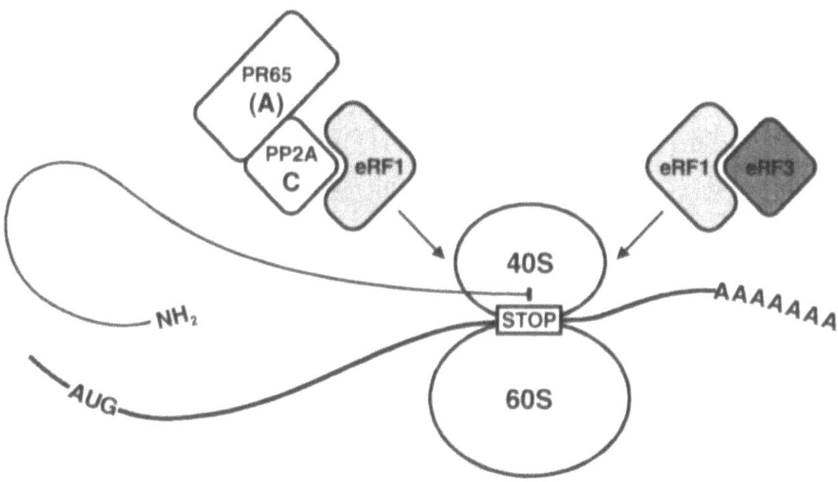


**Figure 3.** eRF1 functions in translational termination and targets PP2A to polysomes.

since PP2A can be coimmunoprecipitated with eRF1 from mammalian cells. There is a significant increase in the amount of dimeric PP2A ($PP2A_2$) associated with the polysomes when eRF1 is transiently expressed in COS1 cells, and eRF1-immunoprecipitated from those fractions contains associated $PP2A_2$. Since there are no dramatic effects of PP2A on the polypeptide chain release activity of eRF1 (or vice versa), we postulate that eRF1 also functions to recruit PP2A into polysomes, thus bringing the phosphatase into contact with putative targets among the components of the translational apparatus (Figure 3).

Besides eRF1, there is a growing list of different cellular proteins that are reported to interact with PP2A to mediate specific cellular responses. Two proteins that were initially purified as potent and specific heat-stable inhibitors of PP2A and designated $I_1^{PP2A}$ and $I_2^{PP2A}$ (Lee et al, 1995) were subsequently shown to correspond to PHAP-I (putative histocompatibility leukocyte antigens class II-associated protein) and myeloid leukemia-associated protein SET, respectively (Lee et al, 1996a; Lee et al, 1996b). Other proteins that interact with PP2A include the retinoblastoma protein-related p107-binding protein p59, which associates with the dimeric form of PP2A. This protein shows ~60% sequence identity to the PR72/130 (B″) variable regulatory subunits of PP2A and therefore can be considered a novel isoform of this subunit family. The physiological consequence of p59 association with the dimeric form of PP2A is not fully delineated, but the possible role of this protein could be to direct PP2A to specifically dephosphorylate p107 and thus regulate its growth-inhibitory activity (M. Voorhoeve and R. Bernards, personal communication). Furthermore, *Saccharomyces cerevisiae* protein Tap42 associates with the catalytic subunit of PP2A without apparently requiring the regulatory subunits (Dicomo & Arndt, 1996). The Tap42/PP2A complex formation is regulated by nutrient growth signals and the rapamycin sensitive Tor signalling pathway.

## *Concluding remarks*

A fundamental question in signal transduction is how protein kinases and protein phosphatases are coordinated to phosphorylate/dephosphorylate the correct target proteins rapidly and preferentially at the correct time and cellular location. The targeting hypothesis of Hubbard and Cohen (1993) postulates that 'targeting subunits' direct protein kinases and protein phosphatases to specific subcellular locations where they act on their targets. A substantial amount of evidence exists to prove that several protein kinases and phosphatases with broad substrate specificity are recruited to their targets by specific interacting proteins (reviewed in Coghlan et al, 1995; Faux & Scott 1996a, b). These interactions are crucial for precise functional regulation by favoring accessibility to the specific substrates. With regard to PP2A, this type of regulation was until recently thought to be mediated exclusively through its constituent regulatory subunits. The isolation and characterization of the novel PP2A interacting proteins described here provide a more sophisticated model for understanding how PP2A is able to regulate a multitude of cellular processes.

## *Acknowledgements*

Parts of the work presented here were supported by fellowships from Howard Hughes Medical Institutes (to S.Z. and B.A.H.) and Krebsliga beider Basel (to N.A.). We are grateful to M. Voorhoeve and Dr. R. Bernards (The Netherlands Cancer Institute, Amsterdam) for communicating their unpublished results. We thank Dr. Timothy Myles for critical reading of this manuscript.

### *References*

Andjelković N, Zolnierowicz S, Van Hoof C, Goris J, Hemmings BA (1996) The catalytic subunit of protein phosphatase 2A associates with the translation termination factor eRF1. EMBO J, in press

Barford D (1996) Molecular mechanisms of the protein serine/threonine phosphatases. Trends Biochem Sci 21:407-412

Coghlan VM, Lester LB, Scott JD (1995) A targeting model for reversible phosphorylation. Adv Prot Phosphatases 9:51-61

Csortos C, Zolnierowicz S, Bako E, Durbin SD, DePaoli-Roach AA (1996) High complexity in the expression of the B′ subunit of protein phosphatase 2A(0) - evidence for the existence of at least seven novel isoforms. J Biol Chem 271:2578-2588

Denu JM, Stuckey JA, Saper MA, Dixon, JE (1996) Form and function in protein dephosphorylation. Cell 87:361-364

Dicomo CJ, Arndt K (1996) Nutrients, via the Tor proteins, stimulate the association of Tap42 with type 2A phosphatases. Genes Dev 10:1904-1916

Fauman EB, Saper MA (1996) Structure and function of the protein tyrosine phosphatases. Trends Biochem Sci 21:413-417

Faux MC, Scott JD (1996a) Molecular glue: kinase anchoring and scaffold proteins. Cell 85:9-12

Faux MC, Scott JD (1996b) More on target with protein phosphorylation: conferring specificity by location. Trends Biochem Sci 21:312-315

Frolova L, Le Goff X, Rasmussen HH, Cheperegin S, Drugeon G, Kress M, Arman I, Haenni A-L, Celis JE, Philippe M, Justesen J, Kisselev L (1994) A highly conserved eukaryotic protein family possessing properties of polypeptide chain release factor. Nature 372:701-703

Hardie G, Hanks S, eds. (1995) The protein kinase facts book. (London: Academic Press)

Healy AM, Zolnierowicz S, Stapleton AE, Goebl M, DePaoli-Roach AA, Pringle JR (1991) CDC55, a *Saccharomyces cerevisiae* gene involved in cellular morphogenesis: identification, characterization, and homology to the B subunit of mammalian type 2A protein phosphatase. Mol Cell Biol 11:5767-5780

Hubbard M, Cohen P (1993) On target with a new mechanism for the regulation of protein phosphorylation. Trends Biochem Sci 81:172-177

Kamibayashi C, Mumby MC (1995) Control of protein phosphatase 2A by multiple families of regulatory subunits. Adv Prot Phosphatases 9:195-210

Lee M, Guo H, Damuni Z (1995) Purification and characterization of two potent heat-stable protein inhibitors of protein phosphatase 2A from bovine kidney. Biochemistry 34:1988-1996

Lee M, Makkinje A, Damuni Z (1996a) Molecular identification of $I_1^{PP2A}$, a novel potent heat-stable inhibitor of protein phosphatase 2A. Biochemistry 35:6998-7002

Lee M, Makkinje A, Damuni Z (1996b) The myeloid leukemia associated protein SET is a potent inhibitor of protein phosphatase 2A. J Biol Chem 271:11059-11062

Mayer-Jaekel RE, Hemmings BA (1994) Protein phosphatase 2A - a 'ménage à trois'. Trends Cell Biol 4:287-291

Mayer-Jekel RE, Ohkura H, Ferrigno P, Andjelković N, Shiomi K, Uemura T, Glover DM, Hemmings BA (1994) *Drosophila* mutants in the 55-kDa regulatory subunit of protein phosphatase 2A show strongly reduced ability to dephosphorylate substrates of $p34^{cdc2}$. J Cell Sci 107:2609-2616

Mayer-Jaekel RE, Ohkura H, Gomes R, Sunkel CE, Baumgartner S, Hemmings BA, Glover, DM (1993) The 55 kDa regulatory subunit of protein phosphatase 2A is required for anaphase. Cell 72:621-633

McCright B, Rivers AM, Audlin S, Virshup D (1996) The B56 family of protein phosphatase 2A (PP2A) regulatory subunits encodes differentiation-induced phosphoproteins that target PP2A to both nucleus and cytoplasm. J Biol Chem 271:22081-22089

McCright B, Virshup DM (1995) Identification of a new family of protein phosphatase 2A regulatory subunits. J Biol Chem 270:26123-26128

Mumby MC, Walter G (1993) Protein serine/threonine phosphatases: Structure, regulation and functions in cell growth. Physiol Rev 73:673-699

Nakamura Y, Ito K, Isaksson LA (1996) Emerging understanding of translation termination. Cell 87:147-150

Okamoto K, Kamibayashi C, Serrano M, Prives C, Mumby MC, Beach D (1996) p53-dependent association between cyclin G and the B′ subunit of protein phosphatase 2A. Mol Cel Biol 16:6593-6602

Shiomi K, Takeichi M, Nishida Y, Nishi Y, Uemura T (1994) Alternative cell fate choice induced by low-level expression of a regulator of protein phosphatase 2A in the *Drosophila* peripheral nervous system. Development 120:1591-1599

Tanabe O, Nagase T, Murakami T, Nozaki H, Usui H, Nishito Y, Hayashi H, Kagamiyama H, Takeda M (1996) Molecular cloning of a 74-kDa regulatory subunit (B″ or delta) of human protein phosphatase 2A. FEBS Lett 379:107-111

Tehrani MA, Mumby MC, Kamibayashi C (1996) Identification of a novel protein phosphatase 2A regulatory subunit highly expressed in muscle. J Biol Chem 271:5164-5170

Tonks NK, Neel BE (1996) From form to function: signaling by protein tyrosine phosphatases. Cell 87:365-368

Tung HYL, Alemany S, Cohen P (1985) The protein phosphatases involved in cellular regulation. 2. Purification, subunit structure and properties of protein phosphatases - $2A_0$, $2A_1$, and $2A_2$ from rabbit skeletal muscle. Eur J Biochem 148:253-263

Uemura T, Shiomi K, Togashi S, Takeichi M (1993) Mutation of twins encoding a regulator of protein phosphatase 2A leads to pattern duplication in *Drosophila* imaginal disks. Genes Dev 7:429-440

Van Hoof C, Cayla X, Bosch M, Merlevede W, Goris J (1994) PTPA adjusts the phosphotyrosyl phosphatase activity of PP2A. Adv Prot Phosphatases 8:301-330

Wera S, Fernandez A, Lamb NJC, Turowski P, Hemmings-Mieszczak M, Mayer-Jaekel RE, Hemmings BA (1995) Deregulation of translational control of the 65-kDa regulatory subunit of protein phosphatase 2A leads to multinucleated cells. J Biol Chem 270:21374-21381

Wera S, Hemmings BA (1995) Serine/threonine protein phosphatases. Biochem J 311:17-29

Zolnierowicz S, Hemmings BA (1996) Protein phosphatases on the piste. Trends Cell Biol 6:359-362

Zolnierowicz S, Van Hoof C, Andjelković N, Cron P, Stevens I, Merlevede W, Goris J, Hemmings BA (1996) The variable subunit associated with protein phosphatase $2A_0$ defines a novel multimember family of regulatory subunits. Biochem J 317:187-194

Zolnierowicz S, Mayer-Jaekel RE, Hemmings BA (1995) Protein phosphatases 2A - purification, subunit structure, molecular cloning, expression, and immunological analysis. Neuroprotocols 6:11-19

# THE CONSTITUTIVE ACTIVATION OF MET, RON AND SEA GENES INDUCES DIFFERENT BIOLOGICAL RESPONSES

Massimo Mattia Santoro and Giovanni Gaudino

Department of Science & Advanced Technologies, University of Torino, Alessandria 15100, Italy

## 1 Introduction

The MET proto-oncogene, encoding the Hepatocyte Growth Factor receptor, is the prototype of a gene family encoding structurally homologous heterodimeric tyrosine kinase receptors, including human RON (19) and avian SEA (8). It can be converted into an oncogene by rearrangement of the kinase domain with a N-terminal unrelated sequence designated TPR (3). The kinase activity of the encoded hybrid protein (Tpr-Met) is deregulated, since two leucine-zipper motifs present in the Tpr moiety promote its constitutive dimerization (18). This conformation mimics receptor activation following ligand binding.

The product of RON has been identified as the receptor for MSP (Macrophage Stimulating Protein; 6, 24). MSP actually exerts a wide spectrum of biological activities, mainly on epithelial, neuro-endocrine and hemopoietic cells (7, 9). Furthermore, naturally-occurring transforming counterparts of RON have not been identified. On the contrary, both the human homologue of avian Sea and its ligand are still elusive. The oncogenic form of SEA (Sarcoma, Erythroblastosis, Anaemia) has been identified as the transforming component of the Avian Erythroblastosis S13 retrovirus, by fusion of extracellular and transmembrane regions of the viral envelope protein with the SEA tyrosine kinase (21).

It has been demonstrated that ligand-stimulation of Met, Ron and Sea induces cell growth, “scattering”, and tubulogenesis (13). These pleiotropic effects are elicited by receptor activation and phosphorylation of two critical carboxy-terminal tyrosine residues embedded in the sequence, which acts as docking site for multiple SH2-containing cytoplasmic effectors. The multifunctional docking site responsible for Met signalling is conserved in the evolutionary related receptors (9, 16).

## 2 Results and Discussion

### 2.1 Constitutive activation of Met, Ron and Sea tyrosine kinases

We used a recombinant approach to obtain constitutively active Ron and Sea tyrosine kinase designed according to the structure of Tpr-Met (Fig.1A). Tpr-Ron and Tpr-Sea cDNAs were stable expressed in NIH 3T3 fibroblasts. Level of expression and tyrosine phosphorylation of the recombinant chimaeras were analysed in stable transfectants by immunoprecipitation and Western blotting. Their enzymatic activity was examined by *in vitro* autokinase assays. All chimaeras were found to be expressed

NATO ASI Series, Vol. H 102
Interacting Protein Domains
Their Role in Signal and Energy Transduction
Edited by Ludwig Heilmeyer

at comparable levels and highly phosphorylated on tyrosine both *in vivo* and *in vitro* (Table I).

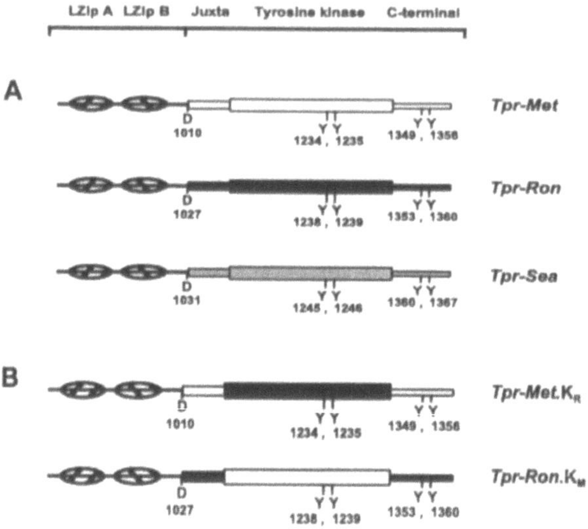


Fig. 1. (A) Schematic representation of the chimaeric proteins containing Tpr and the intracellular domains of Met, Ron and Sea. The leucine zipper motifs (LZipA and LZipB) and the receptor intracellular subdomains are indicated at the top. (B) Schematic representations of the "swapped" chimaeras in which the tyrosine kinase subdomains are exchanged between Tpr-Met and Tpr-Ron. Acronyms on the right identifies the different constructs.

**2.2 Transformation is linked to the functional features of the kinase domains**

Given the strong correlation between the transforming ability and the tyrosine kinase activity of the *MET* oncogene product (3), we assayed the transforming ability of the constitutively activated Met, Ron and Sea kinases. Tpr-Met, Tpr-Ron and Tpr-Sea were tested in a focus forming assay following transfection in NIH3T3 fibroblasts. Cells transfected with Tpr-Sea yielded a two fold higher frequency of foci compared with Tpr-Met. The reason of this behaviour can be explained by the presence of a duplicated Grb-2 binding site in its C-terminal multifunctional docking site ($Y^{1360}$VNL-$X_3$ -$Y^{1367}$VNL). According to this hypothesis, it has been demonstrated that duplication of the Grb-2 binding site in Tpr-Met causes signalling reinforcement along the Ras pathway, and enhances transformation (17). Unexpectedly, Tpr-Ron was completely unable to induce foci of transformation (Table I).

To understand the differences observed in cell transformation we "swapped" the Ron tyrosine kinase subdomain with the corresponding Met region (Fig. 1B). The recombinant proteins were all expressed with the same efficiency, equally capable of autophosphorylation on tyrosine *in vivo*, and displayed comparable kinase activities *in vitro*. The transforming potential of "swapped" Tpr-chimaeras was analysed in focus forming assays and compared with that of Tpr-Met and Tpr-Ron. All Tpr-chimaeras bearing the Ron kinase domain (Tpr-Ron and Tpr-Met.$K_R$) did not induce foci of transformation. Conversely, the chimaeras containing the Met kinase (Tpr-Met and Tpr-Ron.$K_M$) were transforming.

We ruled out that Tpr-Ron signalling involved specific effectors different from those recruited by Tpr-Met, on the basis of the experiments performed with the C-terminal "swapped" chimaeras. Surprisingly, the Ron C-terminal tail was found to be even better than the Met tail in inducing cell transformation. The Ron tail includes the conserved multifunctional docking site ($Y^{1353}$VQL-$X_3$-$Y^{1360}$MNL) that *in vitro* and *in vivo* binds the same set of SH2-containing signal transducers bound by Met ($Y^{1349}$VHV-$X_3$-$Y^{1356}$VNV; 16). Actually, the Ron tail can recruit the Grb-2/Sos complex through $Y^{1360}$MNL (9).

Altogether these experiments show that the transforming ability of Tpr-Ron and Tpr-Met is linked to their kinase domains.

## 2.3 Catalytic efficiency of Met and Ron kinases

There are several reports showing that the oncogenic potential of a tyrosine kinase is dramatically influenced by differences in the catalytic efficiency; as in the case of the EGF receptor (15), of $pp60^{c\text{-}src}$ (10) and the proto-oncogene Neu (11). To evaluate the catalytic efficiency of Tpr-Met and Tpr-Ron we determined the kinetic parameters for tyrosine autophosphorylation and for the exogenous substrate MBP (Myelin Basic Protein) phosphorylation.

The apparent Michaelis-Menten constant [$K_M$ (app)] of Tpr-Met and Tpr-Ron for MBP was in the same order of magnitude ($1.76 \pm 0.5$ μM and $1.79 \pm 0.3$ μM, respectively). On the contrary, there is a strong difference in $V_{max}$ between Met and Ron kinases ($1.15 \pm 0.07$ μM and $0.24 \pm 0.01$ μM, respectively ).

The reported data show that the catalytic efficiency of Tpr-Ron - expressed as a ratio between the $V_{max}$ and the $K_M$ (MBP) - is five times lower than the $V_{max}$ of Tpr-Met (0.13 vs. 0.65 pmol/min, respectively). This suggests that catalytic efficiency is the parameter that discriminates the oncogenic potential of the two kinases.

## 2.4 Invasive phenotype evoked by the Tpr-chimaeras

We next investigated cell motility, and invasiveness. NIH3T3 fibroblasts expressing Tpr-Ron - despite the low efficiency of its kinase - migrated through polycarbonate filters and displayed invasive migration through the artificial basement membrane. Cell migration and matrix invasion induced by Tpr-Ron were comparable to those induced by transfection of Tpr-Met (Table I). Tpr-Ron, in spite of its weak kinase, fulfils the requirements for activating cell migration and matrix invasion, and provides a naturally occurring example of dissociation between the two arms of the biological response triggered by the Met family of receptors.

In contrast with the above, cells expressing Tpr-Sea, that displays an higher transforming ability than Tpr-Met, did not elicit a fully invasive phenotype and displayed only a modest increase in cell motility as well as in matrix invasion. Tpr-Sea docking site has two identical Y*VNL sequences, both binding Grb-2 at high affinity (22). This may prevent recruitment of the necessary amount of PI 3-Kinase for promoting motility and invasion, as demonstrated by a Tpr-Met mutant which binds two Grb-2 molecules but is lacking for binding to PI 3-Kinase. This mutant transformed host cells with higher efficiency, but was unable to trigger matrix invasion and metastasis, indicating that concomitant activation of the two pathways is necessary for the fully malignant phenotype (Giordano *et al.*, 1996, submitted for publication).

### 2.5 Cell polarisation induced by Tpr-Chimaeras

MDCK epithelial cell line is a sensitive target for signals controlling polarised growth. These cells, when seeded in 3D collagen gels and stimulated with HGF, migrate, proliferate, and polarise into collagen matrices. This complex regulation results in the formation of branched tubular structures (14). MDCK cells expressing recombinant Tpr-Ron formed cysts developing few spikes that evolved into long and unbranching tubules. On the other hand, uncontrolled activation of Tpr-Met in these epithelial cells boosts cell proliferation, as shown by the formation of larger spherical cysts, but fails to activate the differentiative program. Also the clones expressing recombinant Tpr-Sea grew as larger spherical cysts, which never formed tubular structures (Table I). Tpr-Ron appears to be able to drive part of the morphogenetic program, inducing linear tubulogenesis but not branching, as occurred in the case of HGF stimulation in presence of TGF-β and vitronectin (20). This suggests that number and morphology of the tubule structures are influenced by the combination of both tyrosine kinase signalling and ECM receptors (2).

The Tpr-Ron.$K_M$ chimaera did not induce tubules in MDCK cells, but formed large cysts as well, whereas the counterpart construct, Tpr-Met.$K_R$, led to unbranching morphogenesis as did Tpr-Ron. A potential explanation for the behaviour of the Met kinase-based constructs could be that the high level of signalling conveyed by the Met kinase, optimal to induce unrestrained proliferation, interferes with the accomplishment of the morphogenic program. On the other hand, the low signalling threshold attained by the Ron kinase seems permissive and adequate to activate at least part of the morphogenic program. This can be explained by differential activation of critical genes due to a lower dosage of transcriptional activators induced by Ron (1).

These data demonstrate that constitutive activation of Ron and Met kinases differentially induces the morphogenic program, independently from the nature of the transducing multifunctional docking site.

Table I: Biochemical and biological characterisation of the Tpr-chimaeras.

| Chimaera | Kinase activity* | Density arrested growth | Transforming ability | Migration and invasion | Cell polarisation |
|---|---|---|---|---|---|
| Tpr-Met | ++ | - | + | + | - |
| Tpr-Ron | ++ | + | - | + | + |
| Tpr-Sea | ++ | - | ++ | +/- | - |
| Tpr-Met.$K_R$ | ++ | n.d. | - | + | + |
| Tpr-Ron.$K_M$ | ++ | n.d. | ++ | ++ | - |

* Measured by *in vivo* tyrosine phosphorylation and *in vitro* autophosphorylation

### 2.6 MAP kinase activation by Tpr-Met, Tpr-Ron and Tpr-Sea

Cell transformation requires a strong mitogenic signal for which MAP kinase phosphorylation is a mandatory step (4). MAP kinase activation in pooled stable NIH

3T3 transfectants was analysed by phosphorylation of MBP exogenous substrate after specific immunoprecipitation. In cells expressing Tpr-Sea and Tpr-Met, the MAP kinase was activated seven and six fold over the background respectively, while expression of Tpr-Ron resulted only in a modest increase of MAP kinase activity (Fig. 2).

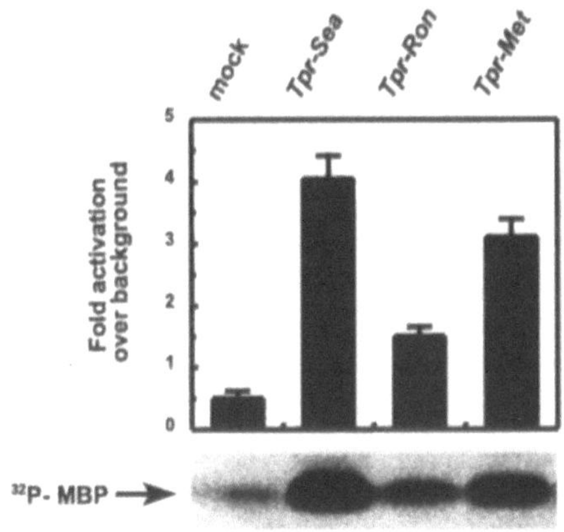


Fig. 2. MAP kinase activation by Tpr-Sea, Tpr-Ron and Tpr-Met. The assay was performed by measuring the amount of $^{32}P$ transferred to myelin basic protein (MBP) by MAP kinase immunoprecipitated with anti-p42$^{ERK2}$ antibodies. MAP kinase activation is expressed as fold increase over the background, in triplicate determinations (bars = S.D.).

MAP kinase stimulation, however, was significant and correlated with induction of cell proliferation. Cells expressing Tpr-Met or Tpr-Sea acquired a transformed behaviour characterised by unrestrained proliferation. Cells expressing Tpr-Ron were able to grow in low serum, but their growth was arrested when saturation density was reached (contact-arrested condition)(Table I).

Then we conclude that the Ron-dependent activation of the MAP kinase pathway does not lead to cell transformation, as occurs in the case of Tpr-Met and Tpr-Sea. A possible explanation is that Met and Sea kinases activate the MAP kinase pathway above a given threshold, that is not reached by the weak intensity of the Ron kinase signal, due to the relative low catalytic efficiency of its kinase. It has been demonstrated that a quantitative difference in MAP kinase activation is translated into a qualitative difference in transcription factors activation, leading to specific gene expression (12). The proliferative *vs* differentiative/morphogenetic responses are modulated by the intensity and duration of the MAP kinase activation, that results in the phosphorylation of transcription factors at various levels (5, 23). According to this interpretation, the higher threshold of MAP kinase activity induced by Tpr-Met and Tpr-Sea causes a higher nuclear concentration of transcription factors that can induce the expression of specific genes associated to cell transformation. On the contrary, the lower threshold of MAP kinase activity attained by Tpr-Ron signalling maintains a lower nuclear level of transcriptional activators, leading to expression of other critical different genes, associated to cell invasiveness and morphogenesis but not to cell transformation.

## 3 References

1. Barros, E., O., Santos, K., Matsumoto, T., Nakamura, and S.K. Nigam. 1995. Proc. Natl. Acad. Sci. USA 92: 4412-4416.
2. Clark, P. 1994. J. Cell Science 107: 1265-1275.
3. Cooper, C.S., M., Park, D.G., Blair, M.A., Tainsky, K., Huebner, C.M., Croce and G. Vande Woude. 1984. Nature 311: 29-33.
4. Cowley, S., H., Paterson, P., Kemp and C. J. Marshall. 1994. Cell 77: 841-852.
5. Dikic, I., J., Schlessinger and I. Lax. 1994. Curr. Biol. 4: 702-708.
6. Gaudino, G., A., Follenzi, L., Naldini, C., Collesi, M., Santoro, K.A., Gallo, P.J., Godowski and P.M. Comoglio. 1994. EMBO J. 13: 3524-3532.
7. Gaudino, G., V., Avantaggiato, A., Follenzi, D., Acampora, A., Simeone and P.M. Comoglio. 1995. Oncogene 11: 2627-2637.
8. Huff, J.L., M.A., Jelinek, C.A., Borgman, T.J., Lansing and T.J. Parsons. 1993. Proc. Natl. Acad. Sci. USA 90: 6140-6144.
9. Iwama, A., N., Yamaguchi and T. Suda. 1996. EMBO Journal 15: 5866-5875.
10. Kato, J.Y., T., Takeya, C., Grandori, H., Iba, J.B., Levy and H. Hanafusa. 1986. Mol. Cell. Biol. 6: 4155-4160.
11. Le Vea, C.M., J.N., Myers, W.C., Dougall, X., Qian and M.I. Greene. 1993. Receptor 3: 293-309.
12. Marshall, C.J. 1995. Cell 80: 179-185.
13. Medico, E., A.M., Mongiovì, J., Huff, M.A., Jelinek, A., Follenzi, G., Gaudino, J.T., Parsons and P.M. Comoglio. 1996. Mol. Biol. Cell 7: 495-504.
14. Montesano, R., K., Matsumoto, T., Nakamura and L. Orci. 1991. Cell 67: 901-908.
15. Nair,N., R.J., Davis and H.L. Robinson. 1992. Mol. Cell. Biol. 12: 2010-2016.
16. Ponzetto, C., A., Bardelli, Z., Zhen, F., Maina, P., Dalla Zonca, S., Giordano, A., Graziani, G., Panayotou and P.M. Comoglio. 1994. Cell 77: 261-271.
17. Ponzetto, C., Z., Zhen, E., Audero, F., Maina, A., Bardelli, M.L., Basile, S., Giordano,R., Narsimhan and P.M. Comoglio. 1996. J. Biol. Chem. 271:14119-14123.
18. Rodrigues,G.A. and M. Park. 1993. Mol. Cell. Biol. 13: 6711-6722.
19. Ronsin, C., F., Muscatelli, M.G., Mattei and R. Breathnach. 1993. Oncogene 8: 1195-1202.
20. Santos,O.F.P. and Nigam,S.K. 1993. Dev. Biol. 160: 293-302.
21. Smith, D.R., P.K., Vogt and M.J. Hayman. 1989. Proc. Natl. Acad. Sci. USA 86: 5291-5295.
22. Songyang, Z., S.E., Shoelson, J., Mcglade, P., Olivier, T., Pawson, X.R., Bustelo, M., Barbacid, H., Sabe, H., Hanafusa, T., Yi, R., Ren, D., Baltimore, S., Ratnofsky, R.A., Feldman and L.C. Cantley. 1994. Mol. Cell. Biol. 14: 2777-2785.
23. Traverse, S., K., Seedorf, H., Paterson, C., Marshall, P., Cohen and A. Ullrich. 1994. Curr. Biol. 4: 694-701.
24. Wang, M.H., C., Ronsin, M.C., Gesnel, L., Coupey, A., Skeel, E.J., Leonard and R. Breathnach. 1994. Science 266: 117-119.

# NUCLEOSIDE DIPHOSPHATE KINASE : EFFECT OF THE P100S MUTATION ON ACTIVITY AND QUATERNARY STRUCTURE

**Sébastien Mesnildrey and Michel Véron**

Unité de Régulation Enzymatique des Activités Cellulaires,
CNRS URA 1149, Institut Pasteur, 75724, Paris Cedex 15, France

## INTRODUCTION

Nucleoside Diphosphate Kinase (NDP kinase) is an ubiquitous enzyme which catalyses the phosphate exchange between a triphospho- and a diphospho- nucleoside by a ping-pong mechanism involving the formation of a phospho-histidine intermediate [1,2]. The enzyme is not specific and can use both ribo- and deoxyribonucleotides as well as purines and pyrimidines as subtrates.

NDP kinase is made of small identical polypeptides of 150 amino-acids with a highly conserved sequence throughout evolution. The structure of NDP kinase from several organisms has been solved at high resolution, showing an identical fold of the monomer with a $\beta\alpha\beta\beta\beta\alpha$ motif. Eukaryotic NDP kinases are hexamers with one three fold axis and three two fold axis [3-5] while NDP kinases from prokaryotes are tetramers [6,7].

There are several indications that NDP kinase is involved in a variety of cellular processes. Thus, in *Drosophila* a null mutation of the gene encoding NDP kinase (*awd*) leads to severe developmental defects [8]. In human, *nm23-H1* encoding NDPK-A shows properties of a metastasis suppressor gene [9,10], while NDPK-B, another isozyme of NDP kinase, was shown to bind to the *c-myc* promoter and to activate its transcription [11]. This activity was maintained in a catalitycally inactive mutant indicating that NDP kinase may be bifonctionnal [12] and there are indications that the oligomeric state of protein binding to DNA may be a dimer [13].

The structure points to the particular importance of the so-called Kpn-loop involved both in the trimer interface of hexameric NDP kinases and in the active site. Pro-100 from this loop (*Dictyostelium* numbering) makes an important contact between subunits with its carbonyl oxygen H-bonding ε-NH2 of Lys-35 in an adjacent subunit [3]. Another contact is the amino group of the C-terminal Glu-155' which H-bonds to Asp-115 in the adjacent subunit.

In order to investigate the oligomeric structure of NDP kinase, we substituted Pro-100 by a serine or a glycine by site directed mutagenesis and we combined these mutations with deletions of 1 to 5 residues at the C-terminus [14]. Although they were purified as active hexamers, all of the single mutants were less stable than the wild type

NATO ASI Series, Vol. H 102
Interacting Protein Domains
Their Role in Signal and Energy Transduction
Edited by Ludwig Heilmeyer

NDP kinase in presence of denaruting agents. In contrast, the double mutant proteins were dimers with only 1% of the activity of the wild-type enzyme [14].

We report here the analysis of the dissociation and unfolding properties of the P100S mutant NDP kinase from *Dictyostelium*.

## RESULTS

The P100S mutant protein of *Dictyostelium* NDP kinase was expressed in *E.coli*. The purified protein was 30% active as compared to wild type NDP kinase. We determined the stability of the activity of the P100S mutant and of the wild type NDP kinases upon urea treatment. The wild type protein is very stable and remains fully active after 24 hr incubation in 5M urea. In contrast, the P100S mutant protein is much less stable, retaining less than 0.5% activity after incubation in 2M urea (Fig. 1). We also studied the renaturation of the protein after previous complete denaturation. For this, the protein was first incubated overnight with 8 M urea, and then diluted to different urea concentrations. The protein was further incubated for 24 hr at 20°C and the activity was measured. The activity of wild type NDP kinase was fully restored at 3M urea indicating a refolding of the protein to the native structure. In contrast, no activity of the P100S mutant NDP kinase could be recovered at any urea concentration indicating that the refolding of the P100S mutant NDP kinase is incomplete (Fig. 1).

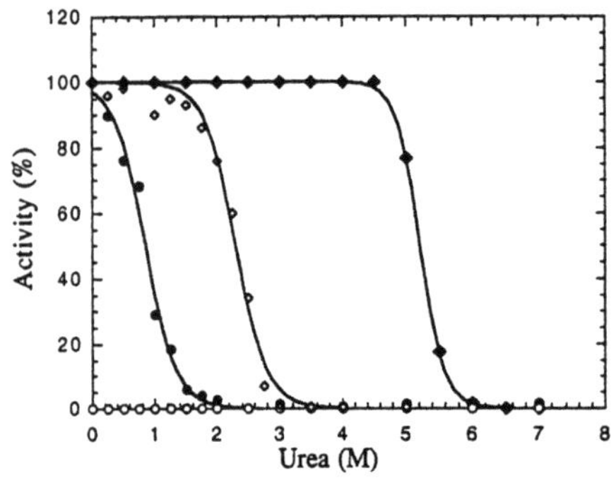


Figure. 1. **Inactivation and reactivation of wild type and mutant NDP kinases in urea.** For denaturation experiment (filled symbols), 400 µg/ml of wild type NDP kinase (squares) or P100S mutant NDP kinase (circles) were incubated at the indicated urea concentration for 24 hr. For the renaturation experiment (open symbols), the protein was first denaturated in 8M urea for 5 hr. After dilution at various urea concentrations, the protein was further incubated for 24 hr. The residual activity was measured at 20°C by a coupled assay using 0.2mM dTDP as phosphate acceptor in the presence of 1mM ATP. 100% corresponds to the specific activity of native proteins.

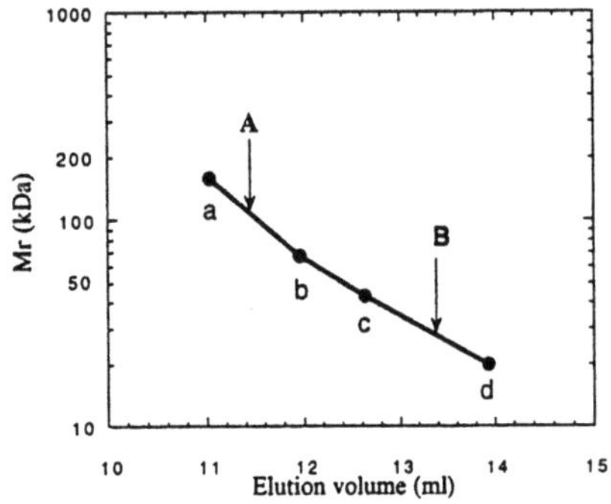


Figure. 2. **Size exclusion chromatography of the wild type and mutant NDP kinase.** Purified wild type and P100S NDP kinase (20 µg to 100 µg) were loaded on Superose 12 column equilibrated in buffer T or in the same buffer containing 2M urea. The elution of the proteins was followed by a UV-detector at 280 nm. Arrow **A** corresponds to the elution volume of wild type NDP kinase in both buffers and of P100S mutant protein in buffer T. Arrow **B** corresponds to the elution of P100S in buffer T containing 2M urea. The P100S-N150stop mutant, shown to be a dimer, is eluting at the same position as B. The proteins used as marker are IgG (a) (Mr 160,000), bovine serum albumine (b) (Mr 67,000), ovalbumin (c) (Mr 43,000), trypsine (d) (Mr 20,000).

The oligomeric state of the mutant protein P100S was determined at two different urea concentrations by size-exclusion chromatography on Superose 12 column (Pharmacia). The column was first calibrated with protein markers in buffer T (50 mM Tris HCl pH 7.5, 150mM NaCl) or in the same buffer containing 2M urea. NDP kinase (20 μg to 100 μg) was applied onto the column and eluted with a Pharmacia FPLC system at 0.3 ml/min. In absence of urea, the P100S mutant NDP kinase eluted at the same volume as the wild type enzyme (arrow A in Fig. 2), indicating that the native P100S enzyme is an hexamer. When the protein was previously incubated for 24 hr in buffer T containing 2M urea and loaded on the Superose 12 column equilibrated with the same buffer, it was eluted later (Arrow B in Fig. 2), indicating that the P100S protein was not a hexamer under these conditions. As shown in Fig.2, the elution was identical to that of the P100S-N150stop double mutant NDP kinase, previously shown to be a dimer [14]. We conclude that the P100S mutant protein dissociates into dimers in the presence of 2M urea.

## DISCUSSION

The residue Pro-100 is totally conserved in all eukaryotic NDP kinases. It H-bonds with Lys-35 from an adjacent subunit. We have shown that P100S mutant protein is only 30% active compared to wild type NDP kinase. This lower activity may be due to an increased flexibility of the Kpn-loop which contains several residues involved in substrate binding including Arg-109 and Val-116. In presence of 2M urea, P100S NDP kinase dissociates into inactive dimers with less than 0.5% activity of the wild type enzyme. This instability suggest the possibilty of a dimer-hexamer equilibrium in the cell.

In *drosophila*, the *Killer of prune* (*kpn*) mutation corresponds to the substitution of Pro-97 (corresponding to Pro-100 in the *Dictyostelium*) by a serine in NDP kinase, leading to a lethal phenotype in the absence of the *prune* gene product [15]. The NDP kinase from the *Kpn* mutant flies was shown to be unstable [16]. Our results with *Dictyostelium* P100S NDP kinase are similar to those obtained with the Kpn mutant from *Drosophila* and indicates that P97S mutant protein also dissociates into dimers. It was proposed that the dominant lethal phenotype of *Kpn* flies was due to the accumulation of dissociated forms of mutated NDP kinases [16]. Timmons *et al.* isolated revertants of *Kpn* flies and found that all revertant mutations lead to catalytically inactive NDP kinases [17]. They hypothetized that in the absence of the *prune* gene product, the accumulation of an unknown compound could become toxic for the cells upon phosphorylation by Kpn NDP kinase. The high activity of the P100S NDP kinase along with an increased flexibility of the Kpn loop may result in the capacity to phosphorylate new substrates. Alternatively, the tendancy of the P100S enzyme to dissociate may allow new substrates to be phosphorylated by the residual activity of the dimeric form.

Our results support the possibility that various oligomeric states of the NDP kinase may carry different biochemical functions. We are currently studying by transfection the effects of the P100S mutation and of the dimeric form of NDP kinase on *dictyostelium* development.

## ACKNOWLEGMENTS

We thank Joël Janin and Anna Karlsson for stimulating discussions. This work was supported by the grants from the Ligue Contre le Cancer, Comité de Paris and from the Association pour la Recherche sur le Cancer.

## REFERENCES

1. Parks, R.E.J. and Agarwal, R.P. (1973) The Enzymes 8, 307-334.
2. Gilles, A.M., Presecan, E., Vonica, A. and Lascu, I. (1991) J. Biol. Chem. 266, 8784-8789.
3. Dumas, C. et al. (1992) EMBO J. 11, 3203-3208.
4. Chiadmi, M., Moréra, S., Lascu, I., Dumas, C., LeBras, G., Veron, M. and Janin, J. (1993) Structure 1, 283-293.
5. Webb, P.A., Perisic, O., Mendola, C.E., Backer, J.M. and Williams, R.L. (1995) J. Mol. Biol. 251, 574-587.
6. Williams, R.L., Oren, D.A., Munoz-Dorado, J., Inouye, S., Inouye, M. and Arnold, E. (1993) J. Mol. Biol. 234, 1230-1247.
7. Giartosio, A., Erent, M., Cervoni, L., Morera, S., Janin, J., Konrad, M. and Lascu, I. (1996) J. Biol. Chem. in press.
8. Dearolf, C.R., Tripoulas, N., Biggs, J. and Shearn, A. (1988) Develop. Biol. 129, 169-178.
9. Steeg, P.S., Bevilacqua, G., Kopper, L., Thorgeirsson, U.P., Talmadge, J.E., Liotta, L.A. and Sobel, M.E. (1988) J. Natn. Cancer Inst. 80, 200-204.
10. MacDonals, N.J., Freije, J.M.P., Stracke, M.L., Manrow, R.E. and Steeg, P.S. (1996) J. Biol. Chem. 271, 25107-25116.
11. Postel, E.H., Berberich, S.J., Flint, S.J. and Ferrone, C.A. (1993) Science 261, 478-480.
12. Postel, E.H. and Ferrone, C.A. (1994) J. Biol. Chem. 269, 8627-8630.
13. Postel, E.H., Weiss, V.H., Beneken, J. and Kirtane, A. (1996) Proc. Natl. Acad. Sci. USA in press.
14. Karlsson, A., Mesnildrey, S., Xu, Y., Morera, S., Janin, J. and Veron, M. (1996) J. Biol. Chem. 271, 19928-19934.
15. Biggs, J., Tripoulas, N., Hersperger, E., Dearolf, C. and Shearn, A. (1988) Genes Dev. 2, 1333-1343.
16. Lascu, I., Chaffotte, A., Limbourg-Bouchon, B. and Veron, M. (1992) J. Biol. Chem. 267, 12775-12781.
17. Timmons, L., Xu, J., Hersperger, G., Deng, X.-F. and Shearn, A. (1995) J. Biol. Chem. 270, 23021-23030.

# Part V

# Regulatory Cascades

# Interaction of adenylyl cyclase type 1 with βγ subunits of heterotrimeric G-proteins

Andrea Hülster[1], Franz-Werner Kluxen[2] and Thomas Pfeuffer[1]

[1]Department of Physiological Chemistry II, P.O. Box 10 10 07, HHU Düsseldorf, D-40001 Düsseldorf, and [2]Merck KGaA, A32-311, D-64271 Darmstadt, Germany

## 1. Introduction

All of the 10 known mammalian adenylyl cyclases (AC) (1) are stimulated by the α subunit of the activating G-protein ($\alpha_s$) but only a certain subset is regulated by G-protein βγ subunits. The isoforms 2, 4, and 7 (?) are conditionally activated by βγ subunits upon costimulation with $\alpha_s$, while AC-1 activity is suppressed by βγ. A physical interaction between βγ and AC is implicated by the work of Taussig (2), Enomoto (3) as well by our own experiments (unpublished results).

A number of other proteins can interact with βγ subunits like phospholipase Cβ, Bruton tyrosin kinase, β-adrenergic receptor kinase, and G-protein gated inwardly rectifying potassium channel. Many of them exhibit a pleckstrin homology domain which contains a βγ binding site. However, no such consensus sequence in AC-1 (and AC-2) has been found. Recently amino acids 956-982 of AC-2 have been shown to be involved in activation by βγ subunits (4). This motif however is unique to AC-2 and is not present in AC-1.

Here we are describing initial experiments to localize the βγ-binding site(s) in AC-1. COS 1 cells coexpressing the cDNAs encoding AC-1, βγ, and selected fragments of AC-1 are monitored for the attenuation of the βγ mediated inhibition by these fragments.

## 2. Material and Methods

*Molecular cloning:* The following AC-1 fragments were amplified by PCR and cloned into both pCDNA3-Flag and pCDNA3-Flag/Gera (see below): 1NT (bp 106-294), 1C1a (bp 814-1510), 1C1b (bp 1510-1939), 1C1T (bp 106-1939), 1C2a (bp 2494-2688), 1C2b1 (bp 2779-3090), 1C2b2 (bp 3154-3504), 1C2ab (bp 2494-3090), 1C2bT (bp 2779-3504), and 1C2T (bp 2494-3504). We constructed two derivatives of the eukaryotic expression vector pCDNA3 to allow expression of the peptides as either (soluble) cytosolic or (geranyl-geranylated) membrane bound fragments. In both vectors the ATG start codon was followed by a tag-sequence (FLAG (Kodak): DYKDDDDK) for detection in western blots with the monoclonal anti-FLAG antibody. For PCR amplification of the soluble fragments, the 3' oligonucleotides contained a stop codon. pCDNA3-Flag/Gera carries in the 3' part of the polylinker a sequence that encodes the amino acids C-V-L-L-Ter. This sequence was previously shown to induce modification of the C-residue by a geranylgeranyl moiety followed by cleavage of the C-V bond (5).

NATO ASI Series, Vol. H 102
Interacting Protein Domains
Their Role in Signal and Energy Transduction
Edited by Ludwig Heilmeyer

*Transfection of COS 1 cells:* COS 1 cells were seeded in 24 multiwell plates and transfected for 2 h by the DEAE-Dextran method using 2.25 µg/ml DNA.
*$^{3}$H-adenine assay:* 40 - 64 hours after transfection cells were labelled with 1 µCi $^{3}$H-adenine for 1 h and after incubation with 1 mM IBMX stimulated by either 10 µM ionophore A23187 ($Ca^{2+}$ stimulation), 10 µM forskolin or 2 mM EGTA (basal) for 20 min. After lysis of the cells by TCA, $^{3}$H-ATP and $^{3}$H-cAMP were determined by sequential chromatography on Dowex (50W-X8) and $Al_2O_3$. $^{32}$P-labelled ATP and cAMP were added as recovery standards.
*Western blot:* Cells of a single well were lysed with 1 % SDS, lyophilized, separated on 10 % Schägger-gels (6) and electroblotted on PVDF-membranes. The blots were sequentially probed with anti-β-antiserum (for detection of β-subunit), BBC-1-antibody (for detection of AC-1) (7), and anti-FLAG-antibody (for detection of AC-1 fragments)(Kodak). Secondary antibodies coupled to alkaline phosphatase and chemoluminescence with CSPD were used for visualization.

## 3. Results and Discussion

The expression of AC-1 in COS 1 cells was monitored by stimulation with the $Ca^{2+}$-ionophore A23187. The activity was usually 3-6 fold compared to vector transfected cells (fig. 1). The synergistic activation of AC-1 by forskolin and A23187 was about 8 fold (not shown). When βγ-subunit cDNAs were transfected together with AC-1 cDNA in 2 or 4 fold excess, the $Ca^{2+}$ stimulated activity was depressed by 38% and 46%. The effect on forskolin stimulated activity was less pronounced while the basal activity was variably affected (fig. 1). When free βγ-subunits were trapped by coexpressed $\alpha_t$-subunits, the inhibition was attenuated (fig. 2) or, with higher amounts of $\alpha_t$ cDNA transfected, reversed to a stimulation. This effect was even more pronounced when only AC-1 and $\alpha_t$ were coexpressed. This is probably due to trapping of endogenous free βγ subunits.

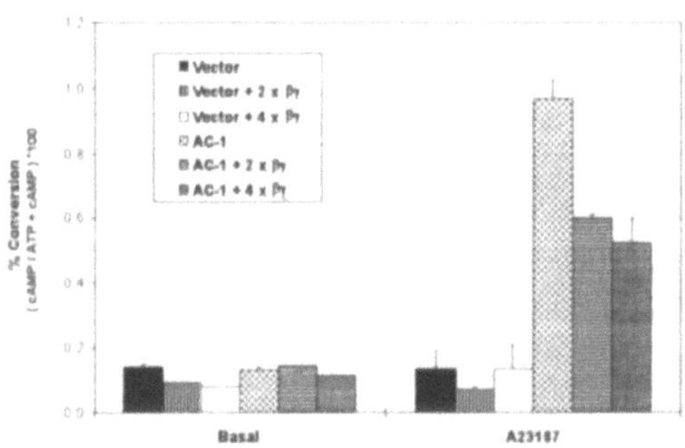


*Fig.1: Inhibition of AC-1 by coexpression of βγ subunits:* Transfections were performed using the following amounts of DNA: 0.25 µg/ml pCDNA3-AC1, 0.5 µg/ml pCDNA3-$\beta_1$, 0.5 µg/ml pCDNA3-$\gamma_2$ (2xβγ) or 1.0 µg/ml of each (4xβγ). In each case total amount of DNA was adjusted to 2.25 µg/ml with pCDNA3. After 3 days, AC-activity was measured.

Fig. 3 shows the position of the AC-1 derived fragments within the AC-1 sequence that were cloned into the pCDNA3 derivatives (while all the fragments were cloned into pCDNA3-Flag, only some of them were cloned into pCDNA3-Flag/Gera). In fig. 4 the expression of the fragments is verified by western blotting. Interestingly, the putative geranylgeranylated fragments were expressed at a higher level than their non isoprenylated counterparts. Of the soluble fragments, all but 1-NT, 1-C2a, 1-C2b1, and 2C2b1 could be detected by the anti-FLAG antibody in Western blots while all of the cloned geranylgernylated fragments were expressed (fig. 4). The fragments bearing the C-terminus were in addition detected by the monoclonal antibody BBC-1, which is directed against the extreme C-terminus of the AC-1 (not shown). The molecular weights of all fragments were of the predicted size.

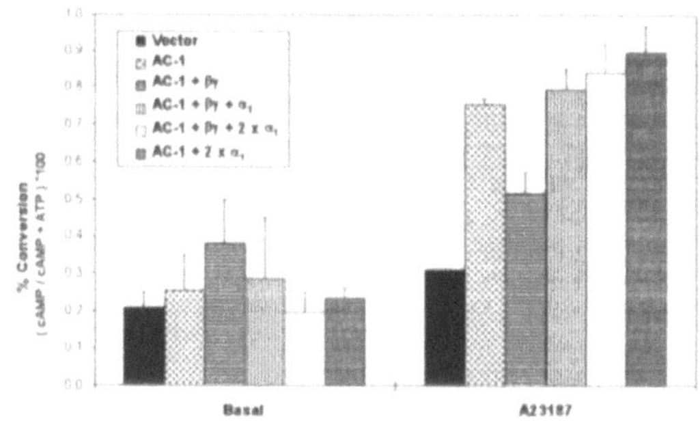


*Fig. 2: Attenuation of $\beta\gamma$ inhibition of AC-1 by $\alpha_t$:* Transfections and determination of adenylyl cyclase activity were performed as outlined in Material and Methods. The amounts of DNA were: AC-1: 0.25 µg/ml, $\beta\gamma$: 0.5 µg/ml each, $\alpha_t$ 0.5 µg/ml, 2x $\alpha_t$ 1µg/ml. The final DNA concentration amounted always to 2.25 µg/ml.

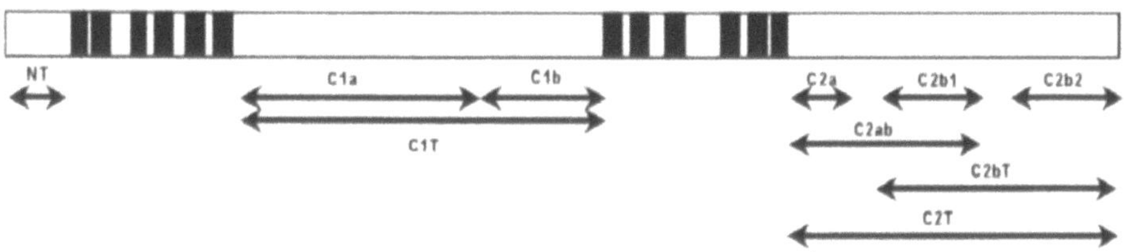


*Fig. 3: Cloning of AC-1 fragments :* A schematic representation of the AC-1 sequence is given. Bars show the position of the putative transmembrane regions and arrows the cloned fragments.

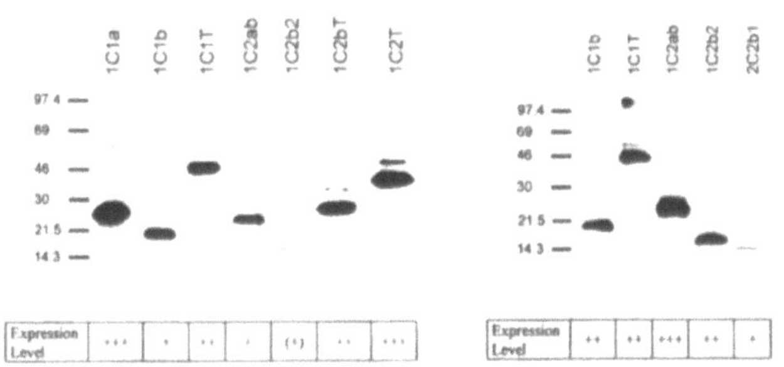


*Fig. 4: Western blot analysis of soluble (left) and geranylgeranylated (right) AC-1 fragments:* cDNAs encoding AC-1 fragments were transfected into COS 1 cells. After 2-3 days the cells were assayed for the expression of the fragments with the anti-FLAG antibody. Molecular weight markers are shown on the left hand sites. Below, the relative expression levels of the AC fragments are indicated.

When the soluble fragments were expressed together with AC-1 and $\beta\gamma$, an additional slight reduction of the AC-1 activity was observed (fig. 5). The same was found with most of the geranylgeranylated fragments (fig. 6). 1C1T, the fragment that encompasses the first intracellular domain, increased the βγ-mediated inhibition of the AC-1 activity. This might be due to a competition for calmodulin, since this domain is reported to contain the CaM binding site of AC-1 (8).

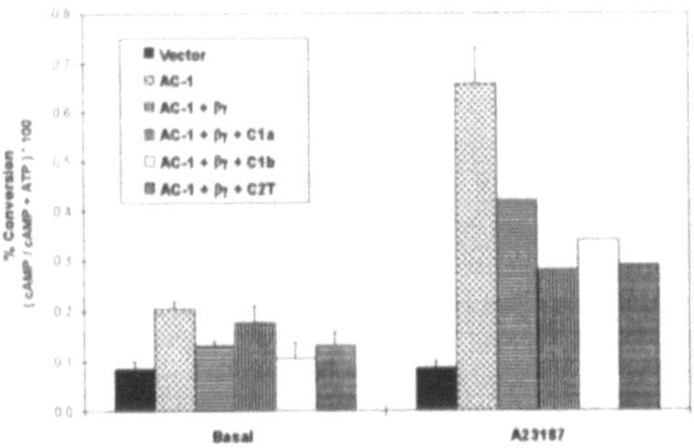


*Fig. 5: Coexpression of AC-1, βγ and soluble AC-1 fragments:* COS 1 cells were transfected with AC-1, βγ, and the indicated fragments while the total amount of DNA was adjusted to 2.25 μg/ml DNA with vector DNA. After 3 days the AC activity was measured with the $^{3}$H-adenine assay.

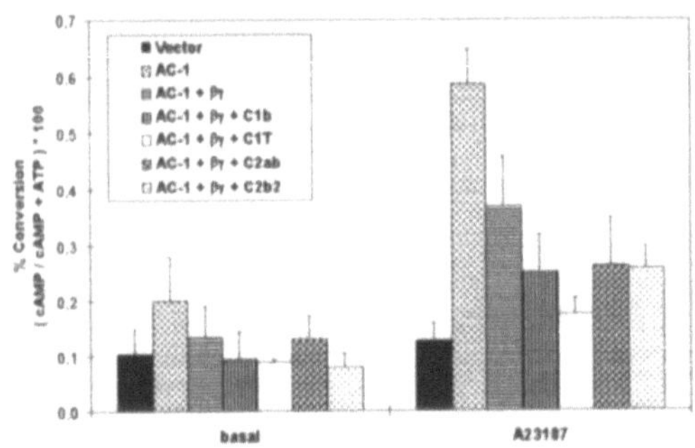


*Fig. 6: Coexpression of AC-1, βγ and geranylgeranylated AC-1 fragments:* COS 1 cells were transfected with AC-1, βγ, and the indicated fragments while the total amount of DNA was adjusted to 2.25 μg/ml DNA with vector DNA. The average (± SEM) of four experiments is shown.

Up to now, none of the AC-1 fragments (in addition to the fragment that contains the AC-2 βγ binding site (4), not shown) did influence the βγ mediated inhibition of the AC-1 activity. Forthcoming studies are designed to achieve higher cellular concentrations while carefully monitoring subcellular distribution of these fragments.

*Acknowledgments:* We are grateful to S. Haeseler for excellent technical assistance. We thank W.-J. Tang for the AC-1 cDNA and P. Gierschik for the $\beta_1$, $\gamma_2$, and $\alpha_t$ cDNAs. The antibodies used in this study were supplied by S. Mollner, this department (BBC-1), and D. Palm, Würzburg (anti-β-antiserum). This work was supported by the Deutsche Forschungsgemeinschaft, grant Pf 80 / 11-1.

## 4. References:

1) Sunahara, R.K. et al. (1996). Annu. Rev. Pharmacol. Toxicol. **36**, 461 - 480
2) Taussig, R. et al. (1993). J. Biol. Chem. **268**, 9 - 12
3) Enomoto, K. et al. (1995). Jpn. J. Pharmacol. **69**, 239 - 250
4) Chen, J. et al. (1995). Science **268**, 1166 - 1169
5) Inglese, J. et al. (1992). Nature **359**, 147 - 150
6) Schägger, H. and von Jagow, G. (1987). Anal. Biochem. **166**, 368 - 379
7) Mollner, S. and Pfeuffer, T. (1988). Eur. J. Biochem. **171**, 265 - 271
8) Vorherr, T. et al. (1993), Biochemistry **32**, 6081 - 6088

# Regulation of G-protein Activation in Retinal Rods by Phosducin

Barry M. Willardson[1], Jon F. Wilkins[2], Tatsuro Yoshida[3] and Mark W. Bitensky[3]

[1] Department of Chemistry and Biochemistry, Brigham Young University, Provo, UT 84602 USA
[2] Department of Biochemistry, University of Wisconsin-Madison, Madison, WI 53706 USA
[3] Department of Biomedical Engineering, Boston University, Boston, MA 02215 USA

## 1. Introduction: Phosducin is a General G-protein Regulator

G-proteins shuttle signaling information between a vast array of seven transmembrane helical receptors to intracellular effectors (1). Effectors determine the concentration of important second messenger molecules in the cell such as cyclic nucleotides, inositol phosphates and $Ca^{2+}$. It is the concentration of these second messengers that determines the ultimate response of the cell to a stimulus, whether it be cell growth, differentiation or death, neuronal signaling, the rate of heart muscle contraction, or a number of other responses.

Phosducin (Pd) appears to play an important role in regulating G-protein activity. Pd was first discovered in retinal rods as a 33 kDa phospho-protein whose phosphorylation state was controlled by light (2). It was phosphorylated in the dark and dephosphorylated in the light. Interest in Pd's function increased when it was shown that Pd bound the βγ subunits of the visual G-protein, $G_t$ and blocked light activation of cGMP phosphodiesterase (3,4). Pd was also found in brain tissue (5). It was shown to inhibit both receptor induced and intrinsic GTPase activity of $G_s$ and the resulting activation of adenylyl cyclase (5). Intrinsic $G_i$ and $G_o$ GTPase activities were also blocked by Pd (5). Recently, studies have shown that when Pd binds Gβγ, it blocks the ability of Gβγ to interact with both Gα and effector enzymes (6-8). Thus when bound to Pd, Gβγ cannot be activated by receptors nor can it activate its effector enzymes. In this manner Pd effectively sequesters Gβγ in an inactive state.

Pd was initially believed to be expressed only in the retina and pineal gland, where it is found in abundance. However, it has also been found in significant amounts in many different tissues and has been shown to bind many different Gβγ isoforms (9,10). Thus, it appears that Pd may be a general regulator of G-protein function.

## 2. The Phosphorylation State of Phosducin Determines its Ability to Sequester Gβγ

Pd is phosphorylated by cAMP dependent protein kinase (PKA). The site of phosphorylation is at Ser 73 (11,12). In retinal rods, this appears to be the principal site of phosphorylation (11). The first report of an effect of phosphorylation of Pd was done using the β2-adrenergic receptor/$G_s$/adenylyl cyclase system in A431 cell membranes (5). Unphosphorylated Pd inhibited isoproterenol stimulation of adenylyl cyclase in this system, but when Pd was phosphorylated by PKA, it had no effect on adenylyl cyclase activity. From these results, we hypothesized that the phosphorylation state of Pd in retinal rods could determine $G_t$ activity. Since Pd was phosphorylated in the dark and dephosphorylated in the light, one could propose a mechanism in which Pd would down regulate $G_t$ activity in the light and contribute to light adaptation. We tested this hypothesis by measuring the ability of Pd or phosphorylated Pd to block $G_t$ binding to its receptor, light activated rhodopsin (Rho*). The results of these experiments are shown in Figure 1. The binding of $^{125}$I-labeled $G_t\alpha$ (0.1 μM), with equimolar $G_t\beta\gamma$, to light

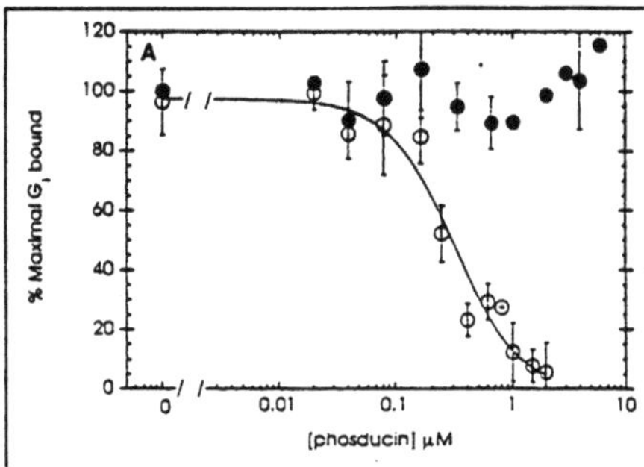


Fig. 1. **Phosphorylation of Pd determines its ability to inhibit $^{125}$I-$G_t$ binding to Rho*.** Light-induced binding of $^{125}$I-$G_t\alpha$ (0.1 μM) and $G_t\beta\gamma$ (0.1 μM) to urea-stripped rod outer segment (ROS) membranes (0.2 μM Rho) was carried out in the presence of unphophorylated (O) or phosphorylated (●) Pd at the concentrations indicated. Reproduced with permission from Ref. 8.

NATO ASI Series, Vol. H 102
Interacting Protein Domains
Their Role in Signal and Energy Transduction
Edited by Ludwig Heilmeyer

activated rhodopsin was measured in the presence of Pd or phosphorylated Pd. At 2 μM concentration, Pd blocked $G_t\alpha$ binding to Rho* completely. Half maximal inhibition was observed at ~0.3 μM Pd, a ratio of phosducin to $G_t$ of 3 to 1. The inhibition was reversed by addition of excess $G_t\beta\gamma$ (data not shown). Since $G_t\beta\gamma$ is required for $G_t\alpha$ to bind effectively to Rho*, we concluded that Pd was blocking $G_t\alpha$ binding to Rho* by sequestering $G_t\beta\gamma$ subunits in an approximately one to one complex. The results were very different with phophorylated Pd. When Pd was phosphorylated by PKA, it had no effect on $G_t\alpha$ binding up to 6 μM concentration. Therefore, we concluded that phosphorylated Pd could not block $G_t\alpha$ binding to Rho* because it could not effectively sequester $G_t\beta\gamma$ from $G_t\alpha$ and Rho*.

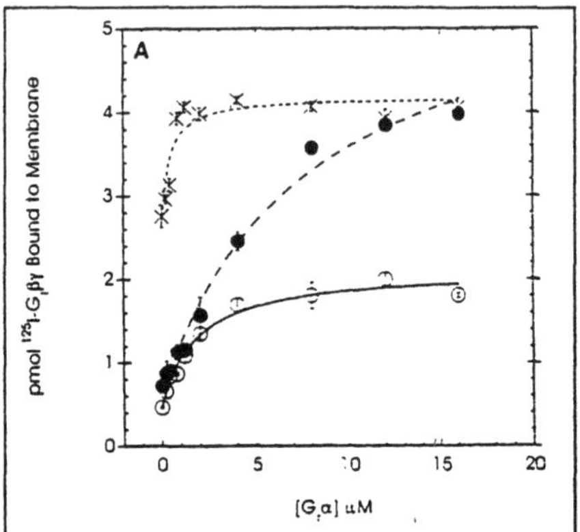


Fig. 2. **$G_t\alpha$ competition with Pd for $^{125}$I-$G_t\beta\gamma$ binding.** Binding of $^{125}$I-$G_t\beta\gamma$ (0.35 μM) to unilluminated urea-stripped ROS membranes (40 μM Rho) was measured in the presence (O) and absence (X) of Pd (2.0 μM) or phosphorylated Pd (●) at the concentrations of $G_t\alpha$ indicated. The total $^{125}$I-$G_t\beta\gamma$ in the 20 μl sample was 7.0 pmol. Reproduced with permission from Ref. 8

These results suggested that Pd was competing with $G_t\alpha$ for binding to $G_t\beta\gamma$. In another experimental format, we looked directly at this competition and also examined effects of phosphorylation of Pd. The experiment exploited the fact that the Pd-$G_t\beta\gamma$ complex is soluble (3), whereas $G_t\beta\gamma$ alone is membrane bound (1). $G_t\beta\gamma$ was labeled with $^{125}$I and its binding to urea-stripped retinal rod cell outer segment membranes was measured (Figure 2). Under these conditions, 2.8 pmol $G_t\beta\gamma$ bound to the membranes. When $G_t\alpha$ was added in the absence of Pd, $G_t\beta\gamma$ binding increased to 4.0 pmol as expected from previous data. When excess Pd was added in the absence of $G_t\alpha$, binding of $G_t\beta\gamma$ to the membrane was decreased to 0.5 pmol bound. Addition of $G_t\alpha$ resulted in a modest increase in $G_t\beta\gamma$ binding to 2.0 pmol bound to the membrane. The results with phosphorylated Pd were very different. Phosphorylated Pd also decreased the binding of $G_t\beta\gamma$ to the membrane, to 0.7 pmol bound. However, when $G_t\alpha$ was added, binding was restored to 4.0 pmol, the amount bound in the absence of phosphorylated Pd. These membrane association data show that Pd competes effectively with $G_t\alpha$ for the binding of $G_t\beta\gamma$, but phosphorylated Pd does not. These results are supported by the crystal structure of the Pd-$G_t\beta\gamma$ complex that has just been reported (13). The binding sites of $G_t\alpha$ and Pd on $G_t\beta\gamma$ overlap considerably, excluding the possibility both could bind simultaneously. The changes in the structure of Pd that occur upon phosphorylation are not yet known.

## 3. Regulation of Phosducin Phosphorylation in Retinal Rods by $Ca^{2+}$

Since phosphorylation of Pd has such dramatic effects on its ability to sequester $G\beta\gamma$, it is important to understand how Pd phosphorylation is controlled. We have examined factors that control Pd phosphorylation in bovine retinal rod cells. In the visual system, light activation of rhodopsin results in $G_t$ activation. $G_t\alpha$-GTP activates cGMP phosphodiesterase, which rapidly hydrolyses cGMP in the rod outer segment (ROS). The decrease in cGMP causes cGMP-gated cation channels in the plasma membrane of the outer segment to close. Channel closure results in hyperpolarization of the rod and triggers neuron activation at the cell's synaptic terminus. Light activation of cGMP phosphodiesterase may also be responsible for the decrease in [cAMP] that occurs in the light response (14). The phosphodiesterase hydrolyzes cAMP in addition to cGMP, albeit with a $K_m$ 100 times higher than cGMP (15). The observed dephosphorylation of Pd in response to light results from this fall in [cAMP]. When dephosphorylated, Pd binds to $G_t\beta\gamma$ and inhibits activation of $G_t$ by Rho*. This may contribute to the process of light adaptation in photoreceptor cells. In the light adapted state, responses to light are substantially decreased. Sequestration of $G_t\beta\gamma$ by dephosphorylated Pd would decrease the number of $G_t$ molecules available for activation, resulting in less $G_t$ activation per Rho*.

The $[Ca^{2+}]$ in ROSs also decreases in response to light because of closure of the cGMP-gated cation channel. $Ca^{2+}$ enters the outer segment through these channels and exits through a $K^+$,$Na^+$/$Ca^{2+}$ exchanger. When the channels close, the influx of $Ca^{2+}$ stops whereas its efflux through the exchanger continues. The result is a decrease in $[Ca^{2+}]$ from ~400 nM in the dark to <100 nM, depending on the intensity and duration of the light exposure (16). This decrease in $[Ca^{2+}]$ has been shown to be responsible for the light adapted state in rods. Two important biochemical effects of decreased $[Ca^{2+}]$ that contribute to the light adaptation process have been described. These include the activation of guanylyl cyclase, which restores [cGMP] (17) and the increased phosphorylation of Rho*, which results in arrestin binding and inactivation of Rho* (18). We wondered whether Pd phosphorylation was in some manner regulated by $[Ca^{2+}]$ in rods. We looked and found no direct effect of $Ca^{2+}$ on Pd phosphorylation in intact ROS preparations. We then asked ourselves whether $[Ca^{2+}]$ could indirectly affect the rate of Pd phosphorylation by influencing the [cAMP] in rods. We were aware that $Ca^{2+}$/calmodulin dependent adenylyl cyclase activity had been described for subtypes 1, 3 and 8 (19-21). Therefore we concluded to examine the $Ca^{2+}$ dependence of rod adenylyl cyclase and its effect on Pd phosphorylation. We found that rod adenylyl cyclase was exquisitely sensitive to $[Ca^{2+}]$ in the physiological range (Figure 3). Activity increased more than 4-fold between 100 and 500 nM $[Ca^{2+}]$. Half maximal activation occurred at 230 nM $[Ca^{2+}]$. The $Ca^{2+}$ dependence was highly cooperative, exhibiting a Hill coefficient of 3.6 +/- 0.5 when the data were fit to the Hill equation. We also showed that calmodulin was necessary for this $Ca^{2+}$ dependence (data not shown). These results demonstrate that ROS adenylyl cyclase can respond with great sensitivity to those changes in $[Ca^{2+}]$ that occur in response to light.

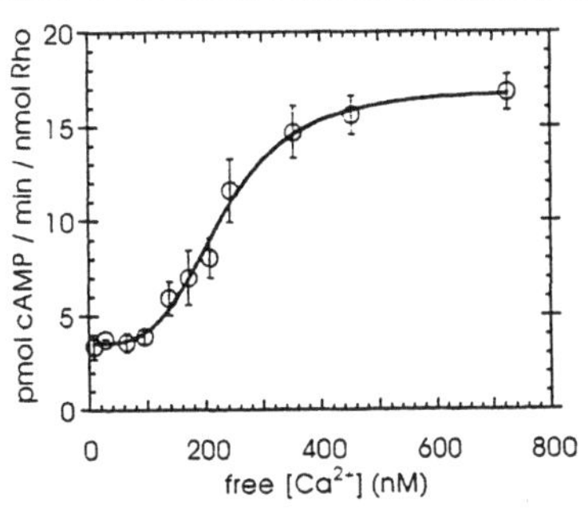


Fig. 3. **Effect of $[Ca^{2+}]$ on ROS adenylyl cyclase activity.** Adenylyl cyclase activity was measured at the indicated concentrations of free $Ca^{2+}$. Free $Ca^{2+}$ was determined using the $Ca^{2+}$ sensitivity dye Fura 2 as described (23). Reproduced with permission from Ref. 23.

We went on to investigate the effect of $Ca^{2+}$/calmodulin dependent adenylyl cyclase activity on Pd phosphorylation. The results are shown in Figure 4. We allowed cAMP to be synthesized at physiologically high and low $[Ca^{2+}]$ in the presence of ATP for 1 min and then measured the initial rate of Pd phosphorylation. We found that the initial rate of Pd phosphorylation was increased by nearly 3-fold at 1100 nM $[Ca^{2+}]$ compared to 8 nM $[Ca^{2+}]$. Without preincubation, no change in the rate of Pd phosphorylation at the two $[Ca^{2+}]$ was observed. Therefore, cAMP accumulation and PKA activation was required to observe an increase in the rate of Pd phosphorylation. This rules out direct activation of a $Ca^{2+}$-dependent Pd kinase because such $Ca^{2+}$ activation does not require a preincubation period. These findings suggest that the light-mediated fall in $Ca^{2+}$ in ROS could lead to a decrease in Pd phosphorylation and sequestration of $G_t\beta\gamma$. Thus, the $Ca^{2+}$ dependence of phosducin phosphorylation may constitute an additional way in which $[Ca^{2+}]$ orchestrates light adaptation in rods.

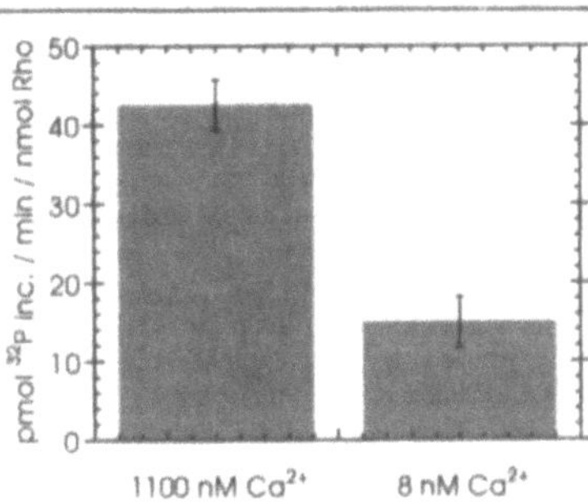


Fig. 4. **Effect of preincubation at high or low $[Ca^{2+}]$ on rates of Pd phosphorylation by ROS PKA.** The initial rate of phosphorylation of Pd by ROS PKA was measured after pre-incubation of the ROS at high (1100 nM) and low (8 nM) $[Ca^{2+}]$. Initial rates were determined as described previously (23). Reproduced with permission from Ref. 23.

A potential mechanism of Pd feedback regulation of $G_t$ activity in photoreceptor cells is shown in Figure 5. After a light stimulus, the decreased rate of cAMP synthesis at low $Ca^{2+}$, coupled with the increase rate of cAMP hydrolysis by activated phosphodiesterase results in a fall in [cAMP]. Rod PKA is inactivated and Pd is dephosphorylated by rod phosphatases acting unopposed by PKA. Dephosphorylated Pd binds to $G_t\beta\gamma$ and blocks dephosphorylated by rod phosphatases acting unopposed by PKA. Dephosphorylated Pd binds to $G_t\beta\gamma$ and blocks further activation of $G_t$ by Rho*. This type of feedback control by Pd may be a general theme by which other G-protein-mediated signals are regulated (22). The specific variations on this theme would, of course, depend on the effectors activated and their effect on [cAMP] in each system.

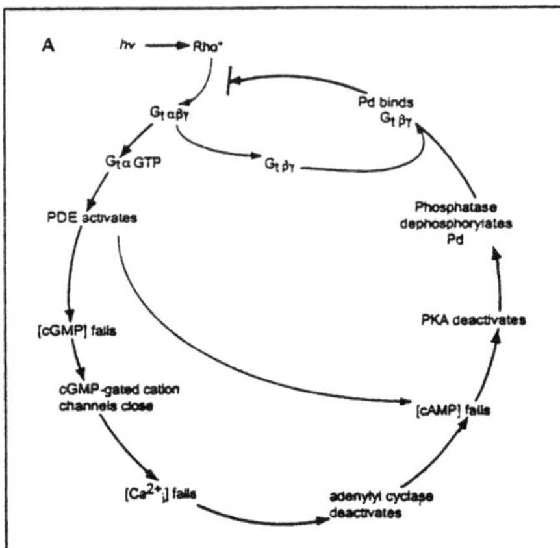


Fig. 5. **Schematic representation of regulation of $G_t$ activity by Pd.** Proposed down-regulation of the light activation cascade by Pd. Reproduced with permission from Ref. 23.

## References


1. Neer, E.J. (1995) *Cell* **80**, 249-257.
2. Lee, R.H., Brown, B.M. and Lolley, R.N. (1984) *Biochemistry* **23**, 1972-1977.
3. Lee, R.H., Lieberman, B.S. and Lolley, R.N. (1987) *Biochemistry* **26**, 3983-3990.
4. Lee, R.H., Ting, T.D., Lieberman, B.S., Tobias, D.E., Lolley, R.N. and Ho, Y.-K. (1992) *J. Biol. Chem.* **267**, 25104-25112.
5. Bauer, P.H., Muller, S., Puzicha, M., Pippig, S., Obermaier, B., Helmreich, E.J.M. and Lohse, M.J. (1992) *Nature (London)* **358**, 73-76.
6. Yoshida, T., Willardson, B.M., Wilkins, J.F., Jensen, G.J., Thornton, B.D. and Bitensky, M.W. (1994) *J. Biol. Chem.* **269**, 24050-25057.
7. Hekman, M., Bauer, P.H., Sohlemann, P. and Lohse, M.J. (1994) *FEBS Let.* **343**, 120-124.
8. Hawes, B.E., Touhara, K., Kurose, H., Lefkowitz, R.J. and Inglese, J. (1994) *J. Biol. Chem.* **269**, 29825-29830.
9. Danner, S. and Lohse, M.J. (1996) *Proc. Natl. Acad. Sci USA* **93**, 10145-10150.
10. Muller, S., Straub, A., Schroder, S., Bauer, P.H. and Lohse, M.J. (1996) *J. Biol. Chem.* **271**, 11781-11786.
11. Lee, R.H., Brown, B.M. and Lolley, R.N. (1990) *J. Biol. Chem.* **265**, 15860-15866.
12. Abe. T., Nakabayashi, H., Tamada, H., Takagi, T., Sukuragi, S., Yamaki, K. and Shinohara T. (1990) *Gene* **91**, 209-215.
13. Gaudet, R., Bohm, A. and Sigler, P.B. (1996) *Cell* **87**, 577-588.
14. Cohen, AI (1982) *J. Neurochem.* **38**, 781-796.
15. Miki, N., Baraban, J.M., Keirns, J.J., Boyce, J.J. and Bitensky, M.W. (1975) *J. Biol. Chem.* **250**, 6320-6327.
16. Gray-Keller, M.P. and Detwiler, P.B. (1994) *Neuron* **13**, 849-861.
17. Koch, K.-W. and Stryer, L. (1988) *Nature (London)* **334**, 64-66.
18. Kawamura, S. (1993) *Nature (London)* **362**, 855-857.
19. Tang, W.-J., Krupinski, J. and Gilman, A.G. (1991) *J. Biol. Chem.* **266**, 8595-8603.
20. Anholt, R.R.H. and Rivers, A.M. (1990) *Biochemistry* **29**, 4049-4054.
21. Cali, J.J., Zwaagstra, J.C., Mons, N., Cooper, D.M.F. and Krupinski, J. (1994) *J. Biol. Chem.* **269** 12190-12195.
22. Schulz, K., Danner, S., Bauer, P., Schroder, S. and Lohse M.J. (1996) *J. Biol. Chem.* **271**, 22546-22551.
23. Willardson. B.M., Wilkins, J.F., Yoshida. T. and Bitensky, M.W. (1996) *Proc. Natl. Acad. Sci. USA* **93**, 1475-1479.

# Selective interactions of the rat μ-opioid receptor and a chimeric μ-opioid receptor expressed in COS-7 cells with multiple G proteins

Georgoussi, Z[1]., Hatzilaris, E[1]., Wise, A[2]. and Milligan, G[2].

[1] Institute of Biology, National Centre for Scientific Research "Demokritos", 15310 Ag. Paraskevi, Athens, Greece.

[2]Molecular Pharmacology group, Division of Biochemistry and Molecular Biology, Institute of Biomedical and Life Sciences, University of Glasgow, Glasgow G12 8QQ, Scotland, UK.

## 1. INTRODUCTION

Agonist activation of all three opioid receptor subtypes μ, δ, and κ has been reported to result in the inhibition of adenylate cyclase and/or the regulation of ion channels by activation of one or more pertussis toxin sensitive guanine nucleotide binding proteins acting as signal transducers (1). cDNA species encoding each of the μ, δ and κ opioid receptors have been reported (2 and references therein). However, the question of which heterotrimeric G protein(s) is involved in a specific opioid function remains still unclear. Therefore expression of these species in heterologous systems would allow a more detail examination of the signalling characteristics of these receptors. In order to explore in detail the molecular identity of the pertussis toxin sensitive G proteins activated by the μ-opioid receptor, we transiently transfected a wild type cDNA corresponding to rat μ-opioid receptor and a chimeric μ-opioid receptor construct composed of the rat μ-opioid receptor and a N-terminal hydrophilic Flag epitope, into COS-7 cells. Moreover, co-transfection of a cDNA encoding $G_{o\alpha}$ was also accomplished. The purpose of the present study was to investigate the specific interactions between the expressed μ-opioid receptors with native G protein(s) or co-transfected $G_{o\alpha}$.

## 2. MATERIALS AND METHODS

The rat μ-opioid receptor cDNA in pRC/CMV was generously provided by Drs. G. Uhl, Baltimore, MD and L. Yu, Bloomington, IN.
*2.1. Generation of the Flag-μ-opioid receptor construct:* Two oligonucleotides were designed as primers for the polymerase chain reaction (Fig.1). The 5' primer (I) in the sense orientation was designed to contain a $BamH_1$ restriction site, an initiation consensus sequence, the Flag sequence, and 27 nucleotides of the rat μ-opioid receptor cDNA. The 3' primer (II), in the antisense orientation was homologous to a 24 oligonucleotide sequence of the rat μ-opioid receptor cDNA containing a unique $BamH_1$ restriction site. A 680 base pair fragment was generated by standard PCR methodology using the rat μ-opioid receptor cDNA and the two oligonucleotide primers (Fig.1B). The PCR fragment was digested using the restriction enzyme $BamH_1$. The μ-opioid

NATO ASI Series, Vol. H 102
Interacting Protein Domains
Their Role in Signal and Energy Transduction
Edited by Ludwig Heilmeyer

receptor cDNA in pBluscript was also digested with $BamH_1$. The $BamH_1$ digested PCR fragment and the linearized plasmid were ligated to produce pBluscript-Flag-µ-opioid receptor cDNA. Transformants were screened by double-stranded DNA sequence analysis. Finally, the Flag-µ-opioid receptor construct in pBluscript was subcloned into $pcDNA_3$ for expression studies.

*2.2. Transfection of the µ-opioid receptor, Flag-µ-opioid receptor chimera into COS-7 cells:* COS-7 cells were cultured in DMEM containing fetal calf serum, 100 U/ml penicillin and 100 µg/ml streptomycin. COS-7 cells were transiently transfected using the DEAE dextran method with either i)pRC/CMV-µ-opioid receptor, ii)$pcDNA_3$-Flag-µ-opioid receptor, or with the above plus a cDNA encoding Goα. Cells (80% confluent) were harvested after 72 hrs, washed twice with PBS and scraped from the plates. Membranes were prepared as described by (3).

*2.3. Binding assays-GTPase activity:* Receptor binding assays and high affinity GTPase activity were performed as described by Georgoussi et al (3-5).

**A) Oligonucleotide primers**

**I.** 5' GCT GGA TCC CCC ACC ATG GAC TAC AAG GAC GAC GAC GAC AAG ATG GAC AGC ACC GGC CCA 3'
BamHI Consensus M D Y K D D D D K
Flag

**II.** 5' AGA GGA TCC AGT TGC AGA CGT TGA 3'

**B) Generation of the Flag-µ-opioid receptor construct**

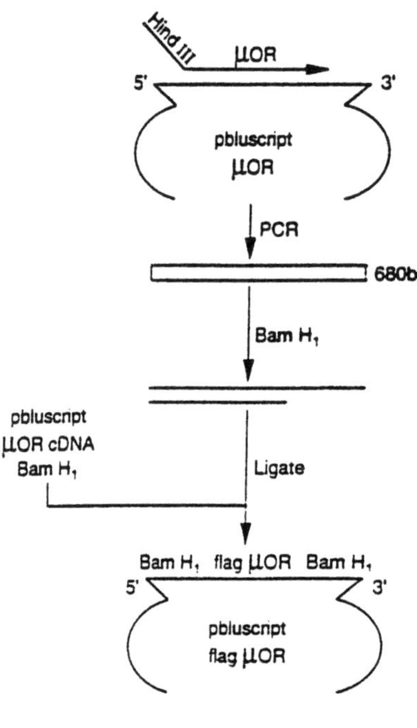


**C) Subcloning**

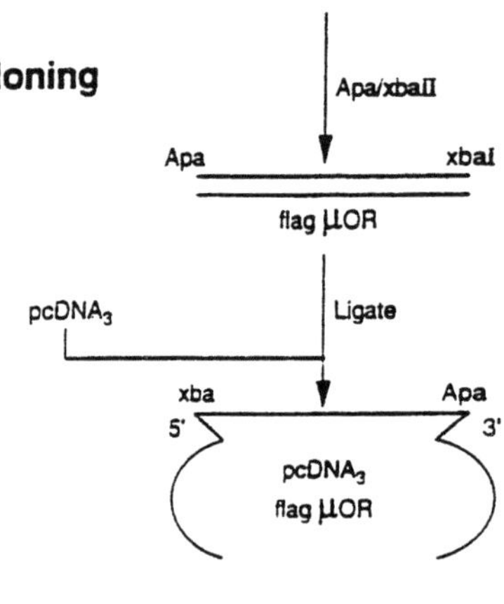


Fig.1: Generation and subcloning of the Flag-µ-opioid receptor construct.

# 3. RESULTS AND DISCUSSION

A chimeric μ-opioid receptor construct was generated with the use of two oligonucleotide primers (Fig.1). Using PCR directed insertional mutagenesis, nucleotides representing a hydrophilic Flag peptide (MDYKDDDDK) was genetically engineered in sequence with the extracellular amino terminal domain of the rat μ opioid receptor. The transformed construct (Flag-μ-OR) chimera and the wild type μ-opioid receptor were transiently transfected into COS-7 cells. As shown in Fig.2, membranes from these cells expressed 3800±80 fmoles per mg of membrane protein, of [$^3$H] Diprenorphine binding for the wild type μ-opioid receptor (μ-OR) and 570±180 fmoles/mg, for the Flag-μ-opioid receptor (Flag-μ-OR). Similarly, co-transfection of $G_{o\alpha}$ resulted also in expression of [$^3$H] Diprenorphine binding, although in lower potency, for both the wild type and the mutated μ-opioid receptor.

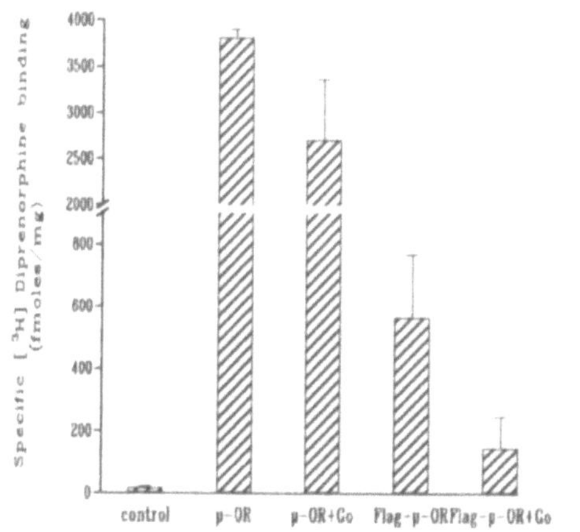


Fig.2: Binding of [$^3$H] Diprenorphine to membranes of COS-7 cells transfected with the wild μ-OR or the Flag-μ-OR chimera and $G_{o\alpha}$.

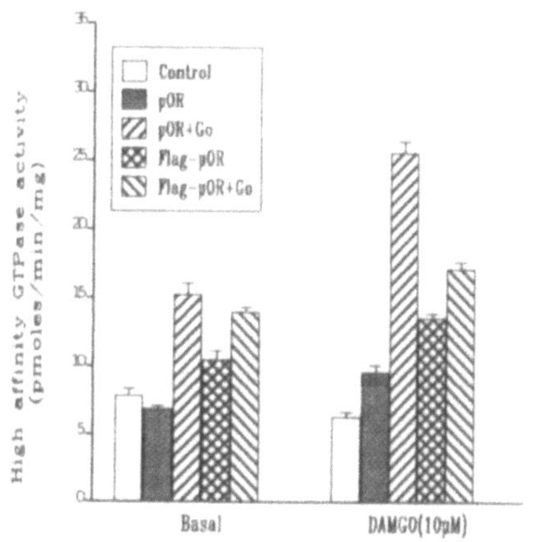


Fig.3: High affinity GTPase activity in membranes of COS-7 cells transfected with the cDNAs of the wild and the transformed μ-opioid receptor in the presence or absence of $G_{o\alpha}$.

To investigate the interactions of the transfected μ-opioid receptor(s) with G proteins we measured the ability of the receptors to stimulate the high affinity GTPase activity of membranes from COS-7 cells co-transfected with μ-opioid receptor(s) and $G_{o\alpha}$. As indicated in Fig.3, membranes from COS-7 cells transfected with the wild type μ-OR and $G_{o\alpha}$ or the Flag-μ-OR and $G_{o\alpha}$, displayed a basal high affinity GTPase activity of 15±0.5 and 13±0.3 pmoles/min/mg of membrane protein, respectively. Addition of the specific μ-opioid agonist DAMGO (10 μM) to membranes from both of these transfected cells resulted to an increase of the GTPase activity to 27±0.5 and 17±0.3 pmoles/min/mg of membrane protein respectively, indicating productive coupling of both μ-opioid receptor(s). Similarly, both the wild μ-OR or the Flag-μ-OR stimulated the high affinity GTPase activity in membranes of these cells (Fig.3). These results indicate that both transfected μ-OR(s) couple effectively to both the endogenous G protein population present in COS-7 cells, as well as to those co-transfected with $G_{o\alpha}$. Antipeptide antiserum OC2 generated against the $G_{o\alpha}$ gene has been characterized (6). Immunoblotting with antiserum OC2 generated against peptides common to polypeptides correspoding to products from both $G_{o1}$ and $G_{o2}$ splice variants, was able to identify a 39 kDa polypeptide in membranes of COS-7 cells co-transfected with $G_{o\alpha}$ in both μ-opioid receptor constructs indicating that $G_{o\alpha}$ was effectively expressed in this cell line (data not shown).

In conclusion, we have succeded to generate a chimeric receptor, comprising a small hydrophillic Flag peptide genetically engineered to the N-terminus of the rat μ-OR. The Flag-μ-OR chimera was fully functional. This was confirmed by assessing the binding characteristics of this receptor and by the effective coupling with the G proteins. Moreover, we have shown that both the wild μ-OR and the Flag-μ-OR successfully interact not only with the endogenous G protein population present in COS-7 cells, but also with $G_{o\alpha}$ following co-transfection of this protein into these cells, suggesting that the functional interactions between these μ-opioid receptors with the endogenous G proteins are not disrupted by the co-expression of $G_{o\alpha}$. These results also suggest that a single opioid receptor subtype can concurrently activate both of these G proteins.

## 4. REFERENCES


1. Childers S.R. Opioid receptor-coupled second messengers. Life Sci.48: 1991-2003 (1991).
2. Uhl G.R., Childers S.R. and Pasternak G. An opiate-receptor gene family reunion. Trends Neurosci. 17:89-101 (1994).
3. Georgoussi Z., Carr C. and Milligan G. Direct measurements of in situ interactions of rat brain opioid receptors with the guanine nucleotide-binding protein Go. Mol.Pharmacol.44:62-69 (1993).
4. Georgoussi Z., Milligan G. and Zioudrou C. Immunoprecipitation of opioid receptor-Go-protein complexes using specific GTP binding-protein antisera. Biochem. J. 306: 71-75 (1995).
5. Merkouris M., Dragatsis I., Megaritis G., Konidakis G., Zioudrou C., Milligan G. and Georgoussi Z. Identification of the critical domains of the δ-opioid receptor involved in G protein coupling using site specific synthetic peptides. Mol.Pharmacol.50:985-993 (1996).
6. McKenzie F.R. and Milligan G. δ-opioid-receptor-mediated inhibition of adenylate cyclase is transduced specifically by the guanine-nucleotide-binding protein $G_{i2}$. Biochem. J. 267: 391- 398 (1990).



## ACKNOWLEDGEMENTS

This study was supported by the European Commission (Z.G) and the European Union Grant Proposal CHRX-CT94-0689 (Z.G, G.M).

# Structural requirements for the EF-Tu-directed kinase

Thomas Plath[1], Charlotte R. Knudsen[2], Nese Bilgin[3], Carsten Lindschau[4], Volker A. Erdmann[1] and Corinna Lippmann[1]

[1]Institut für Biochemie, FU Berlin, Thielallee 63, D-14197 Berlin; Germany
[2]Aarhus University, Division of Biostructural Chemistry, 8000 Aarhus C; Denmark
[3]Institutionen för molekylarbiologi, Uppsala Universitet, BMC, Box 590, S-751 24 Uppsala, Sweden
[4]Franz-Volhard-Klinik am Max-Delbrück-Centrum, Humboldt Universität Berlin, Wiltbergstr. 50, D-13122 Berlin, Germany

## Introduction

The elongation factor Tu is an abundant protein in both prokaryotes and eukaryotes, and plays a central role in protein biosynthesis. It is a multifunctional protein which interacts with RNA, proteins and nucleotides. The function of the protein is the proper alignment of the coding aa-tRNA on the mRNA programmed ribosome. Recently we reported the posttranslational modification by phosphorylation of the prokaryotic elongation factor Tu (EF-Tu) *in vivo* (1, 2). Position of phosphorylation was determined as threonine 382 (in *E. coli*), located in a turn of a β-sheet at the c-terminal end of the molecule. Three-dimensional structure analysis revealed, that the site of phosphorylation is localized in the interface between domain 1 and 3 in the GTP-like conformation (3). The position is absolutely invariant in EF-Tu and even in the eukaryotic counterpart EF-1α (2).

We have investigated the structural requirements of the kinase from *Escherichia coli* by mutants, peptide-phosphorylation, heterologous substrates and mutagenesis.

## Results

The intact structure of the EF-Tu is a prerequisite for the kinase. EF-**Tu** stands for temperature **u**nstable. Thermal denaturation abolishes all activities of the elongation factor including GDP and GTP binding as shown in (Fig. 1). The kinase assay performed as described in (4) after preincubation of the substrate EF-Tu indicated,

NATO ASI Series, Vol. H 102
Interacting Protein Domains
Their Role in Signal and Energy Transduction
Edited by Ludwig Heilmeyer

that the structure of the active protein is necessary for recognition (Fig. 2). The phosphorylation efficiency was drastically reduced after thermal denaturation although the protein was still intact to more than 95 % as visible in the Coomassie stain (not shown).

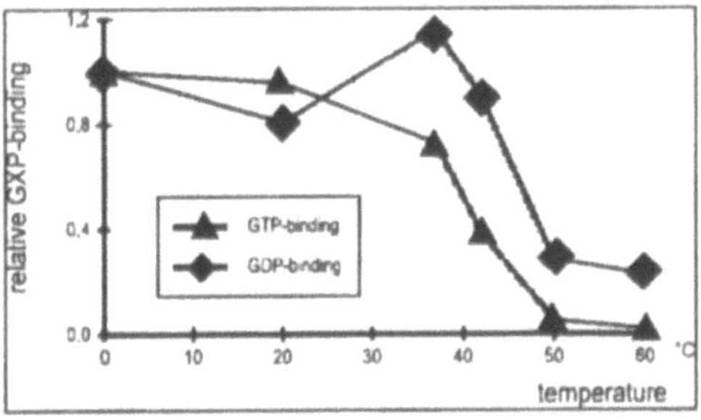


Fig. 1: GDP and GTP binding properties at different temperatures.

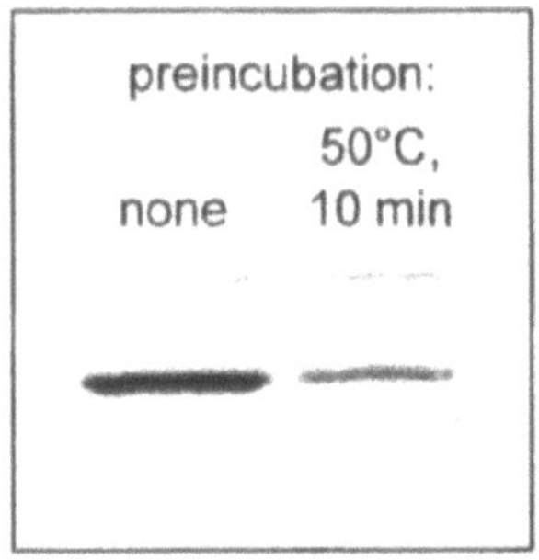


Fig. 2: Denaturation by preincubation of the EF-Tu leads to reduction of phosphorylation efficiency.

To estimate the minimal structure we synthesized a peptide comprising the residues 370-390 as substrate for the kinase. Structural prediction proposed a β-sheet folding with a turn similar to the structure in the native protein. This peptide could be phosphorylated and competed with the intact EF-Tu, but only at a twentyfold molar excess (Fig. 3).

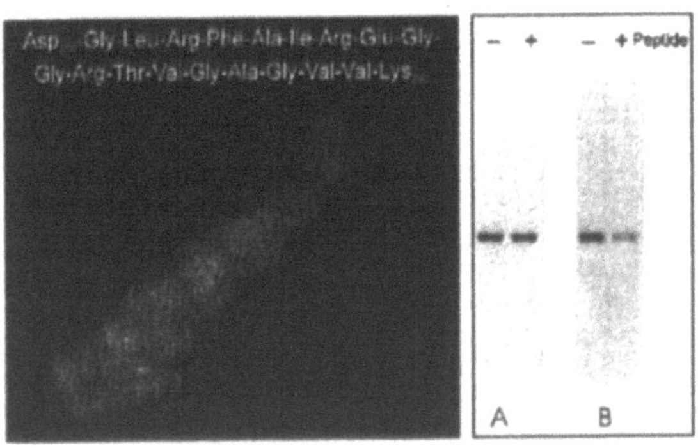


Fig. 3: Proposed structure of the substrate peptide (left) and its effect on *in vitro* phosphorylation of EF-Tu (right). A: Coomassie stain, B: autoradiography. In the assay shown here a twentyfold molar excess of peptide over Tu was used.

The conserved features of the kinase were investigated by a heterologous substrate incubation experiment. The kinase from *E. coli* was able to modify an EF-Tu from thermophilic bacteria (Fig. 4), but not a eukaryotic factor from *Artemia salina.*

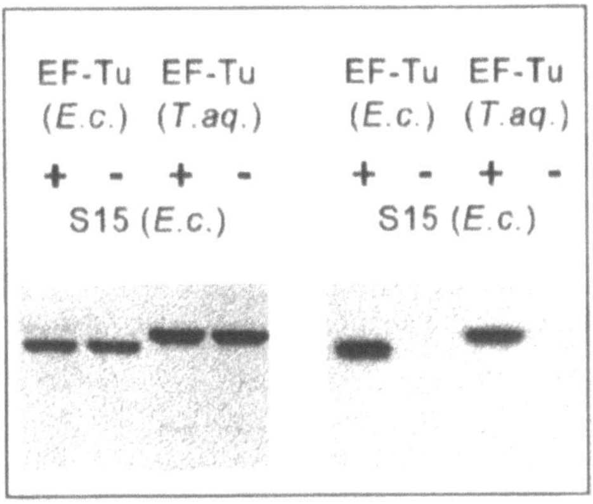


Fig 4: Phosphorylation of EF-Tu from *Thermus aquaticus* by the kinase from *E. coli*. Right panel shows Coomassie stain, left panel autoradiography. The S15 as source for the kinase was prepared as described in (4). Note that optimal temperature for the thermophilic elongation factor is above 60 °C.

The antibiotic kirromycin binds with high affinity to EF-Tu forcing this protein into a conformation that is unable to maintain its proper function during the translational elongation cycle. Kirromycin-resistance mutations are clustered around the site of phosphorylation (e.g. threonine 382 in *E. coli*; 5). We analyzed the effect of the point mutation Ala375→Thr. The mutant strains PM455 and LZ13 (6, 7) expressing only one *tuf* gene with this mutation. Two different approaches were performed: Using the EF-Tu contained in the S15 cell lysate as substrate (Fig. 5, left), we detected an even higher phosphorylation in the mutant strains compared to the wild type strain, independent of the presence of kirromycin, which was shown to inhibit the phosphorylation (4). When we used catalytic amounts of the kinase preparation of the mutants, surprisingly they were not able to phosphorylate the wild type EF-Tu (Fig. 5, right). This indicates that there is a parallel mutation in the kinase, may perhaps be by using the novel phosphoryliable residue as substrate and thereby preventing kirromycin binding.

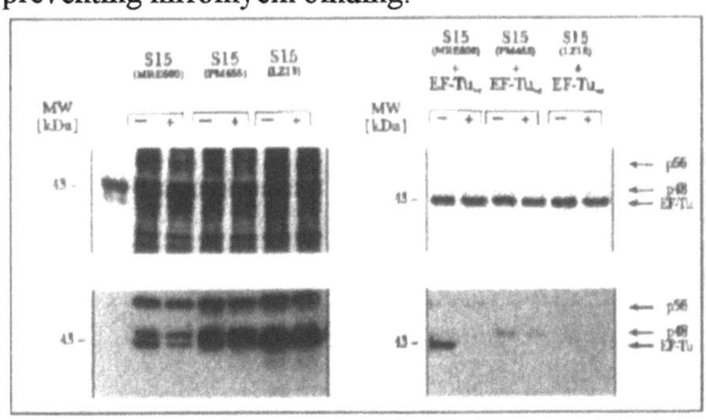


Fig 5: Kirromycin-resistant mutants phosphorylate EF-Tu stronger than the wild-type, but the extracts from the mutant strains are unable to modify the wild-type EF-Tu. Upper panel: Coomassie-stain; Lower panel: autoradiography. The left part shows the phosphorylation with non-catalytic amounts of cell-lysate, using the endogenous EF-Tu as substrate. The right part displays the *in vitro* phosphorylation with catalytic amounts using wild-type EF-Tu from *E. coli* MRE600 as substrate. (+) indicates the presence of kirromycin

These findings were substantiated by recent results, where we can demonstrate that a phosphoryliable residue has to be present. Chromosomal mutants of threonine 382 could only be found when a parallel mutation at 375 was present (Knudsen et al., in preparation). Analysis of plasmid derived mutations at position 382 revealed the importance for aa-tRNA binding, which was only possible with an exchange by serine. GDP and GTP binding was not affected by the amino acid exchange (Plath et al., in preparation). The substrate behavior of the mutated proteins is currently under investigation.

**Summary**

In summary, our results indicate a role of the EF-Tu-phosphorylation in the switch mechanism from the GTP- "on" conformation to the GDP-like "off" form which falls off from the ribosome. The kinase recognition motif is defined by the loop harboring the phosphorylation site, but nevertheless the less conserved three-dimensional environment is necessary for high affinity binding.

**References**


1. Lippmann C., Lindschau C., Buchner K. & Erdmann V.A. (1991) Phosphorylation of Elongation Factor Tu *in vitro* and *in vivo*. in: *Cellular Regulation by Protein Phosphorylation* (Heilmeyer, Jr. L.M.G., Ed.), Nato ASI Serie H 56, 441-445, *Springer-Verlag, Heidelberg*
2. Lippmann C., Lindschau C., Vijgenboom E., Schröder W., Bosch L. & Erdmann V.A. (1993) Prokaryotic elongation factor is phosphorylated *in vivo*. *J.Biol.Chem.*, 268, 601-607
3. Berchtold, H., Reshetnikova, L., Reiser, C. O. A., Schirmer, N. K., Sprinzl, M. & Hilgenfeld, R. (1993) Crystal structure of active elongation factor Tu reveals major domain rearrangements *Nature*, **365**, 126-132
4. Alexander C., Mesters J.R., Kraal B., Bilgin N., Hilgenfeld R., Lindschau C., Erdmann V.A. & Lippmann C. (1995) Phosphorylation of EF-Tu prevents ternary complex formation. *J.Biol.Chem.* 270, 14541-14547
5. Mesters, J. R., Zeef, L. A. H., Hilgenfeld, R., de Graaf, J. M., Kraal, B & Bosch, L. (1994) The structural and functional basis for the kirromycin resistance of mutant EF-Tu species in Escherichia coli *EMBO J.*, **13**, 4877-4885
6. Van der Meide, P. H., Vijgenboom, E., Dicke, M & Bosch, L. (1982) Regulation of the expression of tufA anb tufB, the two genes coding for the elongation factor EF-Tu in Escherichia coli *FEBS Lett.*, **139**, 325-330
7. Zeef, L. A. H & Bosch, L. (1993) A technique for targeted mutagenesis of the EF-Tu chromosomal gene by M13-mediated gene replacement *Mol.Gen.Gen.*, **238**, 252-260

# Effect of the Phytotoxin Fusicoccin on Plant Plasma Membrane $H^+$-ATPase Expressed in Yeast

Lone Baunsgaard[1,2], Anja T. Fuglsang[1,2], Michael G. Palmgren[1]

[1] Department of Plant Physiology, Institute of Molecular Biology, Copenhagen University, Copenhagen, Denmark

[2] Institute of Plant Biochemistry, Royal veterinary and Agricultural University, Copenhagen, Denmark

## 1. Introduction

The plant plasma membrane $H^+$-ATPase is an electrogenic $H^+$-pump transporting $H^+$ from the cytosol to the cell wall. This enzyme has several pivotal roles in the physiology of plants (Palmgren, 1997). The $H^+$-gradient generated across the plasma membrane provides energy for nutrient uptake and osmotic adaptation. The pump plays an obvious role in regulation of cytoplasmic pH. The lowered pH of the cell wall as a result of $H^+$ extrusion seems to be required for plant cell elongation. Furthermore, $H^+$ pump activity controls opening and closure of the stomatal apparatus allowing for water loss and $CO_2$ uptake into leaves.

The fungal phytotoxin fusicoccin promotes $H^+$ extrusion from plant tissues concomitant with an increase in elongation growth. It is well documented that fusicoccin produces a constitutively active form of the plant plasma membrane $H^+$-ATPase (reviewed in Palmgren, 1997). After fusicoccin activation the pump has higher ATP affinity, pH optimum shifts towards neutral pH and $V_{max}$ increases. Fusicoccin does not bind directly to the plasma membrane $H^+$-ATPase. The fusicoccin receptor has been purified and was found to belong to the family of 14-3-3 proteins (reviewed in de Boer, 1997). How the fusicoccin signal is transmitted from a 14-3-3 protein to the $H^+$-ATPase is not known.

The purpose of the present study was to identify the minimal number of components required for fusicoccin activation of the plant plasma membrane $H^+$-ATPase. The *Arabidopsis thaliana* plasma membrane $H^+$-ATPase isoform 2 (AHA2) expressed in yeast is not fully functional unless properly activated (Regenberg *et al.*, 1995). AHA2 was expressed in a yeast devoid of its own $H^+$-ATPase, and the effects of fusicoccin on growth rate and ATPase activity were studied.

## 2. Results and Discussion

### 2.1. Effect of Fusicoccin on the Growth Rate of Yeast Cells Expressing Plant Plasma Membrane $H^+$-ATPase

In the *Saccharomyces cerevesiae* strain employed in these studies (RS72; Cid *et al.*, 1987), the native promoter of the yeast $H^+$-ATPase (*PMA1*) has been replaced by a galactose-dependent (*GAL1*) promoter. Expression of plasmid-born plant $H^+$-ATPase is under the control of the constitutive *PMA1* promoter. When the yeast cells are grown on galactose medium, both the yeast and the plant $H^+$-ATPase are expressed. On glucose medium only the plant pump is expressed and growth of yeast cells is dependent on the plant enzyme.

NATO ASI Series, Vol. H 102
Interacting Protein Domains
Their Role in Signal and Energy Transduction
Edited by Ludwig Heilmeyer

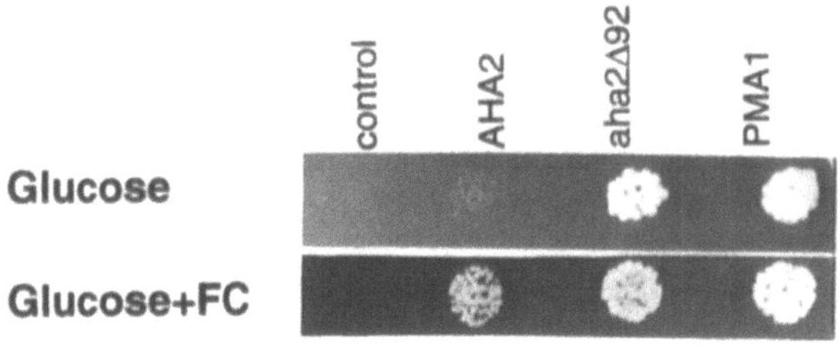


**Figure 1.** Yeast Strains Expressing AHA2 and aha2Δ92: Growth +/- fusicoccin (FC). Yeast drop test (Regenberg *et al., 1995)* on plates containing 2% glucose and 2% glucose + 10μM fusicoccin (pH 6.3). Typical of 3 independent experiments.

Yeast cells expressing AHA2 were barely able to grow on glucose medium (Fig. 1). A similar result was reported by Regenberg *et al.* (1995). A yeast strain carrying an empty vector expressing no $H^+$-ATPase did not grow at all. Strains expressing the plasmid borne yeast *PMA1* cDNA and a strain expressing a constitutively activated truncated mutant of AHA2 (aha2Δ92; Regenberg *et al.*, 1995) showed normal growth (Fig. 1).

Addition of fusicoccin to the growth medium resulted in an increase in growth rate of the yeast cells expressing AHA2 (Fig. 1). Fusicoccin exerted its effect at 5μM and above and worked at all tested pH-values. Fusicoccin did not have any effect on the growth rate of yeast cells expressing either no ATPase, yeast $H^+$-ATPase (PMA1) or constitutively activated plant $H^+$-ATPase (aha2Δ92) (Fig. 1).

These results show that the phytotoxin fusicoccin was able to trigger a growth response in yeast cells transformed with plant plasma membrane $H^+$-ATPase. Since in this particular strain the activation state of the plant $H^+$-ATPase determines the yeast growth rate (Regenberg *et al.*, 1995), the effect of fusicoccin most likely was to increase the activity of the plant enzyme expressed in yeast.

### 2.2. *In vitro* studies of fusicoccin response in *S. cerevesiae*

The plant plasma membrane $H^+$-ATPase exists in at least two conformations: a low activity state and a high affinity state (Baunsgaard *et al.,* 1996). The low activity state has low affinity for ATP ($K_m \approx 1$ mM) and a pH optimum $\approx 6.5$) whereas in the high activity state ATP affinity is increased ($K_m \approx 0.1$ mM) and pH optimum is displaced towards more alkaline pH (optimum $\approx 7.0$).

In order to examine whether fusicoccin in the yeast system influenced the activity state of the plant $H^+$-ATPase, the yeast strain expressing AHA2 was grown on glucose medium containing fusicoccin. Then cells were disrupted and membranes containing AHA2 were prepared. AHA2 is expressed in both the ER and the plasma membrane of transformed yeast cells (Regenberg *et al.* 1995) and accordingly both membranes were isolated. In both membranes the only major ATPase being expressed on glucose medium is AHA2. The dependency of ATP hydrolysis on ATP concentration (Fig. 2A) and pH (data not shown) was measured on freshly prepared membranes. No significant difference was seen between the ATP hydrolyzing activity of membranes prepared from yeast grown with or without fusicoccin in the medium. This suggests that if fusicoccin stimulates the plant plasma membrane $H^+$-ATPase *in situ* as it is expressed in yeast, this effect is not preserved after membrane isolation.

Attempts to activate the plant $H^+$-ATPase *in vitro* by fusicoccin were also made. When ER and plasma membranes from yeast cells expressing AHA2 were incubated in 5μM fusicoccin for 30 min at 30° no change in the pH profile of the enzyme was seen (Fig. 2B). In contrast, Moorhead *et* al. (1996) were able to stimulate plasma membrane $H^+$-ATPase activity in isolated spinach plasma membrane vesicles by incubating them with fusicoccin using the same conditions.

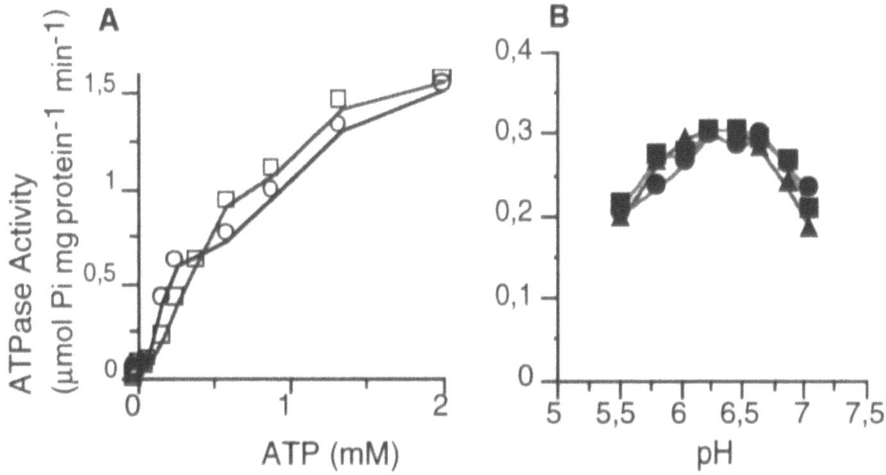


**Figure 2.** ATPase activity: Fusicoccin added *in vivo* and fusicoccin/Raf-peptide *in vitro*. **A**. ATP dependency of ER prepared from AHA2 expressing yeast grown on media minus fusicoccin (O); plus fusicoccin (□). The assay pH was 6.5. Km (ATP) values were estimated to 1.2 for both AHA2 preparations. **B**. pH dependency of plasma membranes at 3 mM ATP (▲); the same + incubation with 5μM FC (■); the same + incubation with phosphoSer259-Raf-1-peptide(●).

### 2.3. Effect of a Peptide Known to Disrupt 14-3-3 Interactions in Other Systems

Several studies have demonstrated that plant plasma membrane $H^+$-ATPase is not a receptor for fusicoccin (de Boer, 1997). Recently fusicoccin binding proteins were purified from plant materials and characterized as 14-3-3 proteins. Two 14-3-3 proteins (BMH1 and BMH2) are found in the yeast *S. cerevisiae*. Perhaps fusicoccin action in this organism involves binding of fusicoccin to one or both of these proteins. It has been proposed (Moorhead *et al.*, 1996) that 14-3-3 proteins bind directly to the plant plasma membrane $H^+$-ATPase, in this way inhibiting the enzyme, and that fusicoccin abolishes this negative interaction. The interaction between 14-3-3 proteins and their targets can be disrupted by the addition of synthetic peptides that represent 14-3-3 binding sites (Muslin, 1996). Thus, Moorhead *et al.* (1996) were able to activate $H^+$-ATPase in spinach plasma membranes by addition of a phosphorylated peptide (phosphoserine259-Raf-1) representing the 14-3-3 binding site of Raf.

Using the same incubation conditions as for the fusicoccin but adding 10 μM phosphoserine259-Raf-1-peptide instead, we were not able to detect any *in vitro* activation of ATP hydrolysis of AHA2 expressed in yeast membranes (Fig. 2B). This would suggest that either the plant $H^+$-ATPase present in purified yeast membranes is not inhibited by 14-3-3 protein or the interaction is not influenced by the peptide.

## 3. Conclusion

We have used a heterologous yeast expression system to characterize the plant plasma membrane $H^+$-ATPase. When activated, the pump is able to replace the endogenous plasma membrane ATPase PMA1 and support yeast growth. Using this system we are attempting to reconstitute into yeast components of plant signal transduction pathways leading to pump stimulation. Our data suggest that the components necessary for fusicoccin mediated activation of the plant plasma membrane $H^+$-ATPase already exist in yeast.

## 4. Materials and methods

*Strains and media. S. cerevisiae* RS72 (MATa ade1-100 his4-519 leu2-3,112; Cid *et al.*, 1987) and its transformants were used throughout this study. Cells were grown in medium with 2% glucose or galactose plus 0.7% yeast nitrogen base without amino acids (Difco) supplemented with adenine (30 μg $ml^{-1}$ ) and L-histidine (30 μg $ml^{-1}$). The liquid synthetic medium was buffered with 50 mM succinic acid adjusted to pH 5.5 with Tris. Solid media contained 2% agar. Fusicoccin (Sigma #F0537) was added to growth media at 5μM (to agar plates to respectively 0μM; 0.1μM; 1μM, 10μM and 50μM after solidification; pH in the plates were respectively 5.7; pH 6.0; pH 6.3, adjusting pH with $KHPO_4$). For membrane preparation of MP136, expressing AHA2, yeast was inoculated in synthetic medium with galactose as sole carbon source and transfered to complex glucose medium (YPD) for 3 cell divisions (16 hours). The culture was concentrated 5 times and further grown in YPD +/- 5μM fusicoccin for 1 more division. ER and plasma membranes isolation as in Regenberg *et al.*, 1995.

*ATPase assay.* ATP hydrolysis was assayed essentially according to Regenberg *et al.*, 1995, with 2 μg membrane protein at 30° C. When indicated, 5μM fusicoccin or 10μM phosphoSerine259-Raf-1-peptide (LSQRQRSTS*TPNVHMV) was added to membranes with expressed AHA2, incubated for 30 min at 30°C and the assay was started by adding 3 mM ATP at different pH values. (* represents the phosphorylated Ser 259 in Raf-1).

### References

Baunsgaard L., Venema K., Axelsen K.B., Villalba J.M., Welling A., Wollenweber B., Palmgren M.G. (1996). Modified plant plasma membrane $H^+$-ATPase with improved transport coupling efficiency identified by mutant selection in yeast. Plant J. **10**, 451-8

Cid A., Perona R., and Serrano, R. (1987). Replacement of the promoter of the yeast plasma membrane ATPase gene by a galactose-dependent promoter and its physiological consequences. Current Genetics **12**, 105-10.

de Boer A. (1997) Trends in Plant Science, in press.

Moorhead G., Douglas P., Morrice N., Scarabel M., Aitken A., Mackintosh C. (1996). Phosphorylated nitrate reductase from spinach leaves is inhibited by 14-3-3 proteins and activated by fusicoccin. Current Biology **6**, 1104-13.

Muslin A.J., Tanner J.W., Allen P.M., Shaw A.S. (1996). Interaction of 14-3-3 with signaling proteins is mediated by the recognition of phosphoserine. Cell **84**, 889-97.

Palmgren, M.G. (1997). Proton gradients and plant growth: Role of the plasma membrane $H^+$-ATPase. Adv. Bot. Res. **28**, in press.

Regenberg B., Villalba J.M., Lanfermeijer, F.C., Palmgren M.G. (1995). Carboxy-terminal deletion analysis of plant plasma membrane $H^+$-ATPase: Yeast as a model system for olute transport across the plant plasma membrane. Plant Cell, **7**, 1655-66.

# Organization and putative regulation of the glucose-specific phosphotransferase system from *Staphylococcus carnosus*.

Ingo Christiansen and Wolfgang Hengstenberg

Department of Microbiology, Ruhr-Universität Bochum, Universitätsstraße 150, D-44780 Bochum, Germany

**1. Introduction.** In bacteria, as in eukaryotes, phosphorylation reactions play a key role in energy and signal transduction. A system which is involved in both processes is the PEP-dependent phosphotransferase system (PTS), an active carbohydrate uptake of anaerob and facultative anaerob bacteria. In contrast to all other known transport systems, the PTS modifies its substrates which become phosphorylated during transport across the cytoplasmic membrane. As phosphoryl donor functions PEP. Thus, for energizing the translocation and activation of the sugar, only one ATP equivalent, under anaerob conditions, PEP is necessary. The phosphogroup is transferred in a reaction cascade of the different components of the PTS which play moreover an important role in regulation and signal transduction (for reviews see Hengstenberg et al., 1993; Lengeler et al., 1994; Saier et al., 1996; Postma et al., 1993).

**2. The PTS as carbohydrate uptake system.** As shown in Fig 1., the phosphogroup is transferred from PEP to the carbohydrate by the different components of the PTS: The first step is an autophosphorylation of a histidyl residue from the approximately 70 kDa protein Enzyme I (EI) at the expense of PEP. P-EI phosphorylates another histidyl residue of the approximately 10 kDa Heatstable Protein (HPr). Both proteins, generally encoded by the *ptsHI* operon, are cytoplasmic, constitutive and are called the general proteins of the PTS. This means they are sugar unspecific and react with all the different sugar specific Enzymes II (EII). These Enzymes II are the membrane integral, inducible permeases which normally consist of 3 or 4 functionally independent domains. EIIA and EIIB are hydrophilic and cytoplasmic and they both contain a phosphorylation site which is a histidyl-residue (IIA) and in most cases a cysteyl-residue (IIB), respectively. IIC is an integral membrane domain and serves as binding site and translocator for the substrate. Depending on the PTS and the bacterial species the domains are expressed independently or as multidomain fusion proteins (Fig. 1).

NATO ASI Series, Vol. H 102
Interacting Protein Domains
Their Role in Signal and Energy Transduction
Edited by Ludwig Heilmeyer

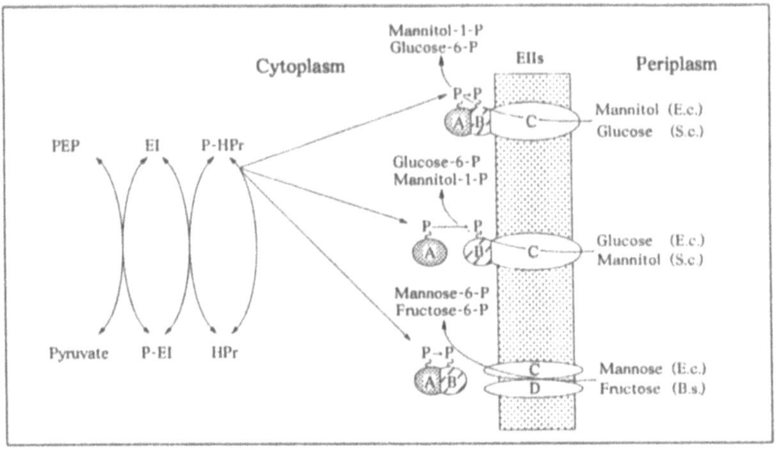


**Fig. 1.** Organization of various PTSs. EI and HPr are the general proteins of all PTSs, the Enzymes II are the sugar specific components of the PTS. Arrows indicate the direction of the biochemical steps. E.c. = *E. coli*, S.c. = *S. carnosus*, B.s. = *B. subtilis*. (Taken from Postma et al., 1993, modified).

**3. The PTS as regulatory system.** In one single bacterium the PTS can comprise up to 30 different catalytic and regulatory proteins or domains. They are building a network by transferring phosphogroups between themselves during the carbohydrate transport activities. The altering activities as consequence of the phosphorylation state of a number of PTS components regulate other metabolic proteins, leading to inducer exclusion or inducer expulsion, Furthermore they are responsible for differential expression of genes and operons encoding metabolic enzymes via mRNA antitermination or regulation of transcription, known as catabolite repression. The PTS proteins are also involved in regulation of chemotaxis proteins leading to a positive reaction towards PTS substrates.

**3.1. Catabolite repression in Gram-negative bacteria.** A regulatory key role in Gram-negative bacteria plays $EIIA^{Glc}$ which is an allosteric effector: nonphosphorylated it prevents uptake and metabolism of alternative carbon sources by inhibiting catabolic enzymes (glycerol kinase) or sugar permeases (lactose or melibiose permease). $P\text{-}EIIA^{Glc}$ is an activator of adenylate cyclase. Its product, cAMP, is a signal for global alteration of gene expression, exerted by binding the catabolite gene activator protein (CAP).

**3.2. Catabolite repression of Gram-positive bacteria.** In Gram-positive bacteria one central regulatory pathway is initiated by HPr. The product of the PEP-dependent reaction, P-His(15)-HPr, phosphorylates EIIs and in some cases regulatory activator proteins, e.g. the LevR protein from *B. subtilis* (Stülke et al., 1995). In contrast to Gram-negative bacteria, HPr can be phosphorylated in an ATP dependent reaction on Ser(46). P-Ser(46)-HPr has a 600-fold reduced PTS activity (Deutscher et al., 1984). The corresponding kinase is regulated negatively by $P_i$ and positively by ATP, 2-phosphoglycerate and fructose-1,6-bisphosphate, indicating good nutriental conditions. As $EIIA^{Glc}$ in Gram-negative bacteria, P-Ser(46)-HPr allosterically controls metabolic enzymes (sugar permeases, sugar-phosphate phosphatase) and interacts with a pleiotropic regulator CcpA (catabolite control protein). This interaction leads to a negative control of gene expression after binding to a short palindromic DNA sequence (CRE).

**3.3. Antitermination.** Often, genes or operons are prevented from expression by a preceding termination stemloop. The transcription is initiated after binding of an antiterminator protein to another short specific imperfect stemloop RAT (RNA target), overlapping the termitator structure. Antiterminators, controlling operons containing sugar specific EIIs and often enzymes for subsequent metabolism of the substrate can be phosphorylated and thus regulated by components of the general or corresponding PTS.

**4. The glucose-specific PTS from *Staphylococcus carnosus*.** Initial studies, using radioactive phosphorylation and in vitro phosphorylation of glucose, on the PTS from the Gram-positive, meat fermenting bacterium *Staphylococcus carnosus* indicated the existence of a glucose-specific Enzyme II fusion protein containg all three domains CBA. Such an $EIICBA^{Glc}$ was unique to that time, so we were interested in cloning the corresponding gene *ptsG* and characterisation of the gene product.

**4.1. Cloning.** Oligonucleotides were derived from the conserved sequences of the two known $IIA^{Glc}$ and $IIB^{Glc}$ from *B. subtilis* and *E. coli* and a 500 bp DNA region of the chromosome of *S. carnosus* was amplified by PCR. Thereby, a specific, homologous probe for cloning the complete *ptsG* gene was obtained.

Using this probe two overlapping *Bcl*I fragments (2k, 4k) and a 7 kbp *Hin*dIII fragment were cloned from an *S. carnosus* gene bank. Selection for positive clones was done by colony hybridisation (2k, 4k) or complementation of glucose fermentation of a $Glc^-$-mutant strain of *E. coli* (see below). The 7 kbp DNA fragment contains two ORFs (Fig. 2) with an identity of 73%, each coding for a EIICBA membrane permease (Christiansen and Hengstenberg, 1996).

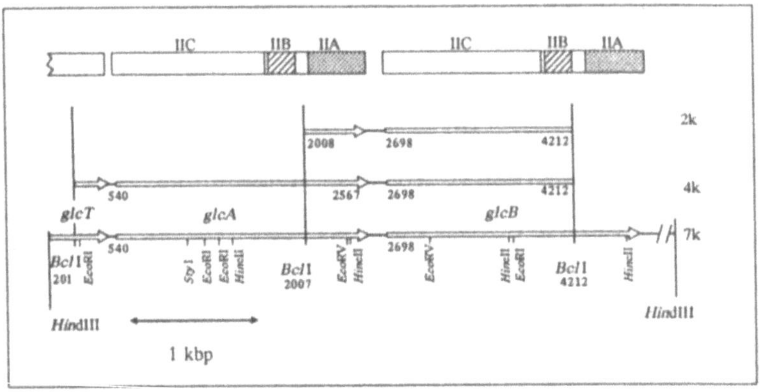


**Fig. 2** Restriction map of the cloned fragments 2k, 4k, and 7k and organization of the two identified ORFs designated as *glcA* and *glcB*, respectively (see below). Furthermore the domains of the encoded EIIs are shown.

**4.2. Expression and purification** In contrast to the *ptsG* gene from *B. subtilis*, whose expression in *E. coli* is toxic for the cells, the two separated ORFs of the 7k fragment can stably be maintained in *E. coli.* Separate overexpression of the two cloned genes in $Glc^-$-mutants of *E. coli* (*E. coli* WA2127Δ*ptsG*::$Cm^r$) restored glucose fermentation as shown on MacConkey agar plates. Membranes of these cells were prepared and the kinetic data of glucose phosphorylation were examined using a coupled fluorescence test. Both gene products phosphorylate glucose in a PEP-dependent reaction with similar $K_M$-values of 12 μM and 18.8 μM. Thus, the gene products were called $EIICBA^{Glc}1$ and $EIICBA^{Glc}2$, respectively. To adapt to the common nomenclature, the genes were called *glcA* and *glcB*.

To permit the purification of the gene products with $Ni^{2+}$-affinity chromatography, gene constructions were made, coding for an EII protein with a C-terminal histidine-hexapeptide fusion. After alkaline extraction of cell membranes, the proteins were solubilised with Triton-X-100 and purified by affinity chromatography and ion-exchange chromatography. From 1 g cell membranes 200 μg $EIICBA^{Glc}1$ and 160 μg $EIICBA^{Glc}2$ could be purified to homogenity (Christiansen and Hengstenberg, 1997).

The substrate specifity of both purified proteins was investigated by inhibition of the in vitro glucose phosphorylation. Although both transport glucose with a similar affinity, they show a different behaviour towards other glucosides (Fig. 3). No inhibition was observed with galactose, fructose, mannose, cellobiose, sucrose, maltose, lactose, melibiose, trehalose, sorbit, and N-acetylglucoseamine.

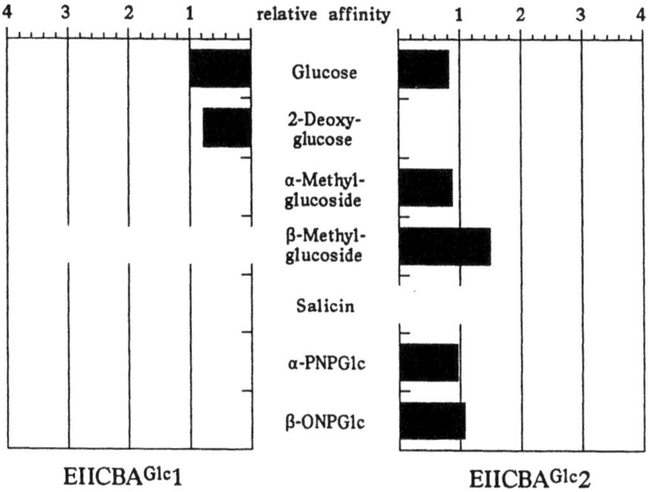


**Fig. 3** Inhibition of glucose phosphorylation via the glucose-specific PTSs of *Staphylococcus carnosus* by different glucosides. Relative comparison of $K_I$-values.

Yet unknown for the PTS is the existence of two closely related ORFs in close vicinity on a chromosome, coding for similar gene products with a common function. This supports the idea, that evolution of the PTS proteins has occured by doubling of an existent gene with subsequent differentiation of the substrate specifity.

**4.3. Elements of regulation** Computer aided analysis of the sequences led to the identification of putative elements of regulation like promoters, ribosome-binding-sites, termination stemloops and CRE-elements, which are involved in catabolite repression of Gram-positive bacteria. Upstream *glcA* the 3′ part of a third ORF was found, the gene product shows homology to antiterminator proteins. Recently, an antiterminator GlcT in *B. subtilis* was identified as well as its target sequence (Jörg Stülke, personal communication), which has a corresponding part in *S. carnosus* (85% identity). All these putative elements in *S. carnosus* lead to a model of gene regulation as it is shown in Fig. 4.

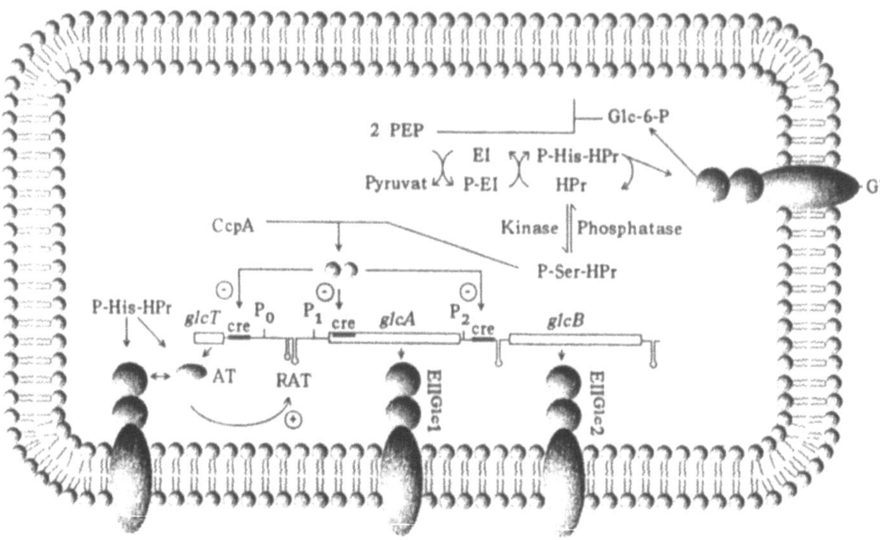


**Fig. 4.** Proposed model for putative pathways in *S. carnosus*, regulating expression of the genes *glcA* and *glcB* encoding glucose-specific membrane permeases. A positive regulation is exerted by an antiterminator (AT), a negative regulation is proposed by catabolite repression via CcpA and CRE (see text).

## 5. Literature


**Christiansen, I. and Hengstenberg, W. 1996.** Cloning and sequencing of two genes from *Staphylococcus carnosus* coding for glucose-specific PTS and their expression in *Escherichia coli* K-12. Mol. Gen. Genet. 250: 375-379.

**Christiansen I. and Hengstenberg, W. 1997.** Staphylococcal phosphoenolpyruvate-dependent phosphotransferase System: Two glucose permeases in *Staphylococcus carnosus* with different glucoside specifity. Protein engineering in vivo? Biochim. Biophys. Acta, submitted.

**Deutscher, J., Kessler, U., Alpert, C.A. and Hengstenberg, W. 1984.** Bacterial phosphoenolpyruvate-dependent phosphotransferase system: P-ser-HPr and its possible regulatory function. Biochemistry 23: 4455-4460.

**Hengstenberg, W., Kohlbrecher, D., Witt, E., Kruse, R., Christiansen, I., Peters, D., Pogge von Strandmann, R., Städtler, P., Koch, B. and Kalbitzer, H.R. 1993.** Structure and function of proteins of the phosphotransferase system and of 6-phospho-ß-glycosidases in Gram-positive bacteria. FEMS Microbiol. Rev.-12: 149-164.

**Lengeler, J.W., Jahreis K. and Wehmeier, U.F. 1994.** Enzymes II of the phosphoenolpyruvate-dependent phosphotransferase systems: their structure and function in carbohydrate transport. Biochim. Biophys. Acta 1188: 1-28.

**Postma, P.W., Lengeler, J.W. and Jacobson, G.R. 1993.** Phosphoenolpyruvate:carbohydrate phosphotransferase systems of Bacteria. Microbiol. Rev. 57: 543-594.

**Saier, M.H., Jr., Chauvaux, S., Cook, G.M., Deutscher, J., Paulsen, I.T., Reizer, J. and Ye, J.J. 1996.** Catabolite repression and inducer control in Gram-positive bacteria. Microbiology: 217-230.

**Stülke, J., Martin-Verstraete, I., Charrier, V., Klier, A., Deutscher, J. and Rapoport, G. 1995.** The HPr protein of the phosphotransferase system links induction and catabolite repression of the *Bacillus subtilis* operon. J. Bacteriol. 177: 6928-6936.

# Part VI

# Regulation of Muscle Contraction

# Single Molecule Myosin Mechanics Measured Using Optical Trapping

A.D. Mehta and J.A. Spudich

Departments of Biochemistry and Developmental Biology, Beckman Center, Stanford University School of Medicine, Stanford, CA 94305, U.S.A.

The idea of using light to manipulate matter was suggested by Hansch and Schawlow in 1975,[1] although the roots likely extend to nineteenth century Maxwell ideas, the understanding that light carries momentum. The optical tweezer, using a single focused laser beam to constrain remotely the position of a dielectric particle in aqueous solution, was demonstrated initially by Ashkin and colleagues in 1986.[2] Light cannot match more direct mechanical probes in crude force, but in this weakness lies its greatest strength: exquisite sensitivity to the tiny forces exerted by single cells and even molecules.

We have used optical tweezers to constrain the position of one micrometer-diameter polystyrene beads, for use as probes to study myosin-induced movement of actin filaments. The reconstitution of motile protein systems *in vitro* that allowed a quantitative assay for velocity was achieved initially by Sheetz and Spudich in 1983, using myosin-coated beads observed to move upon actin tracks on the inner surface of Nitella membranes.[3] In 1985, Spudich et al. demonstrated myosin movement using nothing except purified actin and myosin.[4] The *in vitro* motility assay used to study myosin today was devised by Kron and Spudich in 1986. They decorated a surface with myosin molecules, added fluorescent actin filaments, and observed their directed sliding upon addition of ATP.[5] Although we and others attempted to extract from filament velocities attachment kinetics and the "step size", the movement produced by a single molecule per ATP hydrolyzed, these estimates remained controversial and dependent on model assumptions. While several studies supported a step size around 10 nm,[6-8] others supported estimates well over 100 nm.[9-11]

The optical tweezers allowed us to construct the appropriate geometry to observe single molecules directly, bypassing the assumptions underlying previous estimates of the step size. The laser beams are focused through a microscope objective to diffraction-limited focal waists in a flow cell mounted upon a microscope slide. The polystyrene beads are chemically attached to single actin filaments. By moving the optical tweezers, we can trap the beads, stretch the filament taut, and lower it into closer proximity of myosin molecules,

NATO ASI Series, Vol. H 102
Interacting Protein Domains
Their Role in Signal and Energy Transduction
Edited by Ludwig Heilmeyer

which are attached to silica beads on the surface of a microscope coverslip (figures 1a and 2). We observed bead movement with some stepwise character at high myosin densities.[12] As we reduced the surface myosin density, we observed isolated transient bead deflections, although the bead spent most of its time at the baseline. The myosin binding events were prolonged by reducing the ATP concentration to 10mM or 1mM, since ATP binding precedes the detachment of myosin from the actin filament. These deflections were interpreted to reflect single myosin molecules binding to and pulling upon the actin filament. Myosin-induced actin filament motion shows a Michaelis-Menton-like dependence on ATP concentrations,[5] and we expect the molecular mechanism for generation of movement to remain independent of ATP concentration, aside from change of rate-limiting step.

The results of this initial study by Finer et al. supported a mean stroke distance of about 11 nm with a standard deviation of 2-3 nm, based on tabulation of visible bead deflections from the baseline.[12] We were very excited by these results, because they distinguished clearly between conventional models predicting a stroke near 10 nm and the much higher estimates suggesting less obvious mechanisms. It was clear at that time, however, that the value of 11 nm could be an underestimate or an overestimate due to uncertainties in the methods. For instance, the orientation of the surface-attached myosin with respect to the actin filament axis was random, and if the molecule lacks orientational flexibility, this could artificially decrease the measured stroke. In fact, we suggested the random relative orientation could underlie some of the variance in the bead deflections.[12] Moreover, compliant elements elsewhere in the system, such as the bead/actin interface, could absorb some displacement, again rendering bead movements in the trap an overestimate. We also pointed out that the step size could be smaller than 11 nm, since "displacements smaller than or equal to the level of the Brownian noise... were not scored, and as a result the distribution... may be truncated below ~5 nm."[12] In fact, later experiments first by Molloy et al.[13] and later by Mehta et al.[14] demonstrated this to be the case.

The possible presence of myosin binding events hidden in the background noise, it turns out, is coupled to the issue of variance in the distribution of bead deflections. In 1995, Molloy and colleagues,[13] using an instrument modeled after the Finer/Simmons apparatus, observed a small number of bead deflections in the opposite direction. Although one could conjecture several explanations, including a second filament connecting the beads in the opposite direction or an actual backward stroke following binding, Molloy et al.[13] suggested these backward events represented a visible, low-end subset of a broad distribution, and the 11 nm Finer et al. estimate[11] was based on the visible, high-end subset of the same distribution. In fact, based on thermal diffusion of the trapped actin filament, one expects a data distribution width consistent with this idea.

The Molloy et al. argument[13] becomes clear when one examines the position of the trapped bead with detectors fast enough to track the full frequency range of thermally driven bead movement. An example trace is shown in figure 3. Figure 3b reflects bead motion tracked with our current detectors, with a detector response time of 10.6ms. Figure 3a reflects the analogous trace we would observe with our older detectors, with a 1.4 ms rise time, used by Finer et al.[12] to measure the mean 11 nm deflections. The slow detector trace includes obvious deflections of 10-20 nm or more, but smaller displacements remain indistinguishable from thermally driven bead fluctuation.

The basic argument is summarized in figure 4. The position density of the bead at baseline encompasses 30-40 nm from peak to peak, assuming a trap stiffness in the range used for these experiments. The myosin-induced deflections reflect the end-point of a myosin stroke. Since the beads and attached filament are free to rotate, myosin binding sites should be accessible to surface-attached myosin throughout this range of diffusion. Hence, the start-point of the myosin strokes are also distributed over 30-40 nm, implying that the distribution of end-points should be at least this wide. Since the Finer et al.[12] data is distributed over a 15 nm range, a subset of data must be obscured from view, a possibility Finer et al.[12] considered and Molloy et al.[13] later demonstrated.

The distribution at the baseline is Maxwellian, and thus the bead spends most of its time at the center. Hence, most of the deflections should end at the stroke distance, other errors aside. In the absence of other variance sources, the bead deflections should fit a Gaussian distribution with the shape of the baseline position density and centered about the stroke distance. However, one must first detect all bead deflections, despite the fact that many of them are hidden in the thermal noise.

At the baseline, the optically trapped bead is constrained by its trap on one end and by the second trap, acting through the intervening actin filament, on the other. However, when a relatively noncompliant myosin binds to the actin filament, it effectively clamps the filament and decouples the beads from one another. The bead is now constrained by its trap on one end and by the actin filament to myosin surface attachment on the other. Since this new combination reflects a significant increase in the stiffness constraining bead position, one expects the thermal diffusion level to drop accordingly. This suppression of position noise is detected in figure 3 c-e, and is apparent in the data shown. Molloy et al.[13] used this clamping of bead diffusion to identify displacement events smaller than those scored by Finer et al.,[12] and they estimated a step size around 5 nm. We recently developed a new method of analysis that allows us to detect the full range of myosin binding events,[14] and our data agrees generally with that of Molloy et al.[13] Thus, we arrive at a distribution of shape and size resembling the baseline bead position density. Using the method of maximum likelihood to estimate fit parameters and uncertainties without binning into histograms, we find the distribution center to be about 5 nm, consistent with the Molloy et al.[13] estimate and with straightforward predictions from the crystal structure of the myosin head and neck region (S1).[15] As noted above, this could be a low estimate due to the random relative orientation of the myosin molecule and the actin filament.

Moreover, by tracking simultaneously the position of beads attached to both ends of the actin filament (fig 3 d, e), we have determined that both ends of the filament move by the same amount (fig 5), and at the same time (fig 6). This result is consistent with the conventional swinging crossbridge theory of muscle contraction but inconsistent with the competing actin powerstroke hypothesis, illustrated in figure 1.

A step size around 4-5 nm fits well with mechanical-ratchet based predictions from the S1 crystal structure, but complicates attempts to provide a clean picture of single molecule energetics. Several labs including ours have attempted to measure myosin forces under near-isometric conditions, and the results have varied from 2 to 7pN.[12,13,16] Although these numbers are also subject to potentially significant experimental errors, the 7pN estimate with a 5 nm stroke would provide 17pN.nm of work if the stroke resembles release of an elastic element,[17] and 35pN.nm if the force remains constant throughout the stroke.[7] Bagshaw reports the available work from ATP hydrolysis is 100pN.nm,[18] indicating that the myosin is 17-35% efficient when we take the *high-end* force estimate. However, frog muscles at zero degrees are reported to have 60% efficiency.[19] One would expect single molecules to be at least this efficient and probably more efficient given the likely tendency of molecules in filaments to work against each other sometimes. It remains possible that frog muscle myosin is simply more efficient than the rabbit skeletal proteins used in these experiments. Moreover, the range of force measurements might all be underestimates.

However, force estimates prior to single molecule assays have combined measured muscle filament tension with an estimated fraction of heads attached to arrive at the putative single molecule force.[19] While the range of single molecule displacement measurements remain within the range of previous predictions, a putative unitary force much higher than 7pN will be at odds with the range of such active fraction estimates in the muscle literature, creating a new mystery to solve. Hence, the picture remains incomplete.

The optical tweezer-based assay described here has allowed us to place increasingly precise estimates on mechanical and kinetic properties of single molecules, extending the growing range of approaches used to study motor proteins at this level.[20-26] The technique as described here and elsewhere[14] can be extended to test the kinetic and mechanical properties of any two interacting proteins, even in the absence of nanometer scale movements.

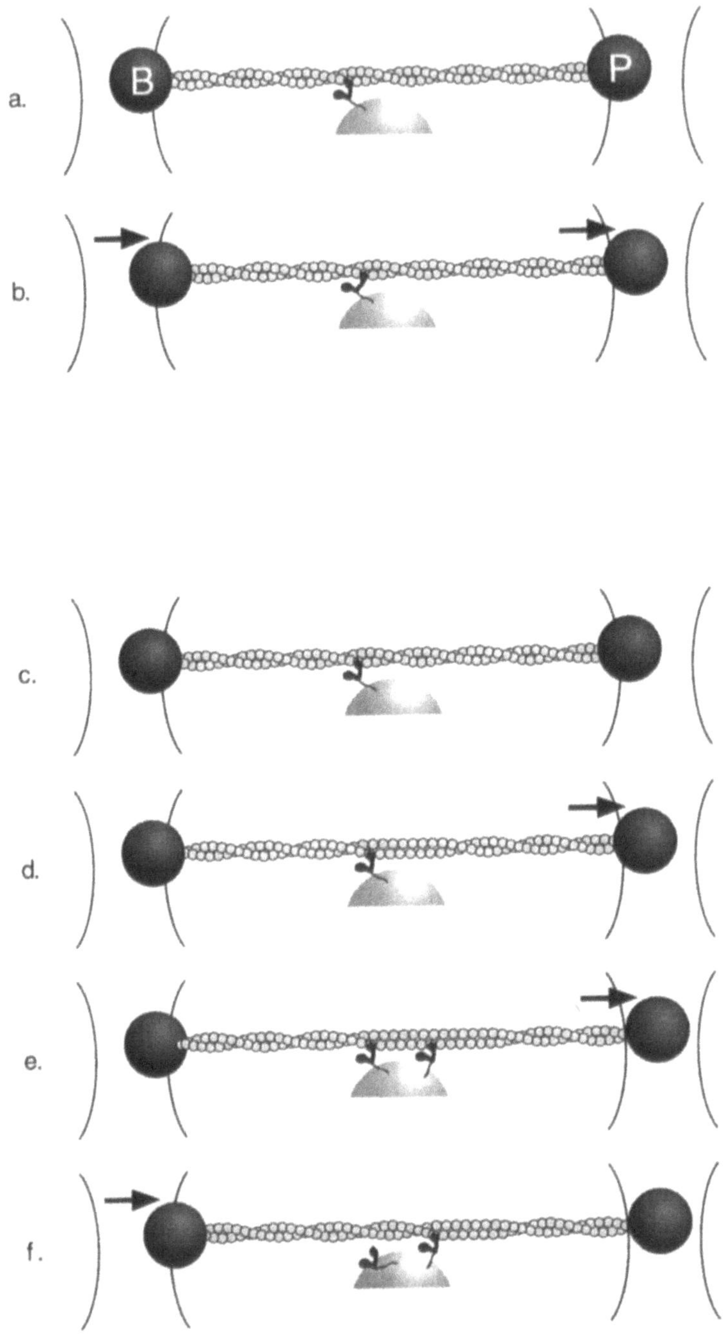


Figure 1

Figure 1. Schematic diagram of a myosin molecule binding to an actin filament held by optical tweezers (*a*). A single actin filament attached to polystyrene beads near each end is suspended in solution by two laser traps. The filament is then stretched taut and moved into close proximity of a silica bead attached to a microscope coverslip. The beads are sparsely decorated by myosin molecules. We then track bead position with nanometer and millisecond resolution by projecting a magnified image of the bead upon a quadrant photodetector.

The swinging crossbridge[27,28] and actin powerstroke models[29,30] make well defined predictions regarding actin filament behavior under the influence of a small number of myosin heads. The actin polymer has a well defined polarity, and is moved through myosin interactions in the direction of its pointed end and away from its barbed end. By the swinging crossbridge model, a single myosin will bind to an actin filament held taut by two optical traps, arresting it within the range of its baseline thermal diffusion and pulling it from this point, in the direction of its pointed end (*a,b*). Both ends should move simultaneously and by the same amount.

The actin powerstroke model predicts that the first myosin binding event arrests the filament somewhere within its range of diffusive motion and then triggers an actin structural change at the pointed end side of the myosin link (*c,d*). The schematic illustrates a helical unwinding suggested by Schutt and Lindberg,[29,30] but the scheme of filament end behavior reflects more general formulations.[31] In our experiment, this would cause bead P, attached to the pointed filament end, to relax into its trap. In the absence of a second myosin binding event, according to the model the ribbon will bind ATP and then return to helix form, restoring bead P to its original position. If one reduces the myosin concentration to the point where most interactions involve only a single molecule, one expects to see transient deflections with a myosin-induced component on one end of the filament only. Net filament movement depends on the attachment of a second myosin, predicted to happen more frequently than the stoichiometry would suggest since the ribbon form is assumed to have higher affinity for myosin. This is consistent with studies suggesting that myosin binding induces a relaxation of the actin filament shape.[32-34] If the preferred solution conformation is a helix form and myosin binding induces the transition to ribbon, thermodynamic arguments suggest the ribbon will have higher affinity for myosin.

A second myosin molecule would bind to the ribbon structure on the pointed end side of the first myosin link, causing further helix to ribbon transitions on the pointed end side of the new attachment and further relaxation of bead P into its trap (*e*). The first crossbridge would then detach, and the part of the ribbon form of the actin filament would return to helix form, pulling bead B, attached to the barbed filament end (*f*). Hence, according to this model, both ends of the filament would move by significantly different amounts, and at different times. Under experimental conditions of 5 uM ATP and given the second order rate constant for ATP binding to actomyosin,[4] one would expect average crossbridge attachment duration to be about 50 ms. The time lag between propagation of displacement from one filament end to the other should therefore be at least 50 ms, probably longer due to the time needed for ATP to bind along the actin filament. Moreover, after the first head detaches, the second head should remain attached for 50 ms on average. Such durations are well within the time resolution of our position detectors. Hence, we expect to see a systematic movement of probe position toward the pointed end as one progresses from the beginning of a myosin binding event if the actin powerstroke model is correct.

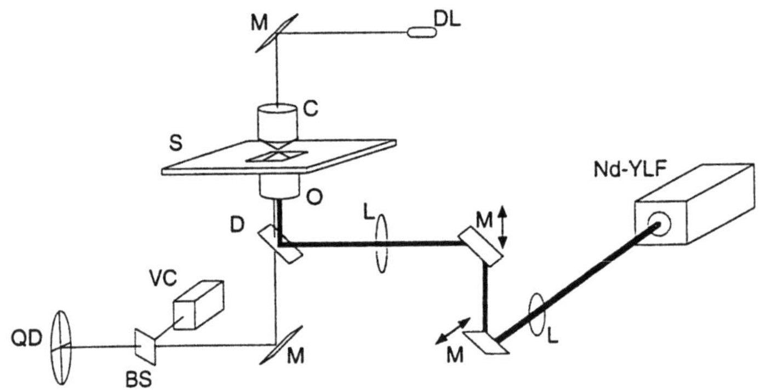


Figure 2. Schematic diagram of the laser trap system. The thick solid line represents the path of the trapping laser beam, and the thin solid line shows the brightfield imaging path. Illumination from a diode laser (DL) is used for visualizing the specimen plane S with a CCD camera (VC). The trapping laser beam is expanded and then focused upon plane S through the microscope objective (O). Moreover, we magnify and project an image of the trapped bead onto a quadrant photodetector (QD). Additional optics include lenses (L), mirrors (M), dichroic filter (D), beamsplitter (BS), and microscope condenser (C). The second trapping beam, fluorescence imaging pathways, and acousto-optic modulators used for rapid trap deflections are not shown. The I-V converters attached to the quadrant detector tend to limit imaging time resolution.

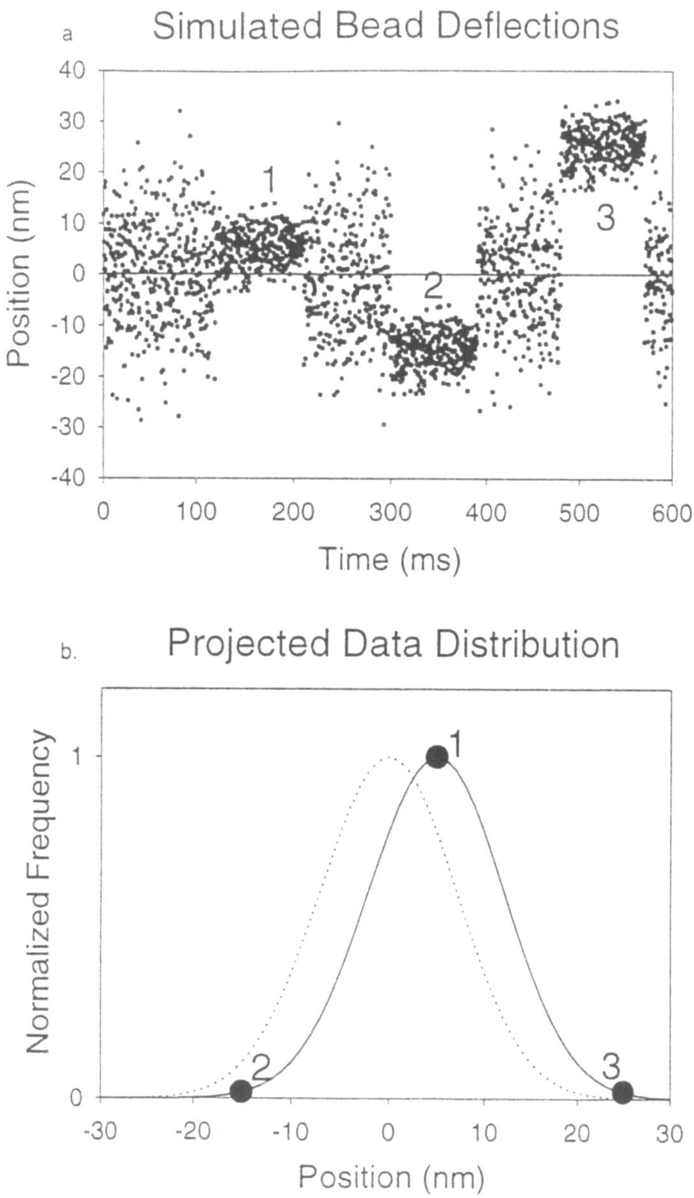


Figure 3. *a* shows a sample trace as it would appear if recorded with the 1.4 ms response time detector used in the early Finer et al.[12] experiments. *b* shows the trace as we actually observe it currently, using new detectors of response time 10.6 ms. The faster detector captures the full frequency range of bead diffusion. Hence, *b* reflects actual bead motion while *a* provides a filtered version. *c* shows a running measure of 15 point (0.75 ms) position variance. Note four regions in which the variance has fallen, indicating that the constraint upon the bead has suddenly become more stiff. The leftmost binding event is not evident in the slow detector data (*a*), and it typifies the myosin-induced bead deflections hidden by background noise in the original Finer et al. study.[12] In *d* and *e* we show simultaneous recordings of beads attached to the barbed and pointed filament ends respectively. Within detector resolution, the beads move by the same distance and at the same time.

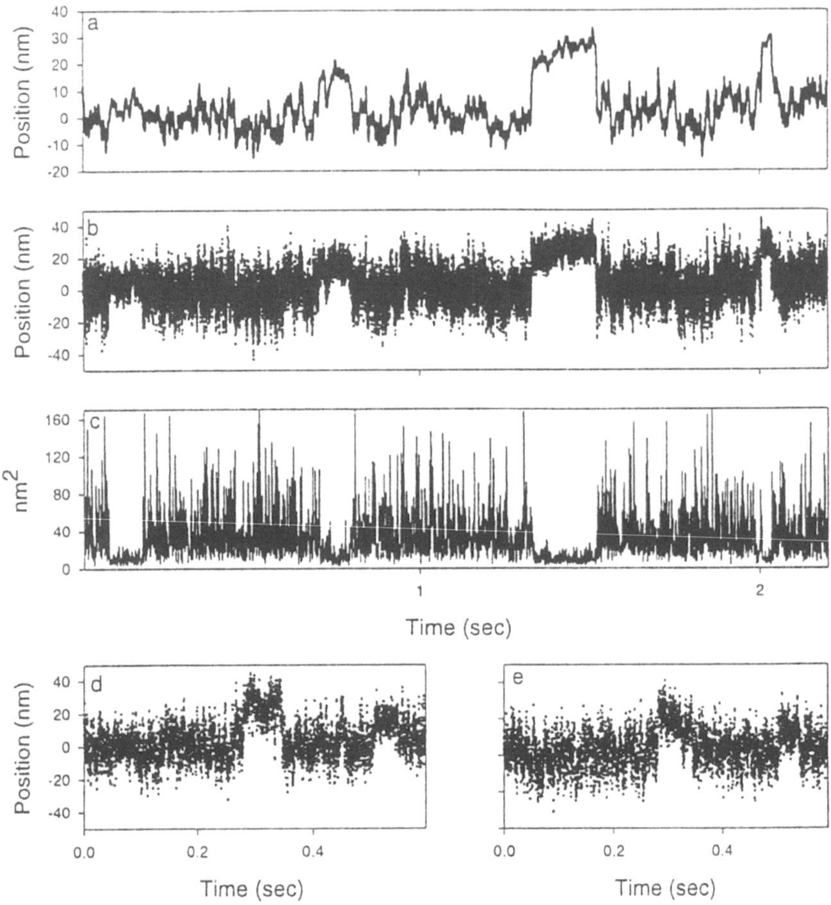


Figure 4. *a* is a schematic illustration of transient arrests of the bead within its range of thermal motion, followed by a 5 nm upward movement. Event 1 reflects myosin binding while the bead is at position 0 nm and pulling the bead to position 5 nm. Since the bead spends most of its time at the center of the distribution, this reflects the most common myosin-induced deflection. In 2, the bead undergoes a thermal fluctuation to position -20 nm. At this instant, myosin binds and pulls the bead up to -15 nm, the low end extreme in the projected data. This illustrates the "backward strokes" reported by Molloy et al.[13] In 3, myosin binds after a thermal fluctuation carries the bead to position 20 nm. The myosin pulls it further up to 25 nm, reflecting the other extreme in the data we anticipate.

*b* illustrates the expected histogram shape characterizing the measured bead deflections. The dotted line represents the baseline density distribution due to thermal motion in the absence of myosin binding. If myosin binds the filament with equal probability anywhere in this range of motion, the distribution of bead deflections should be exactly this baseline density distribution, shifted upward by the working stroke distance. Points 1, 2, and 3 on this expected distribution contain the "data points" in the simulated trace. Much of this distribution sits within 5 nm of the center, exactly in the range obscured from view when one examines filtered records of bead position.

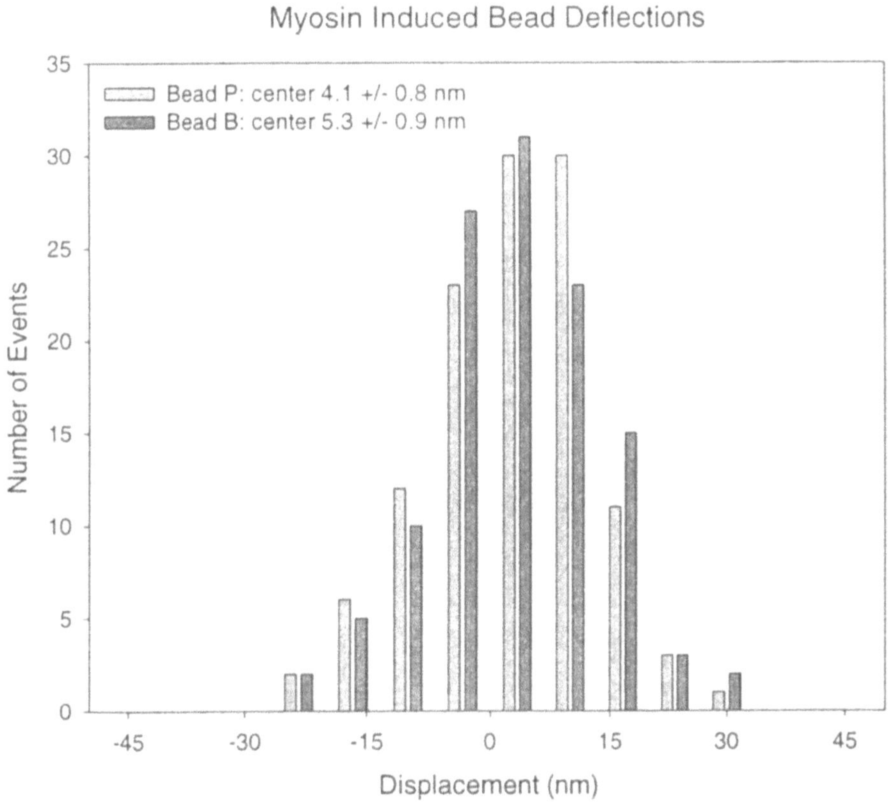


Figure 5 Tabulation of bead deflections, after correction for displacement absorbed by compliant bead/filament connections and elimination of traces affected by low frequency drift. Independent measurements of the stroking distance on either side of the filament yield distributions centered about 4.1 +/- 0.8 nm for bead B and 5.3 +/- 0.9 nm for bead P. The uncertainties are calculated by the method of maximum likelihood, assuming that all errors are statistical, and do not account for possible systematic errors such as that which might be caused by the random relative orientation of surface bound myosin and the actin filament. Both distributions have a variance of 79 nm, the same as the variance of baseline Brownian motion, consistent with the postulate that variance in bead deflection amplitudes is caused by myosin binding to the actin filament anywhere in its range of diffusive motion with equal probability.[13]

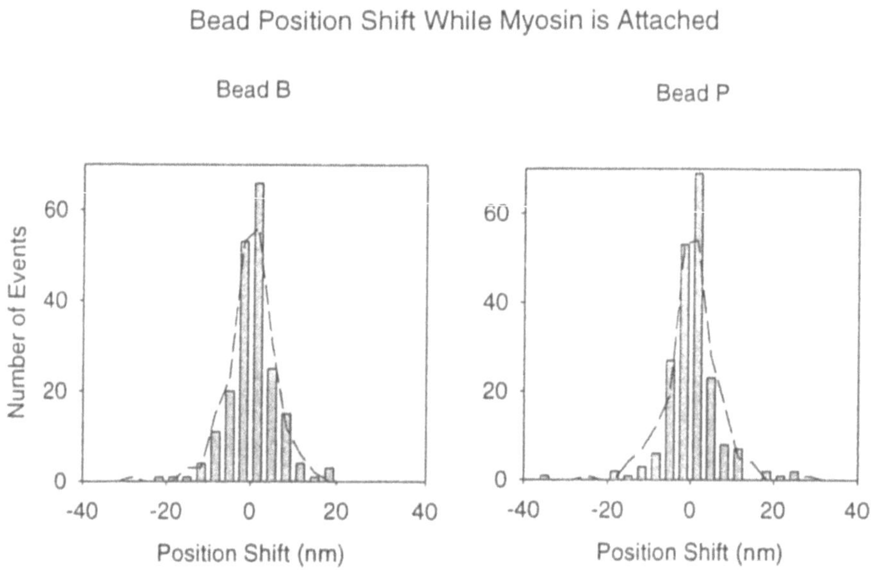


Figure 6. All myosin binding events with attachment times exceeding 60 ms (205 out of 387) were examined to determine if the actin filament undergoes additional displacement while the myosin is attached. The actin powerstroke model predicts that the barbed end of the filament will undergo an additional step once the initially bound myosin releases. On average, the initial myosin will remain attached for 50 ms, and the second myosin for another 50 ms afterwards. Hence, one expects the average event to include a transition and further movement of the barbed end 50 ms after the myosin binding event begins and 50 ms before it ends.
Shown are the differences between bead position during the last 30 ms and first 30 ms of each selected binding event. We see an average shift of only 0.5 nm on each bead, essentially zero within experimental error and far short of that predicted by the actin powerstroke theory. Consistent with the swinging crossbridge mechanism, all movement seems to occur more or less just after the myosin attaches. The dotted line reflects a comparable analysis done to an equal number of 100 ms baseline intervals. The distribution overlap and similarity of width indicate the variance in the distributions are statistical and of no relevance to the actin/myosin interaction.

**References**


1. Hansch, T.W. & Schawlow, A.L. *Opt. Commun.* **13**, 68-??? (1975)
2. Ashkin, A., Dziedzic, J., Bjorkholm, J., & Chu, S. *Opt. Lett.* **11**, 288-??? (1986)
3. Sheetz, M.P. & Spudich, J.A. *Nature* **303**, 31-35 (1983)
4. Spudich, J.A., Kron, S.J., & Sheetz, M.P. *Nature* **315**, 584-586 (1985)
5. Kron, S., and Spudich, J.A. *Proc. Natl. Acad. Sci. USA* **83**, 6272-6276 (1986)
6. Ford, L. E., Huxley, A. F., and Simmons, R. M. J Physiol. **269**, 441-515 (1977)
7. Huxley, A.F., & Simmons, R.M. *Nature* **233**, 533-538 (1971)
8. Uyeda, T. Q. P., Kron, S. J. & Spudich, J. A. *J molec. Biol.* **214**, 699-710 (1990).
9. Harada, Y. & Yanagida, T. *Cell Motil. Cytoskel.* **10**, 71-76 (1988).
10. Harada, Y., Sakurada, K., Aoki, T., Tomas, D. D. & Yanagida, T. *J. molec. Biol.* **216**, 49-68 (1990)
11. Yanagida, T., Arata, T. & Oosawa, F. *Nature* **316**, 366-369 (1985)
12. Finer, J. T., Simmons, R. M., & Spudich, J. A. *Nature* **368**, 113-118 (1994)
13. Molloy, J. E. et. al. *Nature* **378**, 209-212 (1995)
14. Mehta, A. D., Finer, J.T., & Spudich, J. A. in press
15. Rayment, I., et al. *Science* **261**, 50-58 (1993)
16. Ishijima, A., et al. *Biophys. J.* **70**, 383-400 (1996)
17. Finer, J.T., Mehta, A.D., and Spudich, J.A. *Biophysical Journal* **68**, 291s-297s (1995)
18. Bagshaw, C.R. *Muscle Contraction.* Chapman & Hall, London. (1993)
19. Cooke, R. in press
20. Howard, J., Hudspeth, A. J. & Vale, R. D. *Nature* **342**, 154-158 (1989).
21. Kuo, S. C. & Sheetz, M. P. *Science* **260**, 232-234 (1993)
22. Svoboda, K., Schmidt, C.F., Schnapp, B.J., & Block, S.M. *Nature* **365**, 721-727 (1993)
23. Hunt, S. J., Gittes, G. & Howard, J. *Biophys. J.* **67**, 766-781 (1994)
24. Ishijima, A. et. al. *Biochem. Biophys. Res. Commun.* **199**, 1057-1063 (1994)
25. Miyata, H. et. al. *J. Biochem.* **115**, 644-647 (1994)
26. Yin, H. et. al. *Science* **270**, 1653-1657 (1995)
27. Reedy, M. K., Holmes, K. C. & Tregear, R. T. *Nature* **207**, 1276-1280 (1965)
28. Huxley, H. E. *Science* **164**, 1356-1366 (1969)
29. Schutt, C. E. & Lindberg, U. *Proc. Natl. Acad. Sci.* USA **89**, 319-323 (1992)
30. Schutt, C. E. & Rozycki, M. D., Chik, J. K., & Lindberg, U. *Biophys. J.* **68**, 12s-18s (1995)
31. Pollack, G. H. *Biophys. Chem.* **59**, 315-318 (1996)
32. Oosawa, F., Fujima, S., Ishiwata, S., & Mihashi, K. *Cold Spring Harbor Symposia on Quantitative Biology* **XXXVII**, 277-285 (1972)
33. Yanagida, T. & Oosawa, F., *J. Mol. Biol.* **126**, 507-524 (1978)
34. Miki, M. & Kouyama, T. *Biochemistry* **33**, 10171-10177 (1994)


ACKNOWLEDGEMENTS. We gratefully acknowledge K. Eason and B. Simmons for the design of the high-bandwidth I-V conversion circuit used in this experiment.

# Studies on the ATP-binding Site of Actin Using Site-directed Mutagenesis

Herwig Schüler, Elena Korenbaum, Uno Lindberg and Roger Karlsson

Department of Cell Biology, The Wenner-Gren Institute, Stockholm University, S-106 91 Stockholm, Sweden

## Introduction

Hydrolysis of the actin-bound ATP is linked to actin filament turnover. However, neither the role nor the mechanism of the actin ATPase are well established. A novel role for the actin ATPase in muscle contraction has been proposed, making actin the principal force generator (1). The crystal structures of α- (2,3) and β-actin (4) suggest involvement of serine 14 (S14) and aspartic acid 157 (D157) in ATP hydrolysis (2,5). Myslik (5) suggested that the serine hydroxyl, polarized by the aspartic carboxyl, might be transiently phosphorylated under the hydrolysis reaction. According to his analysis (5) of the β-actin structure (4), the geometry of the ATP-site would favor an in-line attack on the ATPγP by the serine hydroxyl. This model was also inspired by the geometric homology of actin to the ATPase domain of Hsc70, which places alcoholic and acidic side chains at similar positions (6).

We have introduced a series of mutations into the β-actin gene, expressed the corresponding mutant proteins in the yeast *S.cerevisiae* and analyzed their properties (7). Here, we compare the actin single mutant S14C and the double mutant S14C,D157A to yeast-expressed wild-type β-actin with respect to polymerizability, ATPase activity, and susceptibility to proteolytic cleavage. Further, the interaction of these actin mutants with myosin was tested in the *in vitro* motility assay and by S1-decoration of mutant actin filaments. Chen & Rubenstein (8) made several mutations of S14 in yeast actin, among them also S14C. However, in the absence of wild type actin, yeast cells expressing these proteins failed to grow; only S14A actin was isolated for analysis. An actin with a replaced D157 has not been analyzed before.

## Methods

Oligonucleotide-directed mutagenesis of the chicken β-actin gene was described earlier (7). The mutant β-actins were produced in *S.cerevisiae* using the temperature-inducible expression system described by Karlsson (9) and isolated by combining affinity chromatography on DNase I-Sepharose with ion-exchange chromatography on hydroxyapatite as in previoius studies (7). Polymerization experiments were carried out at 25°C in 5 mM Tris-HCl, pH 7.6, 0.5 mM ATP, 0.1 mM $CaCl_2$, 0.5 mM DTT, 0.03% $NaN_3$, containing either 2 mM $Mg^{2+}$ or 100 mM $K^+$. To study filament formation, 150 μl of the samples containing 12 μM actin including 2% pyrenyl-labeled β/γ-actin (10) were added to microplate wells. After salt addition the increase in pyrenyl fluorescense at 410 nm (excited at 365 nm) was followed using a

NATO ASI Series, Vol. H 102
Interacting Protein Domains
Their Role in Signal and Energy Transduction
Edited by Ludwig Heilmeyer

Fluoroscan II microplate reader (Labsystems). The critical concentration for actin polymerization ($A_{cc}$) was determined by measuring DNase I inhibition activity in supernatants obtained after ultracentrifugation of polymerized actin samples (11). The actin ATPase was assessed using $\gamma$-$^{32}$P-labeled ATP essentially as described in 12 and 13. Both methods gave the same values. Curve fittings were done using the nonlinear least squares fitter of Microcal Origin. The *in vitro* motility assay (14) was performed in the presence of an oxygen scavenger system (15).

**Results**

We examined the polymerization kinetics of the actin mutants by co-polymerization with pyrenyl-labeled actin. When polymerization was induced with 2 mM $Mg^{2+}$, both S14C and S14C,D157A mutant actin polymerized without a lag phase (Fig. 1A). The rate of elongation in arbitrary units (AU) per minute was 0.28 for the single mutant and 0.22 for the double mutant. The yeast-expressed wild-type β-actin on the other hand showed a brief lag phase before elongation which took place with a rate of 0.12 AU/min. Thus, with the mutants, the polymerization process was approximately 2 to 2.5 times faster than observed for wild-type actin under the same conditions. This fact, however, did not seem to affect the $A_{cc}$ since, for both mutant actins, the $A_{cc}$ values were not significantly different from the value for wild-type actin (0.41±0.12 μM).

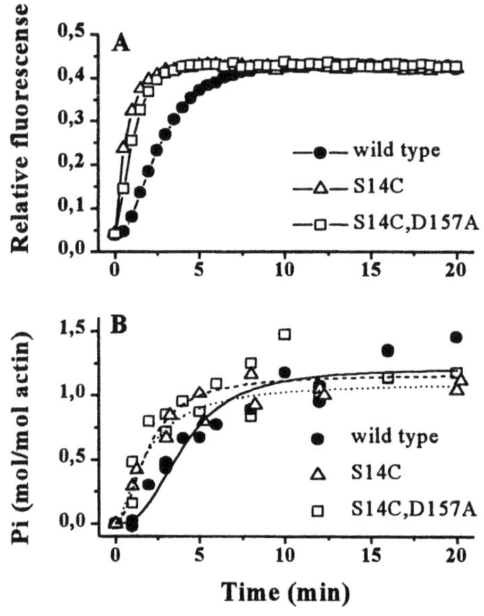


**Fig. 1. A.** Polymerization of wild type and mutant actins containing 2% pyrenyl-labeled β/γ-actin and 2 mM $Mg^{2+}$. The concentration of actin was 12 μM. Filament formation was registered as an increase in pyrenyl fluorescense.

**B.** ATPase activity of wild type and mutant actins during $Mg^{2+}$-induced polymerization. G-actin containing $\gamma$-$^{32}$P-ATP was polymerized as in A. Aliquots were taken at intervals and $^{32}P_i$ content was measured.

Under $K^+/Ca^{2+}$ conditions (Fig. 2), the three actins polymerized differently from what was observed in the presence of $Mg^{2+}$; a lag phase was observed for all three, and the elongation rates distinguished the two mutants from each other. The rate for the double mutant was 0.001 AU/min as compared to 0.0015 and 0.0019 AU/min for

S14C and wild-type actin respectively. This reduced polymerizability was also reflected by a significant increase in $A_{cc}$ (3.8±0.1 μM for the single and 5.1±0.6 μM for the double mutant) compared to that of the wild type (2.1±0.2 μM).

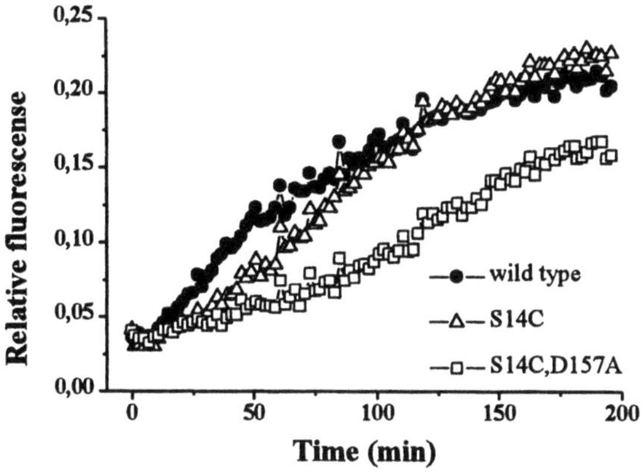


**Fig. 2.** Polymerization of wild type and mutant actins with 2% pyrenyl-labeled β/γ-actin and 100 mM $K^+$. The concentration of actin was 12 μM, and filament formation was measured as in Fig. 1A.

The ATPase activity was monitored during $Mg^{2+}$-induced polymerization. A faster rate of hydrolysis compared to wild-type actin was observed for both mutants, which also could be separated from each other with respect to this quality (Fig. 1B). Thus after replacing these residues, both $Mg^{2+}$-induced filament elongation and ATP hydrolysis rates were increased relative to the wild type.

Myosin S1-decoration of wild-type and mutant actin filaments revealed no difference in myosin binding, and when the interaction of the mutant actins with myosin was examined in the *in vitro* motility assay, the rates of filament movement did not differ significantly (not shown). This shows that myosin interaction and force generation were not dramatically affected by the introduced mutations.

Finally, the wild-type and double mutant actins were incubated separately with subtilisin and trypsin. Both proteases generated cleavage patterns of the double mutant protein that differed from the corresponding wild-type patterns (not shown), suggesting that structural changes of the mutant actin molecule had exposed different cleavage sites.

**Discussion**

We found that mutating actin residues S14 and D157 into C and A, respectively, did not abolish the ATPase activity of the protein. Under $Mg^{2+}$ polymerization, ATP was hydrolyzed faster by the mutants than by wild type actin. Thus residues S14 and D157 in actin do not seem to play a key role in the hydrolysis of the actin-bound nucleotide. Filament formation was affected with both the single and double mutant actin. Additonal experiments are needed to determine whether the absence of a lag-phase in $Mg^{2+}$ polymerization was due to nuclei in the mutant G-actins. It is not yet clear whether the altered elongation rates reflect a change in affinity between monomers or whether they are a result of the changed ATPase activity of the mutants. The opposite effect of $Mg^{2+}$ versus $K^{+}/Ca^{2+}$ on polymerization suggests an altered affinity for the divalent cations. Further, these mutations led to an altered susceptibility of the actin to proteolytic cleavage, suggesting that the overall structure of the actin was slightly changed after the mutagenesis albeit without affecting myosin binding. Therefore these residues appear to stabilize the actin fold and may mediate conformational changes in the ATP binding site that propagate to more distant regions of the molecule. Our observations with the mutant β-actins did not show dramatic deviation from wild-type behaviour. However, replacing S14 in yeast actin is lethal (8), suggesting a crucial role for this residue to maintain proper actin function.

## References


1. Schutt, C.E. & Lindberg, U. (1992), Actin as the generator of force during muscle contraction. PNAS 89, 319-323
2. Kabsch, W. *et al* (1990), Atomic structure of the actin:DNase I complex. Nature 347, 37-44
3. McLaughlin, P.J. *et al* (1993), Structure of gelsolin segment 1-actin complex and the mechanism of filament severing. Nature 364, 685-692
4. Schutt, C.E. *et al* (1993), The structure of crystalline profilin-β-actin. Nature 365, 810-816
5. Myslik, J.C. (1992), The structure of actin in the crystalline profilin:actin complex. Thesis, Princeton University
6. Flaherty, K.M. *et al* (1991), Similarity of the three-dimensional structures of actin and the ATPase fragment of a 70-kDa heat shock cognate protein. PNAS 88, 5041-5045
7. Aspenström, P. *et al* (1993), Mutations in β-actin: influence on polymer formation and on interactions with myosin and profilin. FEBS letters 329, 163-170
8. Chen, X. & Rubenstein, P.A. (1995), A mutation in an ATP-binding loop of *Saccharomyces cerevisiae* actin (S14A) causes a temperature-sensitive phenotype *in vivo* and *in vitro*. J Biol Chem 270, 11406-11414
9. Karlsson, R. (1988), Expression of chicken beta-actin in *Saccharomyces cerevisiae*. Gene 68, 249-257
10. Kouyama, T. & Mihashi, K. (1981), Fluorimetry study of N-(1-pyrenyl)iodoacetamide-labelled F-actin. Eur J Biochem 114, 33-38
11. Blikstad, I. *et al* (1978), Selective assay of monomeric and filamentous actin in cell extracts, using inhibition of deoxyribonuclease I. Cell 15, 935-943
12. Seals, J.R. *et al* (1978), A sensitive and precise isotopic assay of ATPase activity. Anal Biochem 90, 785-795
13. Spudich, J.A. (1974), Biochemical and structural studies of actomyosin-like proteins from non-muscle cells. J Biol Chem 249, 6013-6020
14. Kron, S.J. & Spudich, J.A. (1986), Fluorescent actin filaments move on myosin fixed to a glass surface. PNAS 83, 6272-6276.
15. Kishino, A. & Yanagida, T. (1988), Force measurement by micromanipulation of a single actin filament by glass needles. Nature 334, 74-76.

# Characterisation of the actin-binding protein Insertin

Andreas Teubner, Helmut E. Meyer, Albrecht Wegner

Institute for Physiological Chemistry, Ruhr-University Bochum, POB 102 148, 44780 Bochum, Germany

## Introduction

Insertin is an actin-binding protein which copurifies with vinculin from chicken gizzard. It binds to the barbed ends of actin filaments thereby modulating the polymerisation rate. In contrast to other barbed-end proteins („capping proteins") insertin does not fully suppress the polymerisation, but retards it maximally to one fifth of the normal rate of polymerisation. Thus, insertin is able to permit actin polymerisation while bound to the filament ends and is therefore a good candidate for fine tuning actin polymerisation. Localisation of insertin in focal adhesion plaques (dynamic cell-substrate contacts of migrating cells) underlines this potenial role and raises the question of possible interactions with other focal adhesion proteins.

Tensin is a 200kD protein with similar activities and localisation like insertin and possesses a domain which is nearly 100% identical to insertin (1). It is unknown if these two proteins originate from two different genes, are splice variants of the same gene, or if insertin is a proteolytical product of tensin. Thus, northern blot and southern blot analysis, cDNA library screening and 5'-RACE (Rapid amplification of cDNA ends) as well as mass spectrometry are applied to the insertin protein to reveal the relationship.

NATO ASI Series, Vol. H 102
Interacting Protein Domains
Their Role in Signal and Energy Transduction
Edited by Ludwig Heilmeyer

## Results and Discussion

According to Northern blot analysis few little mRNAs, possibly coding for insertin can be detected in the mRNA pool (Fig.1) To determine the sequence of these mRNA a chicken gizzard cDNA library was screened. Probing with an insertin specific PCR fragment a handful of clones was identified. All clones going beyond the 5'-end of insertin have sequences identical with the tensin protein, so that we were not able to identify the messages seen on the Northern blot by cDNA library screening.Therefore, we are now going to look for these sequences by 5'RACE

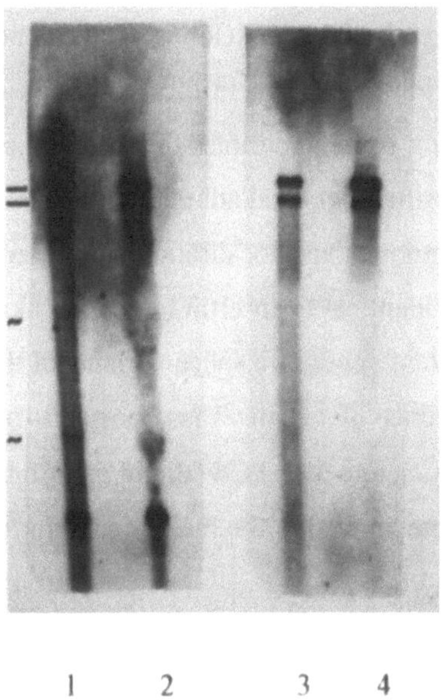


Fig. 1: Northern Blot Analysis of Insertin with Insertin- and Tensin-specific DIG-labelled oligonucleotides reveals multiple smaller bands (~) in addition to the two already described bands of 7 and 11 kbp (-), whereas probes specific for tensin only recognise these large bands. Lane 1 and 3: mRNA of chicken gizzard; lane 2 and 4: dito total RNA; lanes 1 and 2 probed with oligo specific for insertin; lanes 3 and 4 probed with oligos for tensin (C-terminus)

Furthermore we investigated insertin by protein sequencing and mass spectrometry to reveal the relationship between insertin and tensin. During determination of the primary structure of insertin (2) the trypsin-generated arginine at the N-terminus was the first sequencable amino acid. Amino acids more N-terminally located could not be sequenced due to the amount of protein. But mass spectrometry showed that peptides beyond the N-terminal arginine are present and contain only a few amino acids. Therefore, insertin was subjected to MALDI-PSD (**M**atrix-assisted **l**aser **d**esorption **i**onisation followed by **P**ost **S**ource **D**ecay) to determine these amino acids/peptides. The strategy was to generate peptides going beyond the first arginine via GluC-cleavage, then recleaving the isolated peptides wth trypsin to separate the known sequence from the unknown extension (Fig.2).

TENSIN<br>
A Q P T P Q V V Q R<br>
T T T<br>
**S F G T S V G T D P L A K A Y S P G P L V P A A R S T A E P D Y**<br>
G<br>
X X X X X X X X X<br>
INSERTIN

Fig.2: Strategy of the biochemical approach to determine the N-terminus of insertin. In bold the N-terminal amino acid stretch of insertin (2) accomanied by the preceding amino acid of tensin (top) and the preceding sequence in question of insertin (bottom). Insertin was first cleaved with GluC, relevant peptides (according to their mass) were trypsinized and subjected to MALDI-PSD, a method to sequence peptides in the mass spectrometer. G = cleavage site for GluC, T = dito for Trypsin

Finally, we ended up with two peptides with two other N-terminal amino acids. Additionally, Edman sequencing of two other probes resulted in the identification of two peptides with two other N-terminal amino acid around the previously determined N-terminus. Thus, the results are pointing to a ragged N-terminal end of the protein , possibly generated by proteolysis from tensin.

In order to get enough insertin for interaction studies we tried to overcome the poor yield of protein from conventional preparation by overexpressing insertin in E. coli. Since the resulting fusion proteins were immediately degraded in the cell we were not able to generate enough complete protein for interaction studies. Thus, we are now going to reveal these interactions with other focal adhesion site proteins by means of the Two-Hybrid-system

Northern blot analysis suggests that there are at least two different messages for tensin and insertin, supported by Southern blot analysis, pointing to two genes (data not shown) However the amino acid sequence and mass spectrometrical approach rather favourite that insertin is generated by proteolysis of the large protein tensin. To date there is no clear cut answer to the relationship of these two proteins. In the future we have to look for interaction partners for insertin in order to reveal its regulation in the focal adhesion plaques.

References

1. Davis et al., Science, 252, 712-715, 1991.
2. Weigt et al., J. Mol. Biol., 227, 593-595, 1992

# Cardiac Troponin

**Ludwig M.G. Heilmeyer, jr., Karin Lohmann, Silke U. Reiffert, Kornelia Jaquet**

Ruhr-Universität Bochum, Institut für Physiologische Chemie, Abt. Biochemie Supramolekularer Systeme, D-44780 Bochum

Thin filaments of striated muscle contain troponin which confers $Ca^{2+}$-sensitivity to the actomyosin system (for a recent review see: Farah and Reinach, 1995). Cardiac troponin plays a special role since it does not only transduce the $Ca^{2+}$ signal but, maybe more importantly, it also modulates the $Ca^{2+}$ signaling event by being phosphorylated. It is well established that the tropomyosin binding subunit (cTnT) as well as the inhibitory subunit (cTnI) are both phosphoproteins which can be phosphorylated on several sites whereas the $Ca^{2+}$ binding subunit (cTnC) has not been reported to be phosphorylated. Cardiac troponin, therefore, is the merging point of two signal pathways, namely that originating from electrical stimulation and $Ca^{2+}$ entry into the myocyte on the one hand and from epinephrine stimulation, cyclic AMP production and activation of the cyclic AMP-dependent protein kinase on the other hand (Fig. 1). Immediate questions to be answered in this context are: where and to which degree does phosphorylation occur on cardiac troponin, how do incorporated phosphates interact and what are their intermolecular signal pathways and functions.

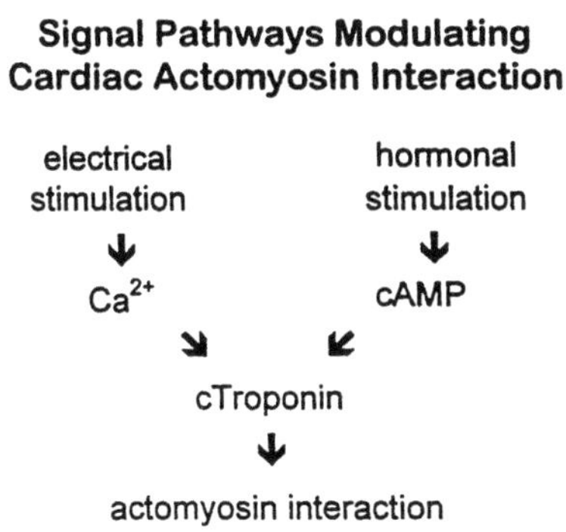


NATO ASI Series, Vol. H 102
Interacting Protein Domains
Their Role in Signal and Energy Transduction
Edited by Ludwig Heilmeyer

Table 1 lists protein kinases and sites involved in phosphorylation of the cardiac holotroponin komplex. It is interesting to see that a variety of protein kinases, like casein kinase (Villar-Palasi & Kumon, 1981), protein kinase C (Noland et al., 1989), phosphorylase kinase (Raggi et al., 1989) and $Ca^{2+}$/calmodulin-dependent protein kinase (Jaquet et al., 1995) can phosphorylate cTnT on a variety of sites which, comparing rabbit and bovine cTnT, reside in similar areas of the whole molecule. The function of none of these phosphorylations is well documented. Serine-1 is phosphorylated as well in skeletal muscle troponin T (Perry & Cole, 1973) which in this tissue, however, seems to be fully phosphorylated all the time (Sperling et al., 1979) whereas serine-1 phosphate in the cardiac system shows turnover and thus variable degrees of phosphorylation. Skeletal muscle troponin I is a poor substrate for protein kinases if integrated into the holocomplex. In contrast, cardiac troponin I is phosphorylated mainly on two adjacent serine residues (serine-23 and -24 in the bovine sequence) located in an N-terminal extension of the sequence added to the skeletal muscle protein apparently for regulatory purposes (Swiderek et al., 1988). Protein kinase A, C and G as well as $Ca^{2+}$/calmodulin-dependent protein kinase all phosphorylate these two adjacent serine residues (Swiderek et al., 1990; compare Table 1).

| Species | Subunit | Protein Kinase | Site |
|---|---|---|---|
| Rabbit<br>Bovine | TnT | Troponin T-Kinase<br>(Casein Kinase) | Ser 1 |
| Rabbit<br>Bovine | TnT | C | Thr 190<br>Ser 194, Thr 190 |
| Rabbit<br>Bovine | TnT | Phosphorylase Kinase | Ser 176 |
| Rabbit<br>Bovine | TnT | Ca/Cm Kinase | ?<br>Thr 190 |
| Rabbit<br>Bovine<br>Human | TnI | A | Ser 22, Ser 23<br>Ser 23, Ser 24<br>Ser 22, Ser 23 |
| Bovine | TnI | C | Ser 43, Ser 45<br>Ser 23, Ser 24 |
| Bovine | TnI | G | Ser 23, Ser 24 |
| Bovine | TnI | Ca/Cm Kinase | Ser 23, Ser 24 |
| Rabbit<br>Bovine | TnC | - | - |

Table 1: Phosphorylation of Cardiac Troponin (Holocomplex)

Protein kinase A, however, seems to be the protein kinase acting *in vivo* since it is present at a high concentration in cardiomyocytes, and it phosphorylates these two residues at a higher rate than the other protein kinases. The N-terminal sequence contains a characteristic motif of three arginine residues followed by two serine residues in rabbit, bovine, man (Mittmann et al., 1992) and rat (Murphy et al., 1991). Also in quail (Hastings et al., 1991) the sequence is homologous and contains the motif two arginine, one lysine, two serine residues (Fig. 2).

| | | | | | |
|---|---|---|---|---|---|
| Bovine | | ADRSGGSTAG | DTVPAPPPV | RRRSS | ANYRAYATE |
| Rabbit | | ADESTDA-AG | EARPAPAPV | RRRSS | ANYRAYATE |
| Man | | ADGSSDA-AR | EPRPAPAPI | RRRSS | -NYRAYATE |
| Rat | M | ADESSDA-AG | EPYPAPAPV | RRRSS | ANYRAYATE |
| Quail | M | AEE------- | E-EPKPPPL | RRKSS | ANYRGYATE |

**Fig. 2 Comparison of cardiac troponin I N-termini.** All sequences contain a proline rich motif - PXPXPX - and a phosphorylation motif - RRR(K)SS - where X designates a hydrophobic aminoacid.

This motif is preceded by a prolin-rich sequence exhibiting the consensus PXPXPX where X are predominantly hydrophobic amino acids. The function of this prolin rich sequence is unknown. The phosphorylation region forms a duplicated minimal recognition motif for the cyclic AMP-dependent protein kinase (Mittmann et al., 1990). This protein kinase recognizes a sequence containing minimally RRXS where X can be any amino acid (Kemp, 1990). If one substitutes X by R and adds a serine C-terminally this minimal recognition motif is duplicated. Indeed, in cardiac TnI four phosphorylation states can be generated by cyclic AMP-dependent protein kinase: the doubly phosphorylated species, two monophosphorylated products carrying phosphate on either serine residue as well as the non phosphorylated form. Phosphorylation of these serines follows a strictly sequential pattern. The distal serine is phosphorylated ca. 12-fold faster than the proximal one (Mittmann et al., 1992). Thus employing a peptide containing this duplicated minimal recognition motif as substrate for protein kinase A a monophosphorylated species appears transiently and finally the bisphosphorylated product accumulates (Fig. 3). Dephosphorylation of the bisphospho species by protein phosphatase 2A occurs as well in a sequential manner (Fig. 4). The dephosphorylation sites have been mapped employing the S-ethylcysteine method as a means to characterize these sites by Edman degradation. The result shows clearly that the distal phosphoserine is dephosphorylated twice as fast as the proximal one (Jaquet et al.,

1995). Thus combining cTnI together with protein kinase A and protein phosphatase 2A a cycling system is created in which the serine residues are consecutively phosphorylated and dephosphorylated finally leading to a steady state which is solely dependent on the activity ratio of protein kinase to protein phosphatase.

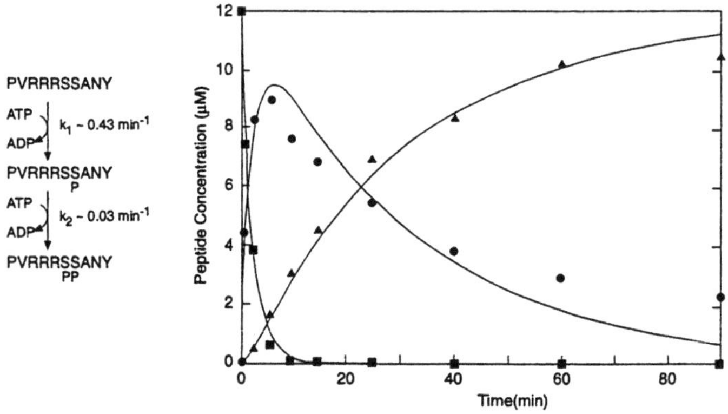


**Fig. 3 Sequential phosphorylation of duplicated recognition motif for protein kinase A present in cTnI.** The amount of each peptide, the nonphosphorylated substrate (■), the monophoposphorylated (●), and the bisphosphorylated form (▲) is determined by [$^{32}$P]phosphate content or by amino acid analysis according to Meyer et al., 1991. The curves calculated with the rate constants k1 and k2 represent the optimal adaptation to the experimental data employing the software Matlab.

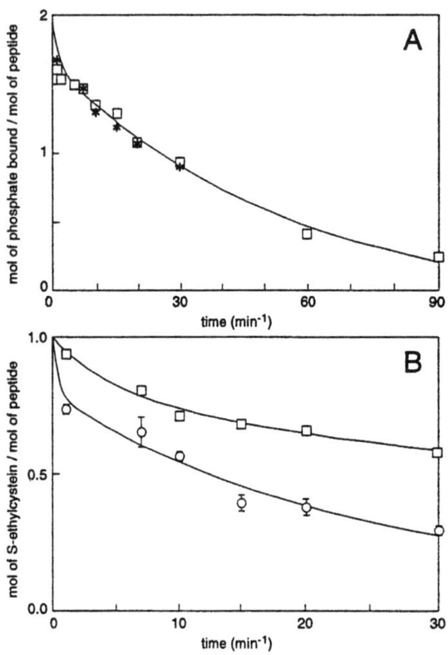


**Fig. 4 Sequential dephosphorylation of cardiac troponin I.** (A) Dephosphorylation of the bisphosphorylated [$^{32}$P]peptide APPPVRRRSS ANYRAYATE (□) was performed with protein phosphatase 2A as described by Thieleczek et al., 1995. The decrease of substrate bound [$^{32}$P]phosphate as function of time was measured. The data resulting from summation of the data shown in B are indicated by *. (B) The amount of phosphoserine-23 (□) and -24 (O) (numbering according to the bovine sequence) present in the peptide at various time points was determined by conversion to S-ethylcysteine prior to solid phase sequencing. Determinations were performed in duplicate. In all experiments the repetitive yield was in the range of 80-84%. The amount of S-ethylcysteine-23, -24 was normalized to 1 nM and corrected for yield.

Fig. 5 shows that in presence of equal activities of protein kinase and protein phosphatase mainly a monophosphorylated species carrying phosphate at the distal serine accumulates, whereas in presence of a 10-fold excess of protein phosphatase over protein kinase activity the major species (70 %) is the dephosphorylated one. Vice versa, at a 10-fold excess of protein kinase over protein phosphatase activity the major species is the bisphosphorylated material (Jaquet et al., 1995). In agreement with these theoretical calculations the freshly isolated cardiac holotroponin complex contains as monophosphorylated form mainly the distally but not the proximally phosphorylated species (Jaquet et al., 1993).

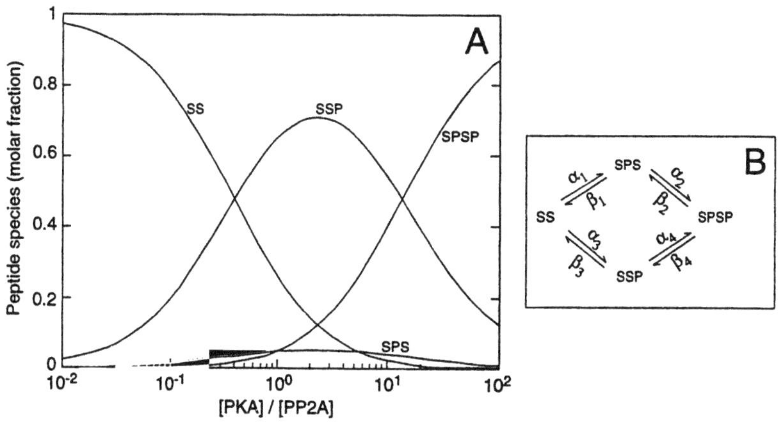


**Fig. 5: Steady state phosphorylation pattern of cTnI as a function of the concentration ratio of active protein kinase A to active protein phosphatase 2A [PKA]/[PP2A].** (A) SS, SPS, SSP and SPSP represent the nonphosphorylated, the monophosphorylated phosphoserine-23 and -24 and the bisphosphorylated species, respectively (numbering according to the bovine sequence). (B) For the phosphorylation of the peptide only the reaction (i = 1, ..., 4) characterized by the rate constatns $\alpha_i$ are relevant. In case of the peptide dephosphorylation only the reactions characterized by $\beta_i$ are involved.

Many proteins have been characterized as phosphoproteins, however, on an atomic level it is very rarely known how the incorporated phosphates affect the target protein. First hand information is obtained from crystallographic studies in which the interaction of incorporated phosphate groups with other amino acid residues of the polypeptide chain can be identified. However, there are only very few examples known. The method of choice to study these kinds of interactions consists in $^{31}$P-NMR spectroscopy. It yields information on the environment and interactions of these phosphate groups and, more interestingly, on the dynamics of these processes. Fig. 6 shows a $^{31}$P-NMR spectrum of the native holotroponin complex freshly isolated from heart which was phosphorylated solely on cTnI (Jaquet et al., 1993). Three signals are observed which have been identified by comparing them with those obtained from homogenously phosphorylated materials. The signal at the highest chemical shift corresponds to the species monophosphorylated at the distal serine. At higher field the monophosphorylated species phosphorylated at the proximal serine is seen. The signal at the highest field corresponds to that of the bisphosphorylated species. Interestingly, the signal at the highest field disappears and only the two signals corresponding to those of each monophosphorylated species are observed if this cTnI is isolated from the holocomplex (Fig. 6).

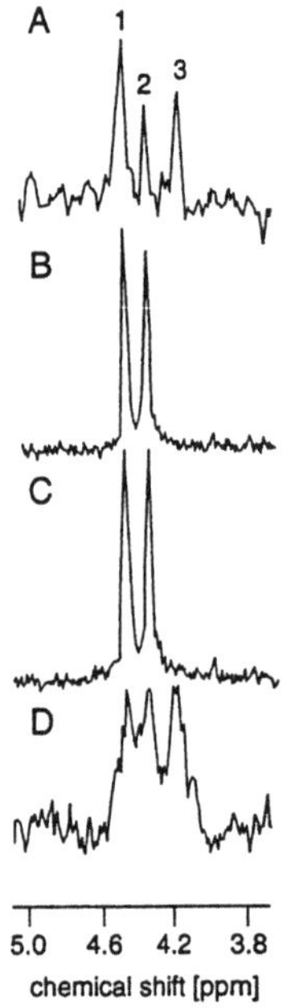


**Fig. 6 $^{31}$P NMR-spectra of reconstituted cTn complexes.** NMR-spectra are recorded with broadband proton decoupling on a 400 MHz Bruker W250C (Jaquet et al., 1993). Proteins are dissolved in $D_2O$ containing 20 mM MOPS, pH 7.2, 0.5 M KCl, 1 mM DTE. In each sample the pH is adjusted such that the chemical shift of inorganic phosphate added as internal standard is constant. All samples contain 1.4 mol of phosphate/mol of cTnI. (A) cTn, 15 mg/ml, 26566 scans, (B) cTnI, 10 mg/ml, 7932 scans, (C) reconstituted cTnI/cTnC complex, 20 mg/ml, 18172 scans, (D) reconstituted cTnI/cTnT complex, 12 mg/ml, 34368 scans, (E) reconstituted cTnI/cTnT/cTnC complex, 15 mg/ml, 42248 scans. 1 can be assigned to serine phosphate-24, 2 to serine phosphate-23 and 3 to serine phosphate-23 and -24 (bisphosphorylated cTnI within the cTn holocomplex; numbering according to the bovine sequence).

These findings indicate clearly that the bisphosphorylated species somehow interacts with (an)other component(s) of the holotroponin complex. Reconstitution experiments with cTnC and cTnT reveal that only the reconstituted holocomplex shows the three signal spectrum. This clearly indicates that bisphosphorylation serves a special function; the monophosphorylated species behave identically in isolated cTnI and in the holocomplex. Two signals of identical chemical shifts as in isolated cTnI are also seen employing a bisphosphorylated tetradecapeptide (Jaquet et al., 1993) comprising the phosphorylation domain. Structural studies have revealed that the phosphorylation motif has a double loop structure which seems to be similarly folded in the dephospho, each monophospho and the bisphospho form (Fig. 7). This double loop structure allows an interaction of the phosphate

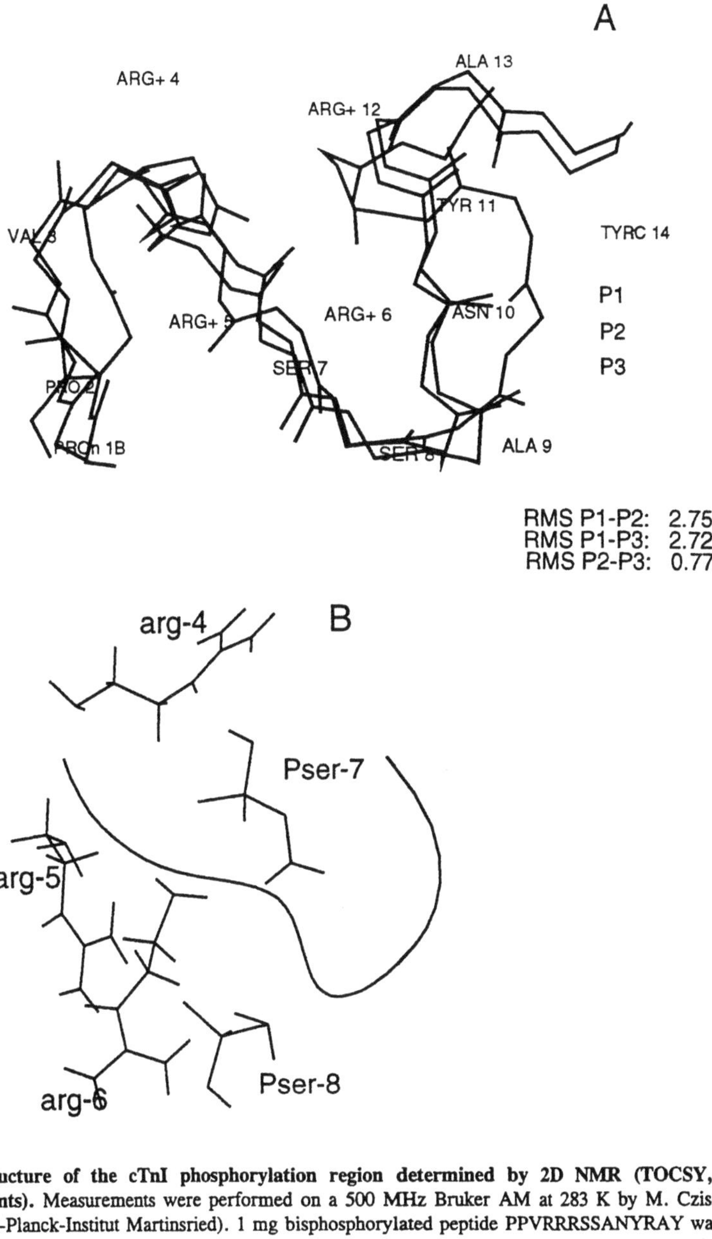


**Fig. 7: Structure of the cTnI phosphorylation region determined by 2D NMR (TOCSY, NOESY measurements).** Measurements were performed on a 500 MHz Bruker AM at 283 K by M. Czisch and T. Holak (Max-Planck-Institut Martinsried). 1 mg bisphosphorylated peptide PPVRRRSSANYRAY was disolved in 95 % $H_2O$, 5 % $D_2O$. pH was adjusted to 6.5. For molecular modeling Insight/Discover (Biosym) was used. Energy minimizations were perfomred in a water box with a dielectric constant of 80 using the conjugate gradient method and the cvff force filed. As constraints NOE distances with upper and lower boundaries of 2 and 4 Å for medium range distances were used. Torsion angles calculated from coupling constants were taken into consideration by selecting structure proposals. (A) Superposition of three selected structures. (B) A model for the phosphorylation domain RRRS(P)S(P)ANY is given. The backbone is indicated by a one-thread ribbon and only Arg and SerP side chains are shown.

group present at the distal serine with the third arginine in the row of the three arginine residues. This is consistent with the findings of Quirk et al. (1995). The distance between this positively charged guanidino group and the phosphate group at the distal serine is ca. 5 nm. The electrical field of this positive charge causes an apparent pKa shift of the phosphoserine which at neutral pH lowers its protonation degree. This fact explains the appearance of the phosphate signal in the $^{31}$P-NMR spectrum at lower fields than the phospho group present at the proximal serine. The phosphate group at the proximal serine shows no such interaction and thus exhibits approximately the same apparent pKa value as free phosphoserine in solution. The appearance of both phosphate groups in only one signal at the highest field employing the bisphosphorylated cTnI holocomplex indicates that both phosphate groups must be present in an acidic environment yielding a higher protonation degree than the monophospho species (Jaquet et al., 1993). An acidic protein being part of the holocomplex is cTnC which in comparison to skeletal muscle troponin C contains a non functional $Ca^{2+}$ binding loop I. Thus, in this non functional $Ca^{2+}$ binding loop I acidic residues like glutamates and aspartates could provide an interaction site for the two phosphates of the cTnI bisphospo species. However, reconstitution of the holotroponin complex with skeletal muscle troponin C instead of cardiac muscle troponin C shows that the same three signal $^{31}$P-NMR spectrum as that of the cardiac troponin holocomplex containing bisphosphorylated cTnI is obtained (Fig. 8). Furthermore, a chimera constructed of the N-terminal sequence of cTnC including the non functional binding loop I and skeletal muscle TnC providing the residual three $Ca^{2+}$ binding loops also reconstitutes a three signal $^{31}$P-NMR spectrum. In addition $Ca^{2+}$ saturation of the troponin complex containing skeletal muscle troponin C does not change the three signal spectrum. These data indicate that binding loop I probably does not provide the interaction site for this bisphosphorylated troponin I N-terminal region. Comparing the sequence of skeletal and cardiac troponin C it is evident that a stretch of acidic residues located in the region between residue 130 and 139 is present in both proteins. These residues are located between the high affinity $Ca^{2+}/Mg^{2+}$ binding loops III and IV. The hypothesis that this region could provide a docking site for the bisphospho troponin I N-terminal region fits with the idea that troponin C and troponin I are arranged in an antiparallel fashion such that the N-terminal domains of troponin C seem to interact with the C-terminal sites in cardiac troponin I and vice versa the C-terminal domain of troponin C with the N-terminal region of troponin I (for review see Farah &

Reinach, 1995). Indeed, molecular modelling studies indicate that the cTnI N-terminal part can be docked onto the C-terminal region of the cTnC with the release of free energy.

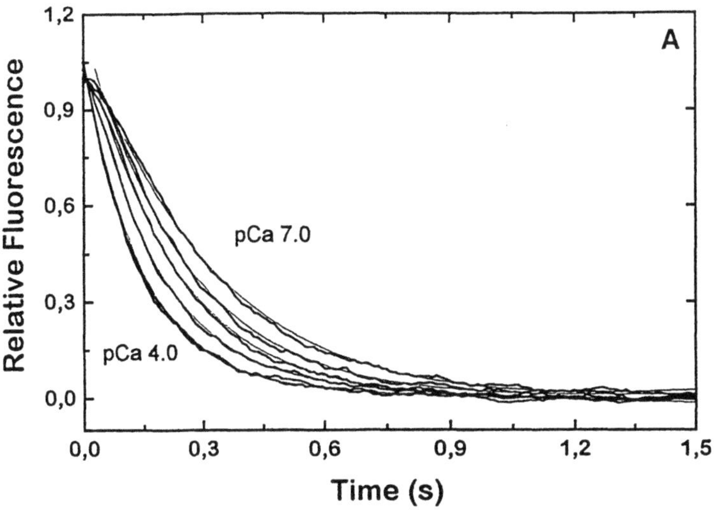


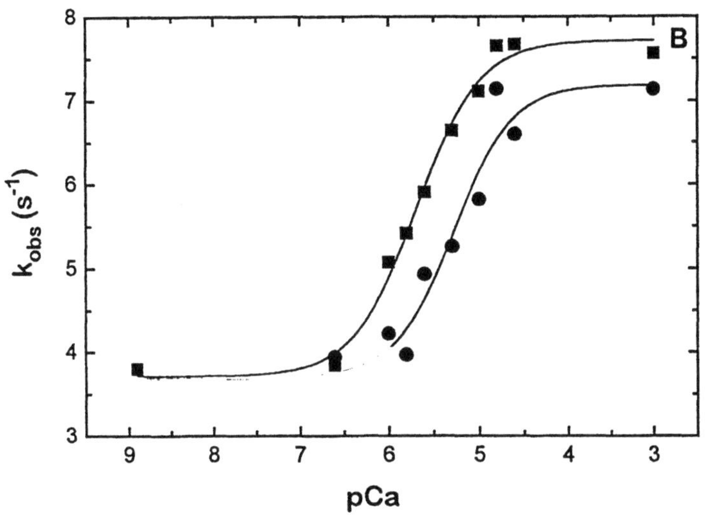


**Fig. 8 Influence of cTn phosphorylation on the calcium-dependent binding of S1 to pyr·actin·Tm·cTn** (A) The fluorescence transient was observed on mixing 5 $\mu$M pyr-actin, 1.45 $\mu$M Tm and 1.45 $\mu$M $cTnTP_0IP_2$ with 1 $\mu$M S1 at different calcium concentrations. For clarity only 5 curves are shwon and the best fit single exponential is superimposed on each data set. The pCa (Robs) values are 4.0 (6.9 $s^{-1}$), 4.5 (7.3 $s^{-1}$), 5.0 (4.4 $s^{-1}$), 6.0 (3.8 $s^{-1}$), 7.0 (2.7 $s^{-1}$). The calcium concentrations were obtained by mixing 2 mM EGTA and 2 mM CaEGTA in appropriate proportions. (B) Robs was plotted as a function of pCa for both $cTnTP_0IP_2$ (•) and $cTnTP_0IP_0$ (■). The data were fitted to the Hill equation. The bst fit parameters are pCa mid point 5.28 and 5.60 and the Hill coefficients are 1.35 and 1.37 for $cTnTP_0IP_2$ and $cTnTP_0IP_0$, respectively. Free $Ca^{2+}$ concentrations were calculated based on the dissociation constants of Sillen and Martell, 1971.

The function of cTnI phosphorylation can be tested by reconstituting thin filaments with a completely non phosphorylated or bisphosphorylated cTnI containing holocomplex and by measuring the $Ca^{2+}$-dependent association of myosin to these filaments (Fig. 9). A ca. 2.5-fold decrease in $Ca^{2+}$ sensitivity is observed upon exchange of the nonphosphorylated against the bisphosphorylated material (Reiffert et al., 1996). This observation is in agreement with the interpretation that bisphosphorylation could be involved in the phenomenon of faster relaxation of cardiac muscle fibres exposed to epinephrine. Under these conditions bisphosphorylation of cTnI occurs *in vivo* (Zhang et al., 1995; Talosi et al., 1993). It is tempting to speculate that force enhancement of cardiac muscle fibres by epinephrine could involve monophosphorylation of cTnI. Therefore, the major problem to be resolved consists in the functional characterization of the monophosphorylated cTnI species which hopefully will be possible in near future.

### *References*


Kemp, B.E. (1990) in: Peptides And Protein Phosphorylation by Bruce. E. Kemp, CRC Press, Inc., Boca Raton, Florida 1990, pp. 14-15

Farah, C.F. & Reinach, F.C. (1995) The Troponin complex and Regulation of Muscle Contraction. FASEB J. 9, 757-767

Hastings, K.E.M., Koppe, R.J., Murmor, E., Bader, D. & Shimada, J. (1991) Structure and Development of Expression of Cardiac Troponin I Isoforms. J. Biol. Chem. 266, 19659-19665

Sperling, J.E., Feldmann, K., Meyer, H., Jahnke, U. & Heilmeyer, L.M.G., jr. (1979) Isolation and characterization of the troponin complexes, $TI_2C$ and $I_2C$. Eur. J. Biochem. 101, 581-592.

Jaquet, K., Fukunaga, K., Miyamoto, E. & Meyer, H.E. (1995) A Site Phosphorylated in Bovine Cardiac Troponin T by Cardiac CaM Kinase II. Biochim. Biophys. Acta 1248, 193-195

Jaquet, K., Korte, K., Schnackerz, K., Vyska, K. & Heilmeyer, L.M.G., jr. (1993) Characterization of the Cardiac Troponin I Phosphorylation Domain by $^{31}P$-Nuclear Magnetic Resonance Spectroscopy. Biochemistry 32, 13873-13878

Jaquet, K., Thieleczek, R. & Heilmeyer, L.M.G., jr. (1995) Pattern Formation on Cardiac Troponin I by Consecutive Phosphorylation and Dephosphorylation. Eur. J. Biochem. 231, 486-490

Meyer, H.E., Hoffmann-Posorske, E. & Heilmeyer, L.M.G., jr. (1991) Determination and location of phosphoserine in proteins and peptides by conversion to S-ethyl-cysteine. In: Methods in Enzymology, Volume 201 *Protein Phosphorylation* (T. Hunter and B.M. Sefton, Hrsg.), pp. 169-185

Mittmann, K., Jaquet, K. & Heilmeyer, L.M.G., jr. (1990) A Common Motif of Two Adjacent Phosphoerines in Bovine, Rabbit and Human Cardiac Troponin I. FEBS Lett. 273, 41-45

Mittmann, K., Jaquet, K. & Heilmeyer, L.M.G., jr. (1992) Ordered Phosphorylation of a Duplicated Minimal Recognition Motif for cAMP-Dependent Protein Kinase Present in Caridac Troponin. FEBS Lett. 302, 133-137

Murphy, A.M., Jones II, L. & Strauss, A.W. (1991) Molecular Cloning of Rat Cardiac Troponin I and Analysis of Troponin I Isoform Expression in Developing Rat Heart. Biochemistry 30, 707-712

Noland, T.A., jr., Raynor, R.L. & Kuo, J.F. (1989) Identification of Sites Phosphorylated in Bovine Cardiac Troponin I and T by Protein Kinase C and Comparative Sites. J. Biol. Chem. 264, 20778-20785

Perry, S.V. & Cole, H.A. (1973) Phosphorylation of the '37000 Component' of the Troponin Complex (Troponin T). Biochem. J. 131, 425-428

Quirk, P.G., Patchell, V.B., Gao, X., Levine, B.A. & Perry, S.V. (1995) Sequential Phosphorylation of Adjacent Serine Residues on the N-Terminal Region of Cardiac Troponin I. Structure-Activity Implications of Ordered Phosphorylation. FEBS Lett. 370, 175-178

Raggi, A., Grand, R.J.A., Moir, A.J.G. & Perry, S.V. (1989) Structure-Function Relationships in Cardiac Troponin T. Biochim. Biophys. Acta 997, 135-143

Reiffert, S., Jaquet, K., Heilmeyer, L.M.G., jr., Ritchie, M.D. & Geeves, M.A. (1996) Bisphosphorylation of Cardiac Troponin I Modulates Calcium Dependent Binding of Myosin Subfragment 1 to Reconstituted Thin Filaments. FEBS Lett. 384, 43-47

Sillen, L.G. and Martell, A.E. (1971) Suppl. no 1, special publ. no 25, The Chemical Society, London

Swiderek, K., Jaquet, K., Meyer, H.E. & Heilmeyer, L.M.G., jr. (1988) Cardiac Troponin I, Isolated from Bovine Heart, Contains Two Adjacent Phosphoserines. A First Example of Phosphoserine Determination by Derivatization to S-Ethylcysteine. Eur. J. Biochem. 176, 335-342

Swiderek, K., Jaquet, K., Meyer, H.E., Schächtele, C., Hofmann, F. & Heilmeyer, L.M.G., jr. (1990) Sites Phosphorylated in Bovine Cardiac Troponin T and I. Characterization by $^{31}$P-NMR Spectroscopy and Phosphorylation by Protein Kinases. Eur. J. Biochem. 190, 575-582

Talosi, L., Edes, I. & Kranias, E.G. (1993) Intracellular Mechanisms Mediating Reversal of ß-Adrenergic Stimulation in Intact Beating Hearts. J. Am. Phys. Soc. H791-H797

Villar-Palasi, C. & Kumon, A. (1981) Purification and Properties of Dog Cardiac Troponin T Kinase. J. Biol. Chem. 256, 7409-7415

Zhang, R., Zhao, J., Mandveno, A. & Potter, J.D. (1995) Cardiac Troponin I Phosphorylation Increases the Role of Cardiac Muscle Relaxation. Circ. Res. 76, 1028-1035

# Studies on the function of the different phosphoforms of cardiac troponin I

Silke U. Reiffert [1], Kornelia Jaquet [2], Ludwig M.G. Heilmeyer Jr.[1], Friedrich W. Herberg[1]

[1] Ruhr-Universität Bochum, Institut für Physiologische Chemie, Abteilung für Biochemie Supramolekularer Systeme, 44780 Bochum, Germany

[2] Herz- und Diabeteszentrum Nordrhein-Westfalen, Georgstraße 11, 32545 Bad Oeynhausen, Germany

## Introduction

Cardiac troponin (cTn) is composed of three subunits: cTnT (tropomyosin binding subunit), cTnI (inhibitory subunit) and cTnC ($Ca^{2+}$-binding subunit). The cAMP-dependent protein kinase (PKA) phosphorylates Ser23 and/or Ser24 in the heart-specific region of bovine troponin I (Swiderek et al. 1988). Ser24 is phosphorylated 12-fold faster (Mittmann et al. 1992) and dephosphorylated by protein phosphatase 2A 2-fold faster (Jaquet et al. 1995) than Ser23. Therefore freshly isolated cTn containes a mixture of non-, two mono- and a bisphosphorylated cTnI species. The bisphosphorylated state was investigated in comparison to the dephospho-state. The $Ca^{2+}$-dependent binding of myosin subfragment S1 to the thin filaments, was reconstituted with pyrene-labelled actin (pyr-actin), tropomyosin (Tm) and cTn in two defined phosphorylation states; one form was completely dephosphorylated on all subunits and the other was bisphosphorylated only in the cTnI subunit. The bisphosphorylated form lead to a rightward shift of the $pCa_{50}$ value of about 0.36 pCa-units concluding that bisphosphorylation is responsible for the decrease in $Ca^{2+}$-affinity of cTnC (Reiffert et al. 1996). Similar results were also obtained by Zhang et al. (1995A) employing skinned fiber experiments.

However, there exist 3 further cTnI forms (one non- and two monophosphorylated) whose functions are not known to date. Zhang et al. (1995B) did not observed any alterations in $Ca^{2+}$-sensitivity of the monophosphorylated compared to dephosphorylated cTnI in skinned fibers.

## Results and Discussion

To investigate the function of the monophosphoforms, human cTnI mutant proteins where one serine residue was replaced by an alanine residue (S22A, S23A) as well as the wildtype cTnI were overexpressed in *E. coli*. The effect of the cTnI-phosphorylation state on the interaction between the cTn-subunits was studied by

NATO ASI Series, Vol. H 102
Interacting Protein Domains
Their Role in Signal and Energy Transduction
Edited by Ludwig Heilmeyer

surface plasmon resonance (SPR) spectroscopy using a BIAcore instrument. Changes in surface concentration due to binding of cTnI to the immobilized cTnC are propotional to changes in the refractive index on the surface resulting in changes in the SPR signal, plotted as Response Units (RU, 1000RU are equivalent to a change of about 1ng/mm$^2$ in surface protein concentration). cTnC derivatized with biotin (NHS-biotin coupled via lysine residues of cTnC) was captured by immobilized streptavidine (SA5-chip). The change in surface concentration resulting from binding of the different phosphorylation states of cTnI subunit to cTnC subunit was measured.

Fig.1A shows interaction of the dephosphorylated cTnI (cTnI-WT-dP) form reaching equilibrium at a higher response level than the bisphosphorylated form (cTnI-WT-bP). This occurs in the presence as well as in the absence of calcium and correlates with the decrease in affinity of bisphosphorylated cTnI. From these measurements could be concluded that the affinity of both, the de- and the bisphosphorylated cTnI form to cTnC is much higher in the presence than in the absence of calcium. In Fig.1B and 1C the phosphorylated mutants (corresponding to the monophosphorylated forms cTnI-S22A-mP, cTnI-S23A-mP) reach equilibrium at a lower response level than the dephosphorylated mutants (cTnI-S22A-dP, cTnI-S23A-dP). However the difference between the mono- and dephosphorylated forms is not that high as between the bisphosphorylated and dephosphorylated forms, displaying an intermediate effect. These results lead to the conclusion that there is an gradual effect of phosphorylation from de- to mono- and then to the bisphosphorylated form.

In Table 1A the calculated affinity constants are presented, showing that the bisphosphorylated cTnI form has the lowest affinity to cTnC. This is 2.3-fold lower than the affinity of dephospho-cTnI to cTnC. These measurements are in good agreement with fluorescence studies by Liao et al. 1994 (see Tabla 1B). The affinity of the monophosphorylated form cTnI-S23A-mP and of the monophosphoform cTnI-S23A-mP is 1.6-fold and 1.4-fold lower, respectively, than that of the dephosphospecies.

These studies demonstrated for the first time a difference between the dephosphorylated and the monophosphorylated species of cTnI. However, no effect of the monophosphorylated cTnI-forms on $Ca^{2+}$-binding was detected using skinned fibers (Zhang et al. 1995 B). Because the difference in $pCa_{50}$ between non- and bisphosphorylated cTnI was only in the range of about 0.3 pCa-Units (Zhang et al. 1995 A, Reiffert et al. 1996) the intermediate effect of the monophosphoforms might not be detectable in skinned fibers experiments.

A standard free-energy $\Delta G^{0'}$ change for the binding of cTnI-WT-dP and cTnI-WT-bP to immobilized cTnC was calculated. The $\Delta\Delta G^{0'}$ of 0.47 kcal/ mol was identical to a value obtained by binding of S1 (myosin subfragment 1) to reconstituted thin filament containing actin tropomyosin and cTn with cTnI-dP and cTnI-bP (Reiffert et al. 1996). In this case bisphosphorylation of the cTnI-subunit led to a shift in calcium sensitivity by 0.36 pCa-units corresponding to 0.47 kcal/ mol. It remains to be clarified if these to events are correlated.

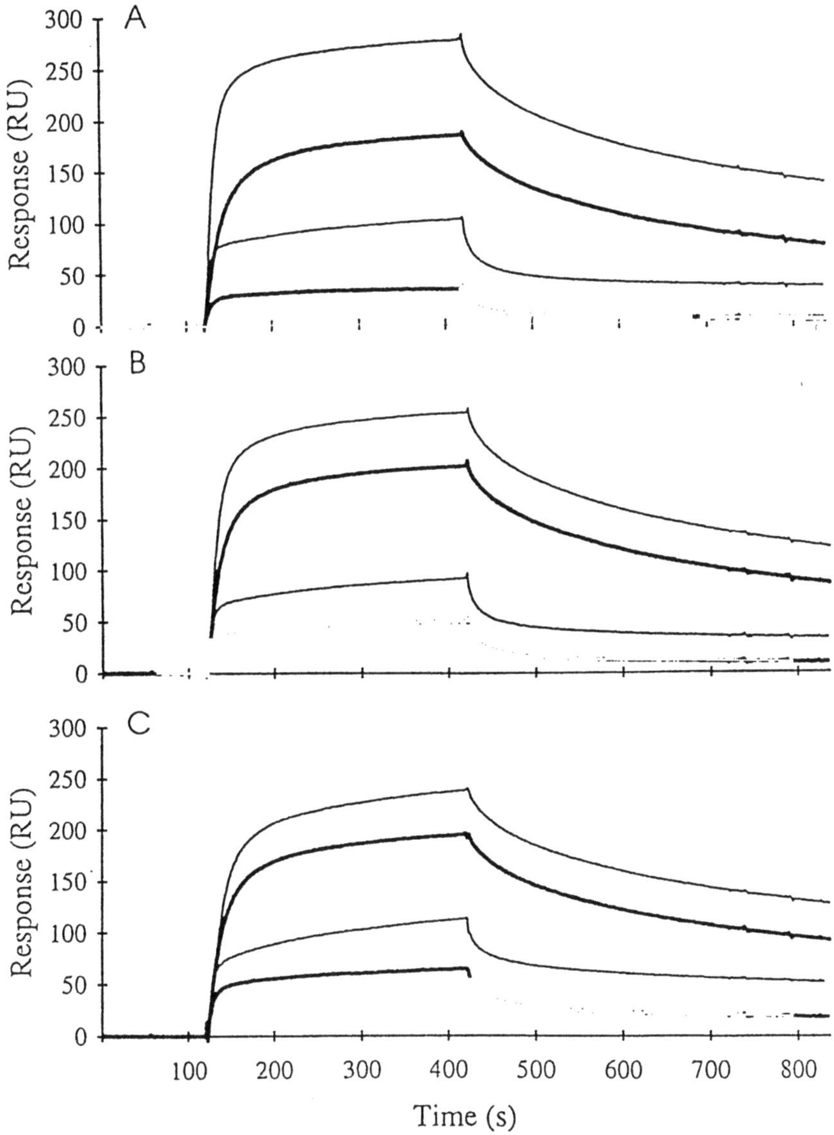


**Fig.1: Interaction of the different phospho- and dephosphospecies of cTnI with immobilized cTnC in presence and absence of Calcium**

1136 RU's of cTnC labelled with 3mol/mol of biotin were immobilized on a streptavidin sensor chip (SA5). For the binding measurements shown above a cTnI-concentration of 0.15µM in 20mM MOPS, 500mM KCl, 5mM $MgCl_2$, 3mM DTT, pH8.0 was used in the presence of calcium [0.1mM $CaCl_2$] (upper curves) and in the absence of calcium [2mM EGTA] (lower curves).

**A)** Wildtype cTnI bisphosphorylated at Ser22 and Ser23 (cTnI-WT-bP; thick line) and nonphosphorylated (cTnI-WT-dP; thin line)

**B)** cTnI with Ser23Ala monophosphorylated at Ser22 (cTnI-S23A-mP; thick line) and nonphosphorylated (cTnI-S23A-dP; thin line)

**C)** cTnI with Ser22Ala monophosphorylated at Ser23 (cTnI-S22A-mP; thick line) and nonphosphorylated (cTnI-S22A-dP; thin line)

**Table 1: Rate and affinity constants for the interaction between immobilized cTnC and cTnI in the presence of $Ca^{2+}$**

A) SPR

| cTnI-species | $k_a$ x $10^5[s^{-1} M^{-1}]$ | $k_d$ x $10^{-2}$ $[s^{-1}]$ | $K_D$ x$10^{-8}$[M] |
|---|---|---|---|
| cTnI-WT-bP | 3.15 | 0.49 | 1.6 |
| cTnI-WT-dP | 6.1 | 0.418 | 0.7 |
| cTnI-S23A-mP | 4.11 | 0.453 | 1.1 |
| cTnI-S23A-dP | 6.1 | 0.427 | 0.7 |
| cTnI-S22A-mP | 3.1 | 0.434 | 1.4 |
| cTnI-S22A-dP | 3.85 | 0.384 | 1 |

B) fluorescence quenching

| cTnI-species | $K_D$ x$10^{-8}$[M] |
|---|---|
| bovine cTnI-bP | 2 |
| bovine cTnI-dP | 0.8 |

A) cTnC was immobilized as described above and the different phospho- and dephosphospecies of cTnI was injected. Phosphate contents: cTnI-WT-bP, 1.93 mol P/mol cTnI; cTnI-S23A-mP, 0.77 mol P/mol cTnI; cTnI-S22A-mP, 1.04 mol P/mol cTnI. Only the constants determined in presence of $Ca^{2+}$ are shown.
**B)** Binding parameters of Liao et al. (1994) using $cTnC_{IAANS}$ fluorescence for interaction with cTnI.

## References


**Jaquet**, K., Thieleczek, R. and Heilmeyer Jr., L.M.G. (1995) Eur. J. Biochem. 231, 486-490

**Jaquet**, K., Korte, K., Schnackerz, K., Vyska, K. and Heilmeyer Jr., L.M.G. (1993) Biochemistry 32, 13873-13878

**Liao**, R., Wang, C.-K. and Cheung, H.C. (1994) Biochemistry 33, 12729-12734

**Mittmann**, K., Jaquet, K. and Heilmeyer Jr., L.M.G. (1992) FEBS Lett. 302, 133-137

**Reiffert**, S., Jaquet, K., Heilmeyer Jr., L.M.G., Ritchie, M.D. and Geeves, M.A. (1996) FEBS Lett. 384, 43-47

**Swiderek**, K., Jaquet, K., Meyer, H.E. and Heilmeyer Jr., L.M.G. (1988) Eur. J. Biochem. 176, 335-342

**Zhang**, R., Zhao, J., Mandveno, A. and Potter, J.D. (1995)A Cir. Res. 76, 1028-1035

**Zhang**, R., Zhao, J. and Potter, J.D. (1995)B J. Biol. Cem. 270, 30773-30780

# INDEX

# INDEX

## I, J



## K



## L



## M



## N

## O



## P

# R

## S



## T



## V



## W



## X, Y, Z

# NATO ASI Series H

Vol. 71: **Protein Synthesis and Targeting in Yeast.**
Edited by A. J. P. Brown, M. F. Tuite, and J. E. G. McCarthy. 425 pages. 1993.

Vol. 72: **Chromosome Segregation and Aneuploidy.**
Edited by B. K. Vig. 425 pages. 1993.

Vol. 73: **Human Apolipoprotein Mutants III. In Diagnosis and Treatment.**
Edited by C. R. Sirtori, G. Franceschini, B. H. Brewer Jr. 302 pages. 1993.

Vol. 74: **Molecular Mechanisms of Membrane Traffic.**
Edited by D. J. Morré, K. E. Howell, and J. J. M. Bergeron. 429 pages. 1993.

Vol. 75: **Cancer Therapy. Differentiation, Immunomodulation and Angiogenesis.**
Edited by N. D'Alessandro, E. Mihich, L. Rausa, H. Tapiero, and T. R. Tritton. 299 pages. 1993.

Vol. 76: **Tyrosine Phosphorylation/Dephosphorylation and Downstream Signalling.**
Edited by L. M. G. Heilmeyer Jr. 388 pages. 1993.

Vol. 77: **Ataxia-Telangiectasia.** Edited by R. A. Gatti, R. B. Painter. 306 pages. 1993.

Vol. 78: **Toxoplasmosis.** Edited by J. E. Smith. 272 pages. 1993.

Vol. 79: **Cellular Mechanisms of Sensory Processing. The Somatosensory System.**
Edited by L. Urban. 514 pages. 1994.

Vol. 80: **Autoimmunity: Experimental Aspects.**
Edited by M. Zouali. 318 pages. 1994.

Vol. 81: **Plant Molecular Biology. Molecular Genetic Analysis of Plant Development and Metabolism.**
Edited by G. Coruzzi, P. Puigdomènech. 579 pages. 1994.

Vol. 82: **Biological Membranes: Structure, Biogenesis and Dynamics.**
Edited by Jos A. F. Op den Kamp. 367 pages. 1994.

Vol. 83: **Molecular Biology of Mitochondrial Transport Systems.**
Edited by M. Forte, M. Colombini. 420 pages. 1994.

Vol. 84: **Biomechanics of Active Movement and Division of Cells.**
Edited by N. Akkaş. 587 pages. 1994.

Vol. 85: **Cellular and Molecular Effects of Mineral and Synthetic Dusts and Fibres.**
Edited by J. M. G. Davis, M.-C. Jaurand. 448 pages. 1994.

Vol. 86: **Biochemical and Cellular Mechanisms of Stress Tolerance in Plants.**
Edited by J. H. Cherry. 616 pages. 1994.

# NATO ASI Series H

Vol. 87: **NMR of Biological Macromolecules.**
Edited by C. I. Stassinopoulou. 616 pages. 1994.

Vol. 88: **Liver Carcinogenesis. The Molecular Pathways.**
Edited by G. G. Skouteris. 502 pages. 1994.

Vol. 89: **Molecular and Cellular Mechanisms of $H^+$ Transport.**
Edited by B. H. Hirst. 504 pages. 1994.

Vol. 90: **Molecular Aspects of Oxidative Drug Metabolizing Enzymes: Their Significance in Environmental Toxicology, Chemical Carcinogenesis and Health.**
Edited by E. Arınç, J. B. Schenkman, E. Hodgson. 623 pages. 1994.

Vol. 91: **Trafficking of Intracellular Membranes: From Molecular Sorting to Membrane Fusion.**
Edited by M. C. Pedroso de Lima, N. Düzgüneş D. Hoekstra. 371 pages. 1995.

Vol. 92: **Signalling Mechanisms – from Transcription Factors to Oxidative Stress.**
Edited by L. Packer, K. Wirtz. 466 pages. 1995.

Vol. 93: **Modulation of Cellular Responses in Toxicity.**
Edited by C. L. Galli, A. M. Goldberg, M. Marinovich. 379 pages. 1995.

Vol. 94: **Gene Technology.**
Edited by A. R. Zander, W. Ostertag, B. V. Afanasiev, F. Grosveld. 556 pages. 1996.

Vol. 95: **Flow and Image Cytometry.**
Edited by A. Jacquemin-Sablon. 241 pages. 1996.

Vol. 96: **Molecular Dynamics of Biomembranes.**
Edited by J. A. F. Op den Kamp. 424 pages. 1996.

Vol. 97: **Post-transcriptional Control of Gene Expression.**
Edited by O. Resnekov, A. von Gabain. 283 pages. 1996.

Vol. 98: **Lactic Acid Bacteria: Current Advances in Metabolism, Genetics and Applications.**
Edited by T. F. Bozoğlu, B. Ray. 412 pages. 1996.

Vol. 99: **Tumor Biology**
**Regulation of Cell Growth, Differentiation and Genetics in Cancer.**
Edited by A. S. Tsiftsoglou, A. C. Sartorelli, D. E. Housman, T. M. Dexter. 342 pages. 1996.

Vol. 100: **Neurotransmitter Release and Uptake.**
Edited by Ş. Pöğün. 344 pages. 1997.

Vol. 101: **Molecular Mechanisms of Signalling and Membrane Transport.**
Edited by K. W. A. Wirtz. 344 pages. 1997.

Vol. 102: **Interacting Protein Domains**
Their Role in Signal and Energy Transduction
Edited by L. Heilmeyer. 296 pages. 1997.

MIX
Papier aus verantwortungsvollen Quellen
Paper from responsible sources
FSC® C105338

If you have any concerns about our products,
you can contact us on
**ProductSafety@springernature.com**

In case Publisher is established outside the EU,
the EU authorized representative is:
**Springer Nature Customer Service Center GmbH**
**Europaplatz 3, 69115 Heidelberg, Germany**

Printed by Libri Plureos GmbH
in Hamburg, Germany